Alkaline Rocks and Carbonatites of the World

Part Two:
Former USSR

Alkaline Rocks and Carbonatites of the World

Part Two:
Former USSR

L.N. Kogarko

Vernadsky Institute, Moscow, Russia

V.A. Kononova

Institute of Geology of Ore Deposits, Petrography, Mineralogy and Geochemistry (IGEM), Moscow, Russia

M.P. Orlova

All-Union Institute of Geology (VSEGEI), St Petersburg, Russia

and

A.R. Woolley

Department of Mineralogy, The Natural History Museum, London, UK

SPRINGER-SCIENCE+BUSINESS MEDIA, B.V.

First edition 1995

© The Natural History Museum, London, 1995
Originally published by Chapman & Hall in 1995

Typeset in 10/10pt Paladium by Colset Private Limited, Singapore

ISBN 978-94-010-9096-4

A catalogue record for this book is available from the British Library

Library of Congress Catalog Card Number: 94-69922

∞ Printed on acid-free text paper, manufactured in accordance with
ANSI/NISO Z39.48-1992(Permanence of Paper).

Contents

Introduction

Although the great diversity of alkaline rocks, with their relatively exotic mineralogies, has always attracted the interest of petrologists, as have the more recently defined carbonatites, it could be argued that little progress has been made over the past 50 years towards formulating a comprehensive petrogenesis of these rocks. It could also be maintained that as the alkaline varieties have the most extreme compositions of all igneous rocks, so an understanding of their genesis is essential if we are to understand fully the workings of the solid earth, while a knowledge of the most extreme products of differentiation must inevitably cast light on rocks of less extreme compositions. The importance of academic research on these rocks is thus clear. There is, however, also a commercial aspect, in so far as they are an increasingly important source of a wide range of industrial raw materials, which has stimulated not only programmes to discover more occurrences, but also to investigate in greater detail those already known.

In recent years many new occurrences of alkaline rocks and carbonatites have come to light to such an extent, indeed, that they can no longer be considered rare. At the beginning of the century the known and described occurrences of alkaline rocks were probably fewer than 100, while only three carbonatites were known. The geological exploration in particular of Africa, the U.S.S.R., Canada and Brazil, much of it inspired by the search for exploitable mineral deposits, has rather changed the picture from the time when Daly (1933, p. 38) estimated that the alkaline rocks comprised only about 0.1% of all the igneous rocks. Although this percentage has probably only increased a few fold, the number of localities to be described in these volumes is over 2000, including more than 300 carbonatites.

That these very numerous occurrences should be listed and briefly described would seem to be justified for several reasons. Firstly, the alkaline rocks are petrographically, chemically and structurally exceptionally diverse, certainly more so than any other group of igneous rocks, so that it is not easy to compare occurrences meaningfully. Secondly, whereas the study of layered basic and ultrabasic rocks, for instance, has greatly benefited from detailed investigations of a few key occurrences, the same has not happened in alkaline rock studies, partly because of this diversity. There are, of course, particular occurrences of considerable petrogenetic importance, as for example the Oldoinyo Lengai natrocarbonatite volcano and the Ilimaussaq complex, but comparative studies on alkaline rocks have in general been hampered by the extraordinary variety encountered and the scattered nature of the literature. Only one book in English has been devoted to the alkaline rocks (Sorensen, 1974) and two to carbonatites (Heinrich, 1966; Tuttle and Gittins, 1966); there are rather more books in Russian and there have been several monographs and papers on the occurrences of single countries, but there is no comprehensive, worldwide survey affording rapid and easy access to the field.

Regional studies of the compositions, form, ages and structural setting of occurrences comprising particular provinces have always seemed fruitful, and the association of alkaline rocks and carbonatites with rifting has long been acknowledged. However, many large, diverse and important provinces are not widely known, including the 20 or so major provinces of the U.S.S.R. and some of the major concentrations of South America. Although alkaline rocks are known to be characteristic of stable, intra-plate environments, it is becoming clear that they also play an important role in the igneous activity concentrated at plate margins. In this respect also, therefore, it is concluded that the collation of all the available regional data can make a significant contribution to understanding these rocks.

Sorensen (1974) in the Preface to 'The Alkaline Rocks' noted that owing to the limitations of space 'it was decided not to include a special section on the mineralogy of alkaline rocks and also not to compile a list of all alkaline complexes of the world corresponding to the valuable catalogues of carbonatites compiled by Tuttle and Gittins (1966) and Heinrich (1966)' (*op. cit.*, 1974, p. ix). It is hoped that the present volume will thus complement 'The Alkaline Rocks', while updating the lists of carbonatites described by Gittins and Heinrich.

Considerable thought has been given to the organization of this series of volumes, and early on it was decided that a mere listing with information in abbreviated form would not be particularly useful. Instead the more discursive style adopted for Currie's 'The Alkaline Rocks of Canada' (1976) and the descriptions of carbonatites compiled by Gittins (Tuttle and Gittins, 1966) were thought more appropriate. What were considered to be the best features of these publications have been adopted, but there are major differences with regard to layout and organization which are aimed at allowing rapid access to descriptions of particular localities, as well as showing, with the aid of numerous maps, the distribution of occurrences. The location of these rocks and their relationship to regional structures, particularly rifts, has always been of interest, and it is thought that this book provides, for the first time for many areas, the data on which meaningful discussion could be undertaken on the relationship between their distribution and temporal, structural and compositional features.

Apart from their scientific interest, the alkaline rocks and carbonatites are of major, and growing, economic importance. They are significant repositories of certain metals and commodities, indeed the only source of some of them, including Nb, the rare earths, Cu, V, phosphate, vermiculite, bauxite, raw materials for the manufacture of ceramics, and potentially for Th, U, diamonds, and many more. The economic potential of these rocks is now widely appreciated, particularly since the commencement of the very lucrative mining of the Palabora carbonatite for copper and a host of valuable by-products, and it is the exploration efforts of mining companies that have led to the discovery of many important new localities. The present volume is likely to be of considerable interest to mineral exploration companies because there appear to be no published reviews of the economic aspects of the alkaline rocks. The economic importance of carbonatites, however, has been usefully documented by Deans (1966, 1978).

Scope of the Catalogue

This series of books includes entries for all occurrences of alkaline igneous rocks and carbonatites that could be traced. The definition of alkaline rocks adopted is generally that of Sorensen (1974, p. 7): that they 'are characterized by the presence of feldspathoids and/or alkali pyroxenes and amphiboles'. The rocks included are, therefore, the nepheline syenites (phonolites) and ijolites (nephelinites), basanites and feldspathoid-bearing gabbroic rocks; peralkaline (i.e. containing alkali pyroxene and/or amphibole) syenite, quartz syenite and granite, together with peralkaline trachyte, comendite and pantellerite. Fenites are also included because of their intimate association with alkaline rocks and carbonatites, as are certain ultramafic and melilite-bearing rocks, including alnoite and uncompahgrite. Analcime is treated as a feldspathoid for purposes of the above definition.

The modal rather than normative composition provides the definitive criteria, so that alkali olivine basalts, for instance, containing normative but not modal nepheline, are excluded. However, sometimes these rocks contain such a high proportion of normative nepheline that it must be present as an occult phase in the glass, or has not been recognized, probably because of the fine grain size. A few highly potassic rocks are included although they would not meet the general criteria.

Carbonatites are taken to be igneous rocks containing >50% modal carbonate (Le Maitre *et al.*, 1989), although it is appreciated that some petrologists, particularly in Russia, consider some, if not all, carbonatites to be metasomatic in origin.

Kimberlites are not included, partly because they are considered to have been dealt with elsewhere (e.g. Dawson, 1980; Mitchell, 1986), and partly because kimberlite pipes do not generally lend themselves to the approach adopted here; further, they are neither alkaline rocks, nor carbonatites, although sometimes comprising part of an alkaline rock or carbonatite association.

ORGANIZATION OF THE CATALOGUE

The catalogue is to consist of four parts:

1 North and South America (including Greenland)
2 Former U.S.S.R.
3 Africa
4 Asia and Europe (excluding the former U.S.S.R.), Australasia, Antarctica and oceanic islands.

Parts 1, 3 and 4 are arranged alphabetically by country but part 2, being devoted solely to the huge area encompassed by the former Soviet Union, has been divided into 23 provinces which are treated individually.

Each section begins with a numbered list of occurrences, which is keyed to an accompanying locality map. The order of the descriptions in each province is geographical with, on the whole, and arbitrarily, the more northerly and westerly occurrences being described first. This simple system is preferred to a genetic one, such as that used by Currie (1976) in his account of Canadian alkaline rocks, because of the difficulty of searching for individual occurrences in such a system and, when there are a large number, locating them on the distribution maps, while the ambiguities in the classification of many occurrences raises further problems.

Occurrence number

Each occurrence has been assigned a province number. For all the other countries covered in this series of books a single national numbering system has been devised. However, for the present volume it was decided that the large number of occurrences in the former U.S.S.R. would make such a numbering system too unwieldy, so separate number sequences are used for each province.

Occurrence name and problems of transliteration from Russian

The preferred names of occurrences are given and these are followed, in brackets, by any synonyms. However, the transliteration from Russian into English of occurrence names presents problems because of the inflected nature of the Russian language, so that endings vary depending on whether the word is used as a noun or adjective. Often the adjectival ending 'skii' has been retained, but this can usually be omitted. Fortunately, this does not generally cause problems when searching the index.

Although occurrence names are often shortened in Russian by omitting the ending, sometimes a shortened form is made by omission of initial syllables. For example, the Yuzhnosakunskii complex is commonly referred to as Sakun and Malomurunskii as Murun. Shortening of names by omission of initial syllables clearly presents problems for anyone searching the index for a particular occurrence, if they are not familiar with all variants. An attempt has been made to alleviate this problem by including shortened forms as synonyms, so that they are cross-indexed.

There are a number of systems for transliterating Russian into English. The 'British System' (British Standard 2979) has been used throughout this book and is illustrated in Table 1. The differences between the British System and that adopted in a number of American publications has caused problems with some references. For instance, the Russian author 'Egorov' is so transliterated using the British System, but his name is given as 'Yegorov' in some journals. In the alphabetical refer-

Table 1 System of transliteration from Russian to English used for locality names

Cyrillic	English	Cyrillic	English
Аа	a	Рр	r
Бб	b	Сс	s
Вв	v	Тт	t
Гг	g	Уу	u
Дд	d	Фф	f
Ее	e	Хх	kh
Ёё	e	Цц	ts
Жж	zh	Уу	ch
Зз	z	Шш	sh
Ии	i	Щщ	shch
Йй	i	Ъъ	'
Кк	k	Ыы	y
Лл	l	Ьь	'
Мм	m	Ээ	e
Нн	n	Юю	yu
Оо	o	Яя	ya
Пп	p		

ence lists all his publications are collected under Egorov, even if in some of the cited references his name is given as Yegorov.

Geographical coordinates

When work on this volume commenced, which was before the major political changes that led to the demise of the U.S.S.R., it was feared that it would not be possible to give geographical coordinates for each occurrence, such information being restricted. However, political changes have led to the easing of such constraints so allowing this information to be given, probably for the first time.

The coordinates are quoted to the nearest minute for the approximate centre of the occurrence, but difficulties have been encountered in determining coordinates for some occurrences, so that they are not always as accurate as we would wish. A range of coordinates is quoted for particularly large occurrences, usually volcanic fields or swarms of dykes or diatremes. Some of these are indicated on the location maps by a numbered rectangle rather than a dot.

Occurrence descriptions

These vary greatly in length and have been tailored as far as possible to reflect the complexity and importance of an occurrence. In some instances, however, very little is known so that the entries contain the only few facts generally available. On the other hand many occurrences are so extensive and complex that the descriptions will only be adequate to give a flavour of the geology. The general aim has been to present a description which is sufficiently full for the overall nature of the occurrence to be appreciated. The emphasis in the descriptions has been given to the main rock types represented, and the form of the occurrence, although often in central intrusive complexes in particular little is known of their three-dimensional geometry. The petrography has been stressed as being the most fundamental characteristic, with a brief outline of the mineralogy, including noteworthy accessory constituents.

Although it would be ideal for the rock names used throughout the catalogue to be consistent, this would be impossible to accomplish in the absence of thin sections of rocks from all the localities cited. In view of this the rock names used are generally those adopted by the authors whose work has been consulted, while general terms such as 'nepheline syenite' and 'ijolite' have been freely applied. However, the relatively full petrographic descriptions which are often given should allow the reader to decide for himself the nature of the rocks described. It is unfortunate that for some occurrences the presence of 'pyroxene' or 'amphibole' is recorded but not its nature, with the result that it is difficult in these cases to decide whether or not the rock is peralkaline.

There are many rock names used in the Russian literature which may be unfamiliar to non-Russian readers, e.g. sviatonossite, synnyrite, but definitions of most of these will be found in the 'Glossary of Terms' in the IUGS 'blue book' (Le Maitre et al., 1989) and Sorensen (1974). For Russian readers the book by Efremova and Stafeev (1985) gives a comprehensive account of the classification system for igneous rocks adopted by the Terminological Commission of the U.S.S.R.

More specialized work, such as mineralogical and petrochemical investigations, trace element, isotope and fluid inclusion studies, are often cited in the occurrence descriptions, but in the former U.S.S.R., as elsewhere, it is apparent how such detailed work has generally been concentrated on a relatively few localities.

Economic notes

Notes on economic aspects of occurrences are given, whether to potential deposits or mineral concentrations being worked at present. It was feared that we would not be able to include information of this kind but again the change in political climate has eased restrictions in this area. However, information such as production statistics, tenors of ore and determined ore reserves is still very difficult to come by for individual occurrences in the former U.S.S.R. and we have only occasionally been able to include such notes, although it is believed that the economic information given, though often sparse, is the fullest so far compiled.

Ages

Geological ages and methods used in obtaining them are quoted when these are available. If more than one method has been used then this is indicated and the results quoted.

References

The references cited are those used for compiling the entries, with preference generally having been given to more recent work. Inevitably in this volume an overwhelming proportion of the primary literature is in Russian, with a little in Ukrainian and other languages. Although some of this literature will be available through a good library, much of it will not. The principal reason for this is that many of the accounts of Soviet geology were published as monographs by the official Soviet publishing houses, the various academies of science and some universities, and the print runs were generally rather short – perhaps 500. Even many Soviet institutions were unable to obtain copies so that some of these works are very rare indeed outside the former U.S.S.R.

The Russian co-authors of this book, working in large geological institutes in Moscow (IGEM and the Vernadsky Institute) and St Petersburg (VSEGEI), were fortunate in having access to large libraries, but also, having worked on alkaline rocks and carbonatites for many years, have large literature collections of their own. Further, between them they have visited, researched upon and mapped many of the occurrences described here and so have been able to give accounts of many of them from first-hand knowledge.

In addition to the primary Russian literature there is a large body of material translated into English. Many papers in *Doklady Akademii Nauk SSSR* are published in English in *Doklady Earth Science Sections*, while material from *Geokhimiya* is given in translation in *Geochemistry International*. Translations from a wide spectrum of journals will be found in *International Geology Review*. When possible the English translation of a paper has been cited rather than the original Russian one. A search has been made of the available translated material for the last 20 years or so and numerous references to this are given. However, the greater proportion of the literature is available only in Russian.

A few important books have been translated into English, notably 'The Lovozero Alkali Massif' (Vlasov et al., 1966) and 'The geochemistry of the Lovozero Alkaline Massif' (Gerasimovsky et al., 1968). The book 'Principal provinces and formations of alkaline rocks'

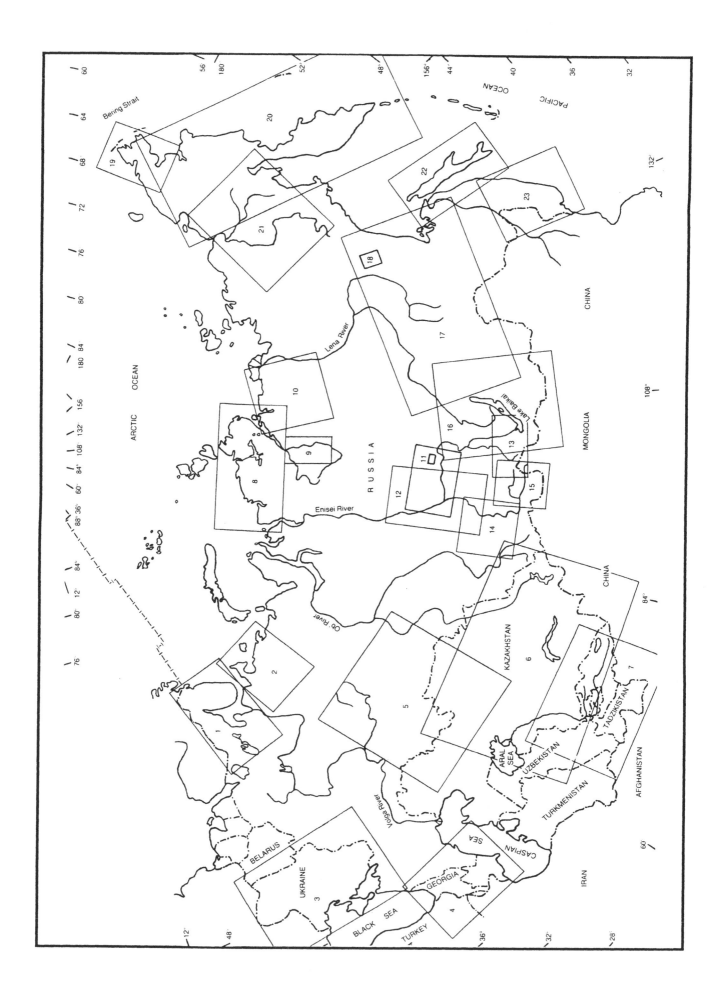

Fig. 1. Map of the former U.S.S.R. showing the distribution of the provinces of alkaline rocks and carbonatites. The numbered outlined areas correspond to the principal locality maps for each of the 23 provinces. The provinces, and numbers of distribution maps, are as follows:

1 Kola and Karelia (Fig. 2)
2 Kanin-Timan (Fig. 38)
3 Ukraine (Fig. 41)
4 Caucasus (Armenia, Azerbaïan, Georgia) (Fig. 50)
5 Urals (Fig. 54)
6 Kazakhstan (Fig. 60)
7 Central Asia (Uzbekistan, Kirgystan, Tadzikistan) (Fig. 89)
8 Taimyr (Fig. 105)
9 Maimecha-Kotui (Figs 109 and 110)
10 Anabar (Figs 105 and 131)
11 Chadobetskaya (Fig. 140)
12 Enisei (Fig. 142)
13 East Sayan (Fig. 146)
14 Kuznetsk-Minusinsk (Fig. 152)
15 East Tuva (Fig. 168)
16 Baikal (Fig. 181)
17 Aldan (Fig. 220)
18 Sette-Daban (Fig. 220)
19 Chukotka (Fig. 241)
20 Kamchatka-Anadyr' (Fig. 243)
21 Omolon (Fig. 246)
22 Sakhalin (Fig. 248)
23 Primorye (Fig. 250).

(Borodin, 1974), which describes numerous occurrences of alkaline rocks in the U.S.S.R., and elsewhere, is probably the only previously published book adopting a similar treatment to the present volumes and as such was of great importance when formulating the present work. Because no English translation was available the Natural History Museum, London, generously financed translation of the U.S.S.R. section of the book by Mrs H. Szabo. Although fewer than half the occurrences now described are to be found in Borodin's book, his descriptions were a great help to the British co-author.

The references in English, in both the author lists at the end of the occurrence descriptions and the full reference lists (to be found at the end of each section), are indicated by an initial asterisk.

Geological and locality maps

These volumes are characterised by the numerous, but generally simplified, geological maps they contain. This approach reflects the conviction that a map, however simple, gives a much better idea of the nature of an occurrence than an equivalent amount of text. There are maps for about two-thirds of the occurrences cited.

For the majority of maps in Russian publications the key to the shading is given in the caption. This style makes for difficult perusal of even the simplest maps and was not considered acceptable. The more normal system of text adjacent to shaded boxes on the map has, therefore, been adopted.

Some attempt has been made to standardize shading throughout, but this has not always proved possible because of the range of rock types involved, and because of the difficulties of showing clearly small areas, for which only a limited range of tints are suitable.

Locality maps, on which the occurrences are indicated by their occurrence number, are given for each province and the positions of the provinces indicated on a general locality map for the U.S.S.R. as a whole (Fig. 1). In some areas the close proximity of occurrences precludes them from being shown clearly on the province maps. In these instances a small rectangle and run of numbers is substituted for the usual dot and single number. A larger-scale location map will then be found in the text at the beginning of the numbered sequence, for example in the central areas of the Maimecha-Kotui and Aldan provinces. Some geographical features are indicated on most locality maps as an aid to the reader using other geological or topographical maps.

Locality index

This lists all occurrences cited, including synonyms, as well as localities referred to in the text. Entries are cited by province and locality number as well as by page. The reader may, however, prefer to turn to the relevant province, which is easily located from the Contents page, and to scan the full province list which is given at the beginning of each section.

ALKALINE ROCKS AND CARBONATITES OF THE FORMER U.S.S.R.

When the series of books describing all the occurrences of alkaline rocks and carbonatites of the world was planned it was intended that each of the four volumes should be of approximately the same length. This necessitated one volume being devoted solely to the Soviet Union, because it was thought that about one quarter of occurrences were located in that country. So the reason-

ing behind the present volume was essentially statistical and political rather than geological. After the project was underway, however, major political changes led to the demise of the U.S.S.R. and the birth of some 15 (at the time of writing) independent states. This necessitated a decision as to whether to retain the original plan of covering the geographical area formerly encompassed by the U.S.S.R. and, if this unit was retained, finding a new title for the book.

A good case can be made for treating all the alkaline rocks and carbonatites of the former U.S.S.R. as a whole. Firstly, because the great majority occur within Russia, and so qualify as a national unit anyway, and secondly because almost all the geological research within the U.S.S.R. was written in Russian and undertaken at institutions founded during the Soviet era. The language, in particular, is a strong unifying factor. It was decided, therefore, to retain the original plan, although the organization of the book would differ from the sister volumes, and to re-title the book more appropriately. Because it is probable that there are still political changes to come, titles such as 'Commonwealth of Independent States' were rejected and 'former U.S.S.R.' adopted as being a unit which is fixed for all time.

Provinces

The political changes and birth of a number of newly independent countries caused few problems when defining the individual provinces. This is because the majority are within Russia itself, while the provinces of Ukraine and Kazakhstan lie wholly within those countries and the descriptions can be considered as comprehensive accounts of the alkaline rocks of those states. Only the provinces of Central Asia and the Caucasus extend across the new national boundaries. The alkaline rocks of the Central Asia province include occurrences in Uzbekistan, Tadzikistan and Kirgystan while the Caucasus province includes alkaline rocks in Armenia, Azerbaïan and Georgia. The decision was taken to retain these two traditional alkaline provinces, and not to subdivide them into six national groups.

In this volume 23 provinces have been defined, with the Baikal province further divided into five sub-provinces. In the book edited by Borodin (1974) 27 provinces are listed but the amalgamation or separation of occurrences in many areas must be arbitrary. For instance, in the first draft of this volume the provinces of Chukotka, Kamchatka-Anadyr' and Omolon were amalgamated as 'Northeastern U.S.S.R.', but in general there is a close correlation of the provinces as described in Borodin and in the present work.

History and literature

Probably the first alkaline rocks to be distinguished in Russia were the widespread alkali basalts in the vicinity of Lake Baikal, which were described by Krapotkin and Cherskii in the middle of the 19th century. The earliest reports on the miaskites (biotite-nepheline syenites) of the Urals were published by Karpinskii in 1891, and the mariupolites of the Oktyabr'skii (Mariupol) intrusion, Ukraine were recognized and described by Morozewicz in 1902.

The unique Lovozero alkaline massif was described for the first time by Ramsay in 1890 and in numerous subsequent publications. He was the first to draw attention to the layered structure of Lovozero (Ramsay and Hackman, 1894), but he thought that the Lovozero and Khibina intrusions were part of a single complex. There was then a lull in the study of these intrusions until 1921 when systematic studies of the mineralogy and petrography were commenced by Fersman, Vorob'eva, Kupletskii, Gerasimovskii, Eleseev and many others. Among the many outstanding achievements of this work was the discovery of the apatite deposits at Khibina and the loparite ores at Lovozero, while the layered structure of the Lovozero massif was thoroughly established.

As a result of intensive mapping after the Second World War dozens of new occurrences and several new provinces were found in the U.S.S.R., notably in Siberia and the Far East. There was a commensurate increase in the number of publications devoted to alkaline rocks, including a summary of data on the alkaline rocks of the whole country (Vorob'eva, 1969). General papers were published on the Cenozoic alkaline basalts of Siberia and the Far East (Belov, 1963; Gapeeva, 1954), on the peralkaline granites of the Transbaikalia area (Arsen'ev, 1946; Nechaeva, 1976) and on the alkaline rocks of the Kola Peninsula (Gerasimovsky *et al.*, 1974), as well as monographs devoted to particularly important massifs such as Lovozero (Vlasov *et al.*, 1966) and Guli (Egorov *et al.*, 1961). There have also been books devoted to particular groups of alkaline rocks, notably those on melilite-bearing rocks (Egorov, 1969), the jacupirangite-urtite series (Kononova, 1976) and lamproites (Bogatikov *et al.*, 1991).

Apart from the general account of Vorob'eva (1969) the only other publication containing a comprehensive account of the alkaline rocks and carbonatites of the former U.S.S.R. is that of Borodin (1974), referred to previously, a partial translation of which was made available to the British co-author. Borodin's book has a number of similarities to the present one, including numerous maps and full accounts arranged by province. A general book devoted to the alkaline rocks, and covering all aspects, has been edited by Kononova (1984). Of inestimable value is the map of alkaline rocks, carbonatites and kimberlites of the U.S.S.R. compiled and edited by Orlova and Krasnov (1976) which shows not only the distribution of occurrences but also classifies them according to rock association.

Carbonatite problems aroused interest slightly later than in the West with the reviews published by Borodin (1957), Kukharenko (1958) and Ginsburgh *et al.* (1958) representing the beginning of carbonatite studies in the U.S.S.R. By the 1970s some 50 carbonatite occurrences had been discovered and more than 200 papers in this field published, including substantial monographs by Kukharenko *et al.* (1965) devoted to the Kola Peninsula and by Butakova and Egorov (1962) on the Maimecha-Kotui Province. More recently a series of books devoted to carbonatites from all over the former U.S.S.R. have been published by scientists of the Geological Survey (VSEGEI) and the Academy of Sciences. Kapustin's book 'Mineralogy of carbonatites' was published in 1971 and translated into English in 1980. This remains the only book devoted to the subject.

There have been very few descriptions of carbonatites from the former U.S.S.R. published in English, apart from that of Gittins (1966a). However, Gittins (1966a, p. 501) noted in his summaries and bibliographies of carbonatites that 'This section dealing with the U.S.S.R. is the least satisfactory. It is very difficult to obtain exact locations of most of the complexes, and the relative inaccessibility of some of the literature makes it difficult to check individual occurrences.' Nevertheless, he compiled a useful review, including descriptions of seven Kola Peninsula occurrences, and cited the most important literature for the Urals, Maimecha-Kotui, Tuva and Sayan provinces.

Although the views of Russian petrologists on carbonatite genesis and those of western petrologists are nowadays generally similar, this was not always the case. In the first few decades after the Second World War the Russian position was very much a metasomatic one, and this encompassed the generation of the silicate alkaline rocks as well as carbonatites. A number of stages of metasomatism were recognized leading on to four stages in the evolution of carbonatite, namely (1) early calcite (2) late calcite (3) calcite-dolomite and (4) ankerite and ankerite-siderite. This approach continues to be influential and many Russian papers on carbonatites identify early and late varieties and recognize certain minerals as characteristic of each stage. Although similar stages are sometimes identified in western carbonatite descriptions, these are generally based solely on cross-cutting field relationships and may not correspond strictly with the stages defined in the Russian literature. Gittins (1966b) gives a useful review of this aspect of Russian carbonatite work.

Economic aspects

The alkaline rocks and carbonatites contain economic and potentially economic deposits of Fe, Nb, REE, Cu, phosphate, vermiculite, nepheline and many others elements and minerals. Although there are numerous geologically and mineralogically orientated papers on these deposits, there is no full, general account for the U.S.S.R. as a whole and it is still difficult, if not impossible, to obtain economic statistics.

There are a number of economic occurrences associated with alkaline and carbonatite intrusions in the former U.S.S.R. that are unique, perhaps the best known being the Khibina apatite deposits, easily the largest source of igneous phosphate in the world. Nepheline is an important commodity and is used not only in the glass industry, as in Canada and Norway, but also for alumina production. There are two industrial centres for alumina production from nepheline: one is near St Petersburg, which uses nepheline from the Khibina apatite-nepheline ore, and the second is in Siberia, using urtite from the Kiya-Shaltyr massif in the Kuznetsk-Minusinsk province. The problems of nepheline ores have been studied in several areas, for example in the Sayan and Baikal provinces (Konev, 1982), and a general account of these rocks is that of Petrov (1978).

As well as the apatite mines at Khibina, other complexes on the Kola Peninsula have been exploited economically. For instance, magnetite, mica, including vermiculite, apatite and baddeleyite are won from the Kovdor complex, while loparite is mined from Lovozero for niobium. Unfortunately, the vast size of the country and the remoteness of many of the alkaline provinces precludes many potential deposits from being exploited. The Maimecha-Kotui province presents a huge economic potential which is unlikely to be realized in the short term because of its geography. Similarly, overlying the Tomtor carbonatite complex in the Anabar province is a regolith rich in vermiculite, francolite, kaolinite and pyrochlore as well as Sc and REE, while the synnyrites of the Synnyr pluton in northern Baikal comprise a huge potential ore body for the production of alumina and potassium. The unique Seligdar occurrence in the Aldan represents a large apatite deposit which could be exploited, because of the proximity of the Baikal–Amur railway, but, although much preliminary work has been undertaken, it remains untouched, as yet.

An interesting, and unique, deposit is the charoite of the Malomurunskii complex in the Aldan. The relatively high value of this attractive blue decorative stone has allowed commercial exploitation in spite of the isolation of the occurrence. It is likely that other similar small, but potentially lucrative, deposits will be exploited in the future.

Although political changes have forced the adoption of strict new economic criteria by the mining industry of the former U.S.S.R., the enormous number of deposits located in the alkaline and carbonatitic occurrences will probably in the longer term lead to the commercial exploitation of more of them.

References

ARSEN'EV, A.A. 1946. The problem of peralkaline granite from the Transbaikal area. *Izvestiya Akademii Nauk SSSR*,3: 125–7.

BELOV, I.V. 1963. *A trachybasaltic association from the Transbaikal area*. Izdatelstvo Akademii Nauk SSSR, Moscow. 371 pp.

BOGATIKOV, O.A., RYABCHIKOV, I.D. and KONONOVA, V.A. *et al*. 1991. *Lamproites*. Nauka, Moscow. 302 pp.

BORODIN, L.S. 1957. On the type of carbonatitic deposits and their relationship with the massifs of alkaline ultrabasic rocks. *Izvestiya Akademii Nauk SSSR*, 5: 3–16.

BORODIN, L.S. (Ed.) 1974. *Principal provinces and formations of alkaline rocks*. Nauka, Moscow. 376 pp.

BUTAKOVA, E.L. and EGOROV, L.S. 1962. The Maimecha-Kotui formation of alkaline and ultrabasic rocks. *In Petrography of Eastern Siberia*, 1: 417–587. Izdatelstvo Akademii Nauk SSSR.

CURRIE, K.L. 1976. The alkaline rocks of Canada. *Bulletin, Geological Survey of Canada*, 239: 1–228.

DALY, R.A. 1933. *Igneous rocks and the depths of the Earth*. McGraw-Hill, New York. 598 pp.

DAWSON, J.B. 1980. *Kimberlites and their xenoliths*. Springer-Verlag, Berlin. 252 pp.

DEANS, T. 1966. Economic mineralogy of African carbonatites. *In* O.F. Tuttle and J. Gittins (eds), *Carbonatites*. 385–413. Interscience Publishers, New York.

DEANS, T. 1978. Mineral production from carbonatite complexes: a world review. *Proceedings of the First International Symposium on Carbonatites, Pocos de Caldas*: 123–33.

EFREMOVA, S.V. and STAFEEV, K.G. 1985. *Petrochemical methods of studying rocks: handbook*. Nedra, Moscow. 511 pp.

EGOROV, L.S. 1969. *The melilite rocks of the Maimecha-Kotui province*. Nedra, Leningrad. 247 pp.

EGOROV, L.S., GOLDBURT, T.L., SHIKORINA, K.M., EPSTEIN, E.M., ANIKEEVA, L.I. and MIKHAOLOVA, A.F. 1961. *Geology and petrology of magmatic rocks of the Guli intrusion*. State Publishing House of Literature for Mining (Gosgortekhizdat), Moscow. 272 pp.

GAPEEVA, G.M. 1954. On the occurrence of ankaramite on the territory of the U.S.S.R. (Data on the petrology of the Primorye alkaline magmatic province). *Doklady Akademii Nauk SSSR*, 9: 155–6.

GERASIMOVSKY, V.I., VOLKOV, V.P., KOGARKO, L.N. and POLYAKOV, A.I. 1974. Kola Peninsula. *In* H. Sorensen (Ed.), *The alkaline rocks*. 206–21. John Wiley, London.

GERASIMOVSKY, V.K., VOLKOV, V.P., KOGARKO, L.N., POLYAKOV, A.I., SAPRYKINA, T.V. and BALASHOV, Yu.A. 1968. *The geochemistry of the Lovozero alkaline massif. Part 1: Geology and petrology. Part 2: Geochemistry*. Translated by D.A. Brown. Australian National University Press, Canberra. 224 and 369 pp.

GINSBURG, A.I., LAVRENEV, Yu.B., NECHAEVA, E.A. and POZHARITSKAYA, L.K. 1958. Rare-metal carbonatites. *In* A.I. Ginsburg (Ed.), *Geology of the rare-element deposits*, 1: 19. Gosgeoltekhizdat, Moscow.

GITTINS, J. 1966a. Summaries and bibliographies of carbonatite complexes. *In* O.F. Tuttle and J. Gittins (eds), *Carbonatites*. 417–541. Interscience Publishers, New York.

GITTINS, J. 1966b. Russian views on the origin of carbonatite complexes. *In* O.F. Tuttle and J. Gittins (eds), *Carbonatites*. 379–82. Interscience Publishers, New York.

HEINRICH, E.W. 1966. *The geology of carbonatites*. Rand McNally, Chicago, 608 pp.

KAPUSTIN, Yu.L. 1980. *Mineralogy of carbonatites* (translated from Russian). Amerind Publishing, New Delhi. 259 pp.

KONONOVA, V.A. 1976. *The jacupirangite-urtite series of alkaline rocks*. Nauka, Moscow. 214 pp.

KONONOVA, V.A. (Ed.) 1984. *Alkaline rocks*. Nauka, Moscow. 416 pp.

KUKHARENKO, A.A. 1958. The Palaeozoic complex of the ultrabasic and alkaline rocks from the Kola Peninsula and related rare-metal deposits. *Zapiski Vsesoyuznogo Mineralogicheskogo Obshchestva, Moskva*, 87: 304–14.

KUKHARENKO, A.A., ORLOVA, M.P., BULAKH, A.G., BAGDASAROV, E.A., RIMSKAYA-KORSAKOVA, O.M., NEPHEDOV, E.I., IL'INSKII, G.A., SERGEEV, A.S. and ABAKUMOVA, N.B. 1965. *The Caledonian complex of ultrabasic alkaline rocks and carbonatites of the Kola Peninsula and north Karelia*. Nedra, Moscow. 772 pp.

LE BAS, M.J. 1977. *Carbonatite-nephelinite volcanism*. John Wiley, New York. 347 pp.

LE MAITRE, R.W., BATEMAN, P., DUDEK, A., KELLER, J., LAMEYRE, J., LE BAS, M.J., SABINE, P.A., SCHMID, R., SORENSEN, H., STRECKEISEN, A., WOOLLEY, A.R. and ZANETTIN, B. 1989. *A classification of igneous rocks and glossary of terms. Recommendations of the International Union of Geological Sciences Subcommission on the Systematics of Igneous Rocks*. Blackwell, Oxford. 193 pp.

MOROZEWICZ, J. 1902. Ueber Mariupolit, ein extremes Glied der Elaeolithsyenite. *Tschermaks Mineralogische und Petrographische Mitteilungen. Wien*. 21 (2 serie): 238–46.

NECHAEVA, I.A. 1976. *Peralkaline granites and their associations*. Nauka, Moscow. 147 pp.

ORLOVA, M.P. and KRASNOV, V.I. 1976. *Map of the distribution and mineral specialisation of alkaline magmatic formations of the territory of the USSR*. 1:10,000,000. The All-Union Geological Research Institute (VSEGEI), Leningrad.

PETROV, V.P. (Ed.) 1978. *Nepheline raw material*. Nauka, Moscow. 190 pp.

RAMSAY, W. 1890. Geologische Beobachtungen auf der Halbinsel Kola. Nebst einem Anhange. Petrographische Beschreibung der Gesteine des Lujavrurt. *Fennia*, 3: 1–52.

RAMSAY, W. and HACKMAN, V. 1894. Das Nephelinsyenitgebiet auf der Halbinsel Kola. I. *Fennia. Bulletin de la Societé de Geographie de Finlande. Helsingfors*, 11 (2): 1–225.

SORENSEN, H. (Ed.) 1974. *The alkaline rocks*. John Wiley, London. 622 pp.

TUTTLE, O.F. and GITTINS, J. 1966. *Carbonatites*. John Wiley, New York. 591 pp.

VLASOV, K.A., KUZ'MENKO, M.Z. and ES'KOVA, E.M. 1966. *The Lovozero alkali massif*. Translated by D.G. Fry and K. Syers. Oliver & Boyd, Edinburgh and London. 627 pp.

VOROB'EVA, O.A. 1969. Principal features of the distribution and origin of the alkaline rocks. *In* F.V. Chukrov (Ed.), *Problems of the geology of mineral deposits, petrology and mineralogy* 2: 62–81. Nauka, Moscow.

Acknowledgements

Lia Kogarko and Maya Orlova would like to acknowledge the financial assistance of the Russian Federation Foundation (RFF93058463) and an INTAS grant (1010CT930018) during the present work and Lia Kogarko also benefitted from a SOROS grant (MBY000). The financial support of IGEM and the Director of VSEGEI, Academician D.A. Scheglov, during their work on the project, are acknowledged by Victoria Kononova and Maya Orlova.

The authors were helped by many specialists who they would like to acknowledge. In particular we thank L. Borodin for general discussion about our book, which can be considered a successor to his own compilation. The late L. Egorov put his unrivalled knowledge of the Maimecha-Kotui province at our disposal and provided two unpublished maps. S. Krivdik helped considerably by providing a manuscript copy of his then unpublished book on alkaline rocks of the Ukraine. Information on the literature of Armenian alkaline rocks was generously provided by R. Gevorkyan who also checked our MS on these rocks. Similarly, G. Zakaziadze helped with information on the alkaline rocks of Georgia. V. Dustmatov is thanked for directing us to materials on the alkaline rocks of Tadzikistan which otherwise we should have missed. A. Zeitsev kindly provided additional information on some Kola occurrences and read through our account of this province.

We thank Engineer S.E. Smirnova, who drew the maps, and Sally Alexander, who added the text to the locality and geological maps.

There is no definitive way of presenting Russian references in English, but Alan Woolley is most grateful to Zara Frenkiel for making numerous suggestions for improving the presentation and consistency of the references; she was also most helpful with problems of transliteration from Russian. The late Mrs H.P. Sabo made an invaluable translation into English of the USSR section of the book 'Principal provinces and formations of alkaline rocks' (Borodin, 1974). Librarians Eileen Brunton, Ann Lum and Helen Santler at The Natural History Museum, London, cheerfully solved many questions of a bibliographic nature.

ARW must thank his Russian co-authors for submitting most cheerfully to a continuous barrage of questions and requests for more information which was often difficult, if not impossible, to obtain. Not only has he learned during this work a tremendous amount about the alkaline rocks and carbonatites of the former USSR, but he has come to appreciate the wealth of knowledge and experience embodied in his three Russian co-authors.

Lia Kogarko would like to point out that the aims of this book correspond with those of IGCP Project 314 (Alkaline and carbonatitic magmatism).

KOLA AND KARELIA

The Kola and Karelia alkaline province is situated in the eastern part of the Baltic shield and occupies the Kola Peninsula and northern part of Karelia (Fig. 2). In the centre of the peninsula are the large layered, agpaitic complexes of Khibina and Lovozero which have several unique features including, in the former, huge apatite deposits. Lovozero was first described in 1887 by Ramsay, who recognized the igneous layering, although he considered that Lovozero and Khibina were but a single complex. However, it was detailed work in the 1920s by numerous petrologists that revealed the extraordinary nature of these two intrusions. Apart from the apatite mined from Khibina, Lovozero has been worked for loparite and the Kovdor complex is mined for magnetite, apatite, baddeleyite and mica.

A number of general monographs describing the Kola alkaline rocks and carbonatites have been published, the most detailed being that of Kukharenko *et al.* (1965), for parts of which there are translations into English by Brown (1970 and 1980). Other general accounts include those of Semenov (1963), Kukharenko *et al.* (1971) and Kogarko (1977), while monographs devoted to individual occurrences are particularly numerous and are detailed in the descriptions that follow. There are brief reviews in English on the province by Gerasimovsky *et al.* (1974) and Kogarko (1987).

The oldest massifs are the Proterozoic alkaline-gabbro occurrences of Gremyakha-Vyrmes, Elet'ozerskii and Tikshozerskii, which are associated with a zone of deep, north–south-trending infrastructural faults of long duration. The distribution of middle Palaeozoic alkaline-ultramafic intrusions, including carbonatite and nepheline–syenite occurrences, is associated with the formation of the central Kola palaeorift (aulacogen) and cross-cutting faults of approximately east–west trend. The location of the alkaline occurrences of the Turiy peninsula and the alkaline dykes of Kandalaksha Bay and Archipelago is, according to Staritsky *et al.* (1981), controlled by the development of the long-lived Kandalaksha–Onezh palaeorift.

1	Chagvedaiv	19	Ozernaya Varaka
2	Seblyavr	20	Afrikanda
3	Gremyakha-Vyrmes	21	Lesnaya Varaka
4	Ivanovskii	22	Salmagorskii
5	Nizyavrskii, Iokan'skii	23	Ogorodnyi
	and Koyutyngskii	24	Kanozerskii
6	Pachinskii	25	Turiy Peninsula
7	Ponoiskii	26	Kandalakshskii dykes
8	Strel'ninskii and	27	Elovyi Island
	Purnachskii	28	Kandagubskii
9	Pesochnyi	29	Mavragubskii
10	Lavrent'evskii	30	Kovdor
11	White Tundra	31	Sallanlatvi
12	Sakhariokskii	32	Vuoriyarvi
13	Western Keiv	33	Kovdozerskii
14	Kontozerskii	34	Tikshozerskii
15	Kurginskii	35	Elet'ozerskii
16	Lovozero	36	Kostomukshskii dyke
17	Khibina		complex
18	Soustova	37	Elisenvaarskii

Fig. 2. *Distribution of alkaline igneous rocks and carbonatites in the Kola-Karelia province.*

1 CHAGVEDAIV
69°22'N; 32°45'E
Fig. 3

The Chagvedaiv intrusion is situated in Archaean biotite gneisses and takes the form of a stock with an area of 1 km². Eliseev (1958a) considers it to be a multiple intrusion in which albite syenites were emplaced as the first phase, followed by quartz nordmarkites then granosyenites with the last phases pegmatites and quartz veins. Aegirine and riebeckite are present in some rocks.

References Batieva *et al.*, 1985; Eliseev, 1958a; Polkanov, 1938.

2 SEBLYAVR
68°43'N; 32°08'E
Fig. 4

The complex cuts and fenitizes Archaean gneisses and is a downward tapering intrusion of 4 × 5 km. Clinopyroxenites are abundant with a gradual transition to nepheline pyroxenites at the margins of the complex. The pyroxenites consist of diopside–augite, perovskite, titanomagnetite and phlogopite with subordinate amphibole, chlorite, calcite, actinolite and apatite. Amongst the pyroxenites are blocks 10–150 m in diameter of

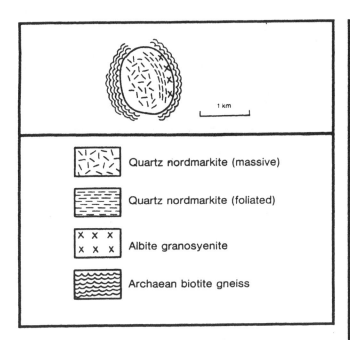

Quartz nordmarkite (massive)

Quartz nordmarkite (foliated)

Albite granosyenite

Archaean biotite gneiss

Fig. 3. Chagvedaiv (after Polkanov, 1938, Fig. 1).

olivinites comprising olivine (Fo_{90}), perovskite and titanomagnetite with minor phlogopite, calcite, serpentine,
clinohumite and tremolite. The olivine blocks have
marginal zones of olivine–pyroxene rocks. Schlieren of
perovskite–titanomagnetite ores are associated with the
ultramafic rocks which are also crossed by numerous
ijolite veins consisting of nepheline and aegirine–augite
with lesser amounts of hastingite, biotite, schorlomite,
titanite, magnetite, pyrite and chalcopyrite. The most
abundant veins are of carbonatite and apatite-rich rocks
(Lapin, 1979), which are arcuate and mainly concentrated in the central part of the complex and define a conical, layered body dipping inwards at 30–60°. They cut
the silicate rocks. The carbonatite veins include
dolomite–phlogopite types, which are massively textured and contain titanomagnetite and perovskite
and subordinate ilmenite, apatite, zirconolite and baddeleyite; dolomite–amphibole carbonatite veins include
richterite, tremolite, titanomagnetite and hedenbergite
with minor phlogopite, apatite, zircon and pyrochlore.
Bulakh and Ivannikov (1984) distinguish an older series
comprising major calcite, phlogopite, actinolite and titanite with subordinate hedenbergite–diopside, alkaline
amphibole, dolomite, titanomagnetite, iddingsite, serpentine, zirconolite, baddeleyite and zircon and calcite-
dolomite and ankerite carbonatites containing baryte
and strontianite with minor quartz, chlorite, pyrite,
galena, sphalerite, ancylite, pyrochlore and zircon. A
younger series consists of veins and pods of carbonate-
baryte rocks. The apatite-rich veins, the complex mineralogy and mineral and rock chemistry, which is
described in some detail by Lapin (1979), have sharp,
discordant contacts but there are also areas of apatite-
rich rocks which pass gradually into the enclosing rocks.
The apatite-rich rocks, which are referred to as 'cama-
phorites' by Lapin, are phoscorites. Alkaline lamprophyres and alkaline syenites cut all the rocks of the
complex described above. The surface of the complex is
covered by a weathered crust in which apatite and vermiculite are concentrated.

Economic High concentrations of apatite have been
observed in carbonatites, with averages of 10%
(Ivanova, 1968; Subbotin and Michaelis, 1986).

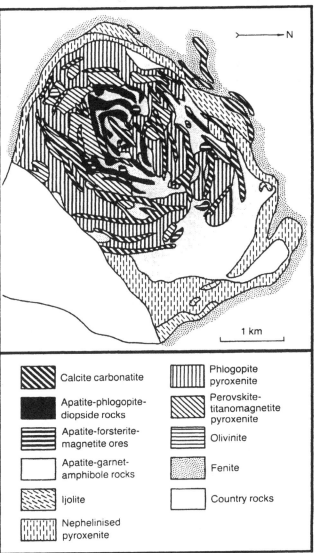

Calcite carbonatite

Apatite-phlogopite-
diopside rocks

Apatite-forsterite-
magnetite ores

Apatite-garnet-
amphibole rocks

Ijolite

Nephelinised
pyroxenite

Phlogopite
pyroxenite

Perovskite-
titanomagnetite
pyroxenite

Olivinite

Fenite

Country rocks

*Fig. 4. Seblyavr (after Subbotin and Michaelis, 1986,
Fig. 1).*

Age K–Ar on phlogopite and biotite from pyroxenite
and ijolite and on phlogopite from carbonatite gave
383–388 Ma (Gerling in Kukharenko *et al.*, 1965).

References Bulakh and Ivannikov, 1984; Ivanova, 1968;
Kukharenko *et al.*, 1965; *Lapin, 1979; Subbotin and
Michaelis, 1986.

---

3 GREMYAKHA-VYRMES

68°38'N; 32°28'E
Fig. 5

This layered multiple-phase complex having an area
of about 100 km² is confined to the Archaean Kola
series. The complex is built up in three phases, each
being represented by a layered, spatially distinct group
of rocks. The earliest phase comprises peridotite-
clinopyroxenite, gabbro, akerites and pulaskites, which
make up the larger part of the south of the complex. The
second phase consists of a series of melteigites, ijolites,
urtites and nepheline syenites which are also concentrated in the southern part of the complex. Igneous
activity was completed by the intrusion of a peralkaline,
but more siliceous, magma which formed the large body

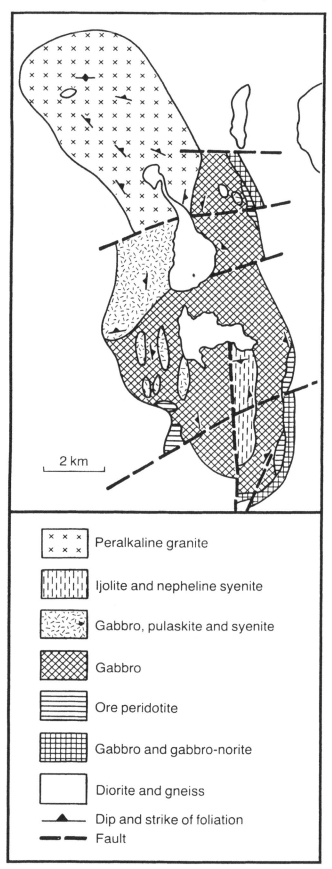

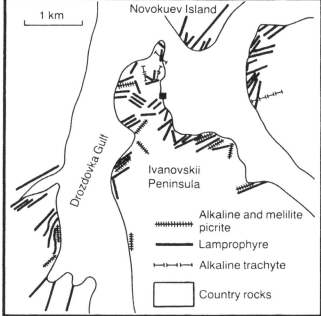

of aegirine granite, nordmarkite and alkaline pulaskite in the northern part of the complex. The melteigite-ijolite–urtite series comprises nepheline, aegirine–augite and biotite with subordinate alkali feldspar, titanite, calcite, magnetite, apatite and zircon. The nepheline-syenite series is somewhat more complex and ranges from foyaite and foyaite pegmatite to peralkaline syenite. The principal minerals are microcline, nepheline, aegirine, albite and minor apatite, calcite, prehnite, pyrochlore and zircon. The peralkaline granites contain arfvedsonite and minor aenigmatite, astrophyllite, chevkinite, orthite and fergusonite. Concentrations of rare earth minerals, zircon and pyrochlore are associated with the ijolite–urtite series, the latter mineralization being of autometasomatic and metasomatic type.

Economic There are ilmenite–titanomagnetite ores in the gabbroic rocks as well as rocks containing high concentrations of apatite (up to 8%) that form layers from 100 to 300 m thick (Kukharenko *et al.*, 1971).

Age K–Ar on amphibole from gabbro gave 1890 Ma and on biotite from alkaline pegmatites 1960 Ma; U–Pb dating of pyrochlore from aegirinites gave 1870 Ma (Kukharenko *et al.*, 1971). Shanin *et al.* (1967) obtained K–Ar dates on nepheline of 1870 ± 40 and 1820 ± 40 Ma, on biotite 1750 ± 40 Ma and on lepidomelane 1830 ± 40 Ma; all minerals from urtite. Pushkarev *et al.* (1987) obtained 2000 ± 40 Ma by U–Pb on apatite from gabbro.

References Kukharenko *et al.*, 1971; Pushkarev *et al.*, 1987; Shanin *et al.*, 1967.

---

4 IVANOVSKII   68°21′N; 38°28′E
                                      Figs 6 and 7

The Ivanovskii complex is located at the northeastern end of the Kola aulocogen on the northeastern coast of the Kola peninsula in the area of the Ivanovka Gulf. It is represented by several clusters of dykes and

Legend for Fig. 5

× × ×	Peralkaline granite
┃┃┃	Ijolite and nepheline syenite
∴	Gabbro, pulaskite and syenite
⊠	Gabbro
☰	Ore peridotite
▦	Gabbro and gabbro-norite
☐	Diorite and gneiss
▲—	Dip and strike of foliation
━ ━	Fault

Fig. 5. Gremyakha-Vyrmes (after Kukharenko et al., 1971, Fig. 6).

Fig. 6. Alkaline dykes in the vicinity of the Drozdovka Gulf, the northern part of the Ivanovskii Peninsula and the Island of Novokuev (after Rusanov et al., 1989, Fig. 1). Black square indicates location of Fig. 7.

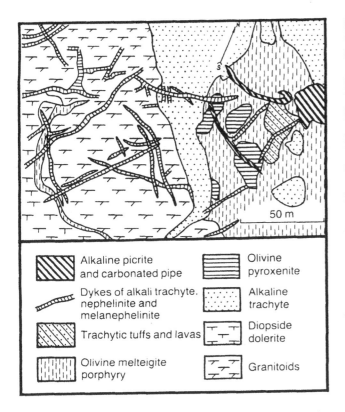

Fig. 7. *Detailed geological map of part of the Ivanovskii Peninsula indicated by black square on Fig. 6. (after Rusanov et al., 1989, Fig. 2).*

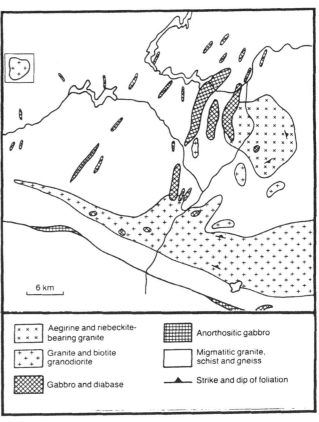

Fig. 8. *Nizyavskii (northwest of centre, in box; intrusion is located 20 km northwest of the map area), Iokan'skii (northeast of centre) and Koyutyngskii (east-west trending; bottom half of map) (after Batieva, 1976, Fig. 4).*

small intrusions as well as tuffs and lavas of alkaline and melilite picrites, lamprophyres, alkaline trachyte, olivine melteigite porphyry, olivine pyroxenite and melteigite. These rocks are concentrated in a northeasterly-trending zone embracing the area of the Drozdovka Gulf, the northeastern extremity of the Ivanovskii Peninsula and the Island of Novokuev. The zone is 6–8 km broad and extends along the deep fault zone of the Kovdor–Kharlov lineament. The extrusive rocks are represented by tuffs, tuffisites, lavas and lava breccias of alkaline trachyte and nepheline trachybasaltoids, which are part of a volcanogenic sedimentary series that is preserved as relics in a linear zone extending over 18 km. The series is subhorizontal and 30–40 m thick. The intrusive rocks comprise small plugs and necks of 0.2–4 km². The largest plug in the area has a circular structure and is composed of olivine melteigite and olivine pyroxenite with zones of eruptive breccia. Small necks of peralkaline trachyte (e.g. 20 × 40 m) and a diatreme (20 × 30 m) of picrite pass gradually into carbonatite. The diatreme fragments are of sandstone, olivine pyroxenite, melteigite porphyry, alkaline trachyte and their tuffs, monchiquite and melanephelinite. The cement of the breccia is of both picritic and carbonatitic composition. Among the dyke series alkaline lamprophyres are the most abundant (77%) with alkaline picrites comprising 13% and alkaline trachytes 10%.

Age The age of the complex has not been determined but the general geology and petrography are similar to those of the Kontozerskii and Turiy complexes and it may be of a similar age.

References Grib *et al.*, 1981; Rusanov *et al.*, 1989.

5 NIZYAVRSKII, IOKAN'SKII and KOYUTYNGSKII

67°55′N; 37°49′E
67°55′N; 38°53′E
67°41′N; 38°24′E
Fig. 8

The outcrops of Nizyarvskii, Iokan'skii and Koyutyngskii are probably all part of a single large body of alkaline granite.

Reference Batieva, 1976.

6 PACHINSKII 67°30′N; 39°24′E

The Pachinskii occurrence lies close to the South Keiv fault, takes a layered form and covers some 120 km². It cuts through Archaean granodiorites, migmatites, gneisses and anorthosites and contains xenoliths of these rocks. It comprises arfvedsonite–aegirine granites.

Reference Batieva, 1976.

7 PONOISKII 67°01′N; 39°12′E
Fig. 9

The Ponoiskii occurrence has an area of about 700 km², only part of which is shown on Fig. 9, and is located in a zone in which northwesterly and northeasterly-trending faults cross. It is formed of a series of circular

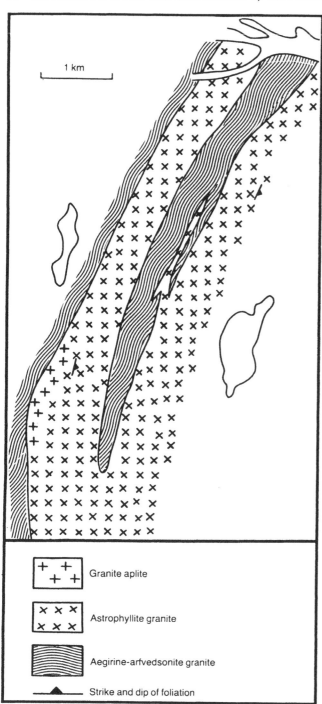

Fig. 9. Ponoiskii (after Batieva, 1976, Fig. 9).

Economic There are concentrations of zircon and rare earth minerals (Batieva, 1976).

Age Pb determinations on zircon gave 2405 Ma (Pushkarev, 1990).

References Batieva, 1976; Batieva *et al.*, 1985; *Ozhogin 1969a and 1969b; Pushkarev, 1990.

---

8 STREL'NINSKII AND PURNACHSKII
66°50′N; 37°52′E
Fig. 10

It is very probable that Strel'ninskii and Purnachskii represent a single body of peralkaline granite. With an area of about 60 km² the intrusion lies in metabasites of Archaean age. It has a concentric, zonal structure, the outermost ring being composed of arfvedsonite–aegirine granites with a clearly defined orientation of the mafic minerals. The middle ring is composed of aegirine-arfvedsonite granites containing aenigmatite and astrophyllite; the mafic minerals are not orientated in this ring. The contacts between the granites of different petrographic facies are sharp, from which it is inferred that they represent separate intrusive phases. The granites sometimes assume a very coarse-grained pegmatitic texture.

Age The lead method on zircon gave 1950 Ma (Pushkarev, 1990).

References Batieva, 1976; Batieva *et al.*, 1985; Pushkarev, 1990.

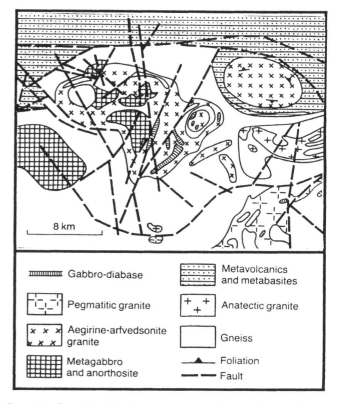

Fig. 10. Strel'ninskii (west of centre) and Purnachskii (oval intrusion in northeast) (after Batieva, 1976, Fig. 11).

ring-dykes and arcuate dykes, situated in the Archaean basement. In the eastern part of the intrusion granites have been emplaced into the nucleus of a dome-shaped anticlinal structure, thus forming a stratiform body. The granites generally contain aegirine and arfvedsonite and are well foliated. The main rock-forming minerals are quartz, microcline, albite, aegirine and alkaline amphibole with the principal accessories zircon, titanite, apatite, monazite and minor astrophyllite and fluorite. A statistical analysis of the rock chemistry of Ponoiskii was undertaken by Ozhogin (1969a) who also studied zircon (Ozhogin, 1969b).

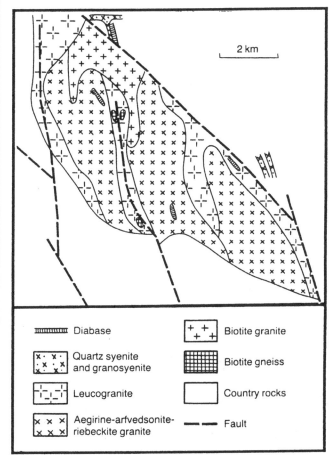

9 PESOCHNYI   67°04′N; 37°07′E
(Kholodnoye Ozero)   Fig. 11

This is an irregularly shaped, poorly exposed massif covering about 6 km². It is intruded into granite gneisses of Archaean age and is composed mainly, in order of decreasing age, of peridotite, pyroxenite, ijolite–urtite, urtite, apatite–amphibole rocks, carbonatites, alkaline and nepheline syenites, augitites and amphibole monchiquites. Ultrabasic rocks compose about 90% of the total area of the complex, of which pyroxenites occupy no less than 80%. The peridotites generally occur in the marginal parts of the intrusion or between pyroxenites which form bodies of irregular form. The contacts between the peridotites and pyroxenites are generally observed to be gradational. The pyroxenites are coarse- and gigantic-grained rocks, which are sometimes enriched in apatite and kaersutite. The alkaline rocks are situated in the central part of the complex and form two separate bodies and some veins. They are represented by ijolite–urtite, urtite and ijolite which almost everywhere are carbonated. The mineralogy of these rocks is aegirine–augite and nepheline with subordinate hastingsite, titanite, apatite, calcite, natrolite and phlogopite. Alkaline pyroxenites are observed among the carbonated alkaline rocks. To the east of the main body of alkaline rocks giant-grained ijolite–urtites and urtites form distinct units. The carbonatites occur as cross-cutting, vein-like bodies with widths of 0.5–1.2 m and as nests and stockworks which, by gradual transition, grade into pyroxenites. The carbonatites consist of

Legend (Fig. 11)

- – – – Monchiquite
- Alkaline and nepheline syenites
- Carbonatite
- Amphibole-calcite and cancrinite-calcite-nepheline rocks
- Apatite-amphibole rocks
- Urtite and ijolite-urtite
- Pyroxenite and olivine pyroxenite
- Peridotite
- Country rocks
- – – – Fault

Fig. 11. Pesochnyi (after Kukharenko et al., 1965, Fig. 3).

Legend (Fig. 12)

- Diabase
- Quartz syenite and granosyenite
- Leucogranite
- Aegirine-arfvedsonite-riebeckite granite
- Biotite granite
- Biotite gneiss
- Country rocks
- – – – Fault

Fig. 12. Lavrent'evskii (after Batieva, 1976, Fig. 10).

calcite, aegirine–diopside and apatite with minor amphibole, titanite, zircon, baryte, ankerite, dolomite, pyrrhotite, galena and francolite. The peridotites, pyroxenites, ijolite–urtites, urtites, carbonated rocks and carbonatites are cut by dykes of nepheline syenite, augitite and amphibole monchiquite.

References Kirnarsky, 1962; Kukharenko *et al.*, 1965.

---

10   LAVRENT'EVSKII 67°16′N; 37°01′E

Fig. 12

Lavrent'evskii has an area of 40 km² and is emplaced in Archaean plagiogranites and gneisses. It comprises two principal phases: an early phase is represented by aegirine–arfvedsonite granites, and a later one mainly by lepidomelane granites. Contacts with the host rocks are intrusive. The mafic minerals define a distinct linear orientation. Extensive apophyses extend into the country rocks and pegmatites containing amazonite, quartz veins and zones of metasomatic alteration are developed throughout the intrusion.

Economic Zircon and rare-earth mineral concentrations occur (Batieva, 1976).

Age The Pb method on zircon gave 2560 Ma (Pushkarev, 1990).

References Batieva, 1976; Batieva *et al.*, 1985; Pushkarev, 1990.

---

11   WHITE TUNDRA 67°28′N; 35°50′E

Fig. 13

The White Tundra occurrence is situated within basic rocks of the Pansk massif and lower Archaean granitoids. The intrusion has an area of 240 km² and is apparently related to a northwesterly-trending fault. The intrusion consists essentially of alkaline granites which have an undoubted intrusive relationship to the country rocks and which contain xenoliths of basic rocks and granodiorite derived from them. The most typical granites are peralkaline, porphyritic rocks with aegirine and arfvedsonite. The main rock-forming minerals are quartz, microcline, albite, aegirine, riebeckite, arfvedsonite, katophorite, aenigmatite and lepidomelane with accessory zircon, titanite, apatite, monazite, fluorite and xenotime.

Economic There are concentrations of rare-earth minerals and zircon (Batieva, 1976).

Age The Pb method on zircon gave 2430 Ma (Pushkarev, 1990).

References Batieva, 1976; Batieva and Bel'kov, 1984.

---

12 SAKHARIOKSKII 67°40′N; 36°12′E

Figs 13 and 14

Sakhariokskii is situated in the central part of the Kola Peninsula along the upper reaches of the Uzkaya River, which is a tributary of the Sakharnaya River. The intrusion is some 6 km in length and 2 km wide and emplaced in Archaean granite and peralkaline granite of Western Keiv (Fig. 13 and locality 13). The southern contact is tectonic and poorly exposed. The rocks of the occurrence are syenites and nepheline syenites and cover about 4 km², of which the nepheline syenites extend over 1.7 km². It has the form of a steeply dipping dyke-like body, the hanging wall of which is composed of nepheline syenite and the footwall of syenite. The contact between the two types of syenite is sharp. According

to Batieva and Bel'kov (1984) the two rock types were not intruded simultaneously but the nepheline syenites were only emplaced after crystallization of the syenites, which they nephelinized. The syenites comprise albite, microcline, biotite and ferrohastingsite with accessory orthite, apatite, fluorite, calcite and zircon. The nepheline syenites vary from leucocratic to melanocratic varieties, the former being composed dominantly of albite, microcline and nepheline and the latter enriched in aegirine, aegirine–augite, amphibole and biotite. Within the nepheline syenites are bands or layers up to tens of metres wide with concentrations of pyroxene, amphibole and biotite. Accessories in the nepheline syenites include pyrochlore, britholite, fluorite, galena and orthite.

Age Pb, U and Th isotope determinations on britholite gave 2185–2060 Ma and on orthite 1910 Ma, while K–Ar determinations on nepheline gave 1740–2260 Ma (Batieva and Bel'kov, 1984). U–Pb dating of syenites gave 1810 ± 50 Ma (Vinogradov *et al.*, 1985).

References Batieva, 1976; Batieva and Bel'kov, 1984; Vinogradov *et al.*, 1985.

---

13 WESTERN KEIV 67°47′N; 36°30′E

Fig. 13

This is the largest granite intrusion on the Kola Peninsula, with an area of 1300 km². The enclosing rocks are Archaean granite gneisses. The intrusion is divided into northern and southern parts by an east–west line of gneisses. The granites contain aegirine and arfvedsonite. The southern part of the massif is characterized by a more complicated structure, there being a large number of dykes. The dykes comprise aegirine–augite–ferrohastingsite granites, granosyenites, quartz syenites and natrolite syenites and these alkaline dykes cut earlier dykes of picrite and diabase. In the alkaline granites of the Western Keiv linear structures are apparent as manifested by orientated dark-coloured minerals and there are huge xenoliths of country rocks. Erosional 'windows' reveal granodiorite, oligoclase granite, plagioclase microcline granite and metagabbro. The main rock-forming minerals of the aegirine–arfvedsonite granites are quartz, microcline, albite, alkaline amphiboles, aegirine and lepidomelane with accessory titanite, zircon, orthite, apatite, fluorite, magnetite and sulphides. The main rock-forming minerals of the granosyenites and quartz syenites are plagioclase (albite, oligoclase), microcline, quartz and aegirine–augite with the accessories titanite, zircon, apatite, fluorite, magnetite, sulphides, chevkinite, astrophyllite, pyrochlore, monazite, britholite and molybdenite.

Economic A zircon and rare-earth mineralization is present and there is a deposit of amazonite which is mined (Batieva *et al.*, 1985).

Age A Sm–Nd isochron gave an age of 2100 ± 50 Ma and the lead method on zircon and titanite 2400 ± 50 Ma (Batieva *et al.*, 1985).

References Batieva, 1976; Batieva *et al.*, 1985.

---

14 KONTOZERSKII (Kontozero) 68°08′N; 36°07′E

Fig. 15

Morphologically the Kontozerskii complex forms a caldera. It lies within the Lovozerskii graben, which is a part of the Kola aulacogen. The caldera has an area of about 55 km² and, as was pointed out by Kirichenko (1970), is a unique structure without an analogue in the

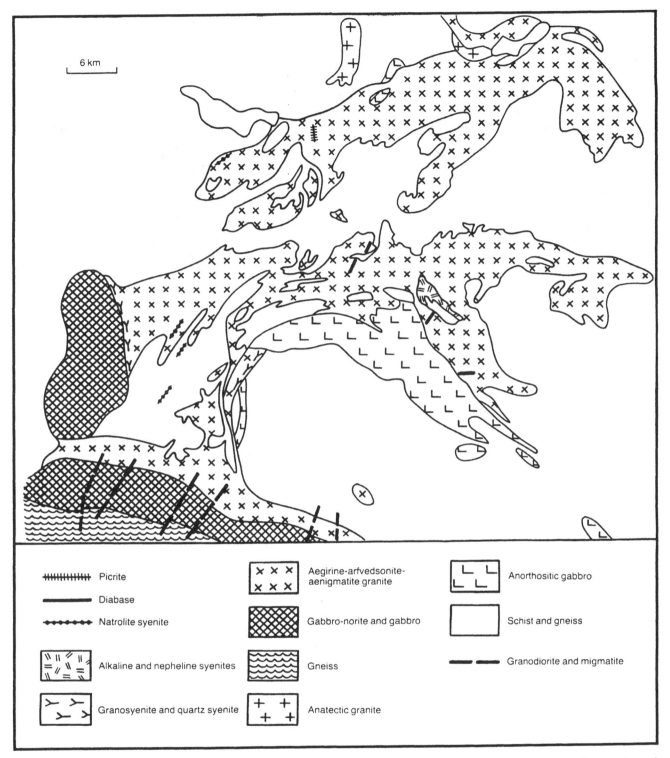

Fig. 13. Western Keiv (north of map), White Tundra (southern part of map) and Sakhariokskii (small intrusion of syenite and nepheline syenite right of centre) (after Batieva, 1976, Fig. 5).

adjacent regions of Russia. The caldera is circumscribed by arcuate fractures along which are breccia zones. The thickness of the sedimentary and volcanogenic rocks preserved within the caldera is over 2 km (Pyatenko and Osokin, 1988) and there are two distinct series, although Pyatenko and Osokin (1988) distinguish three. The first, and lower, one, which can be correlated with the

Devonian rock series of the Lovozero complex, is represented by picrites, alnoites, augitites, nephelinites and limburgites which are interlayered with tuff breccias, tuffs, conglomerates and shales. The second, and upper series, is made up of sheets of lava breccia, lavas and tuffs of melilitite, melanephelinite, melilitic picrites and carbonatites which are intercalated with tuffs, shales,

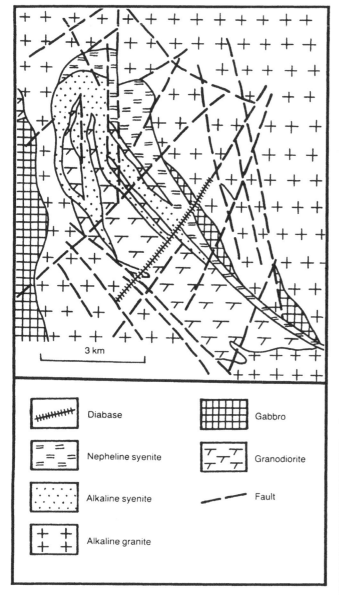

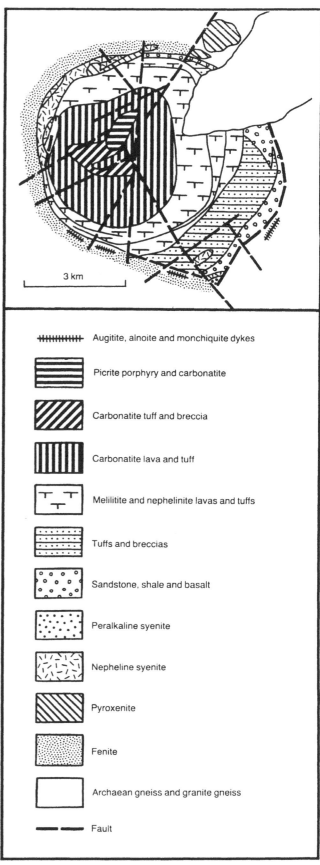

Fig. 14. *Sakhariokskii (after Batieva, 1976, Fig. 12).*

tuffaceous sandstones and limestones. The thickness of individual sheets varies between 0.1 and 0.5 m up to 20 m. About 10% of the succession is composed of carbonatite tuffs and lavas which are intercalated with bedded carbonate tuff siltstones, limestones and dolomites. The picrite–alnoite series are porphyritic rocks with phenocrysts of olivine, pyroxene and biotite set in a groundmass of olivine, augite–diopside, biotite, melilite, titanomagnetite and perovskite. Limburgites have phenocrysts of augite and serpentinized olivine with a groundmass of augite, magnetite, perovskite, serpentine, chlorite and glass. Augite phenocrysts 3–10 mm in diameter occur in the augitites and are set amongst small grains of augite (0.1–1 mm), magnetite, biotite and glass, the last occupying 40–50% by volume; rarely in the groundmass there is analcime and apatite. The nephelinites are composed of nepheline, aegirine–augite, melanite, biotite, an opaque phase and some carbonate. Melilite picrites and melilitites are strongly carbonated rocks containing phenocrysts of altered melilite and olivine in a groundmass of augite aggregates, melilite, apatite, magnetite and carbonated glass. The volcanic

Fig. 15. *Kontozerskii (after Pyatenko and Saprikina, 1981, Fig. 1).*

Legend for Fig. 15:

- Augitite, alnoite and monchiquite dykes
- Picrite porphyry and carbonatite
- Carbonatite tuff and breccia
- Carbonatite lava and tuff
- Melilitite and nephelinite lavas and tuffs
- Tuffs and breccias
- Sandstone, shale and basalt
- Peralkaline syenite
- Nepheline syenite
- Pyroxenite
- Fenite
- Archaean gneiss and granite gneiss
- Fault

Legend for Fig. 14:

- Diabase
- Gabbro
- Nepheline syenite
- Granodiorite
- Alkaline syenite
- Fault
- Alkaline granite

carbonatites are lava breccias, lavas and tuffs with calcite carbonatite predominating (Pyatenko and Saprykina, 1976 and 1981). They contain calcite phenocrysts together with olivine and biotite and fine-grained magnetite and ilmenite. Less common dolomite-calcite carbonatites contain a little albite, zeolite, chlorite and pyrrhotite. Carbonatites also occur in necks, which cut the stratified series, and are formed after picrite breccias, which they replace and cement. Pyatenko and Osokin (1988) give analyses of the minerals of the carbonatites, including rare elements, and chemical data, including rare earths, for numerous rocks. Intrusive rocks are subordinate to volcanogenic sedimentary formations. They extend along the western margin of the Kontozerskii depression, where they form an upland area of 4×0.8 km. The intrusive rocks comprise pyroxenites, melteigites, nepheline syenites and alkaline syenites. The pyroxenites consist of augite–diopside, biotite, titanomagnetite, perovskite and minor melanite, titanite and apatite. The principal minerals in the melteigites are nepheline and zoned pyroxene with some biotite, titanite, apatite, titanomagnetite and amphibole. The nepheline syenites comprise K-feldspar, nepheline, aegirine and sodalite with apatite, analcime, albite, biotite, titanite and alkali amphibole. Finally, the alkaline syenites contain both plagioclase (30–40%) and alkali feldspar (60–70%), chlorite, which has probably replaced aegirine, and accessory zircon. The volcanic and sedimentary rocks adjacent to the contact are fenitized.

Age The sedimentary carbonate rocks were estimated, on the basis of pollen analysis, to be Carboniferous (Pyatenko and Saprykina, 1981).

References Kirichenko, 1970; *Pyatenko and Osokin, 1988; *Pyatenko and Saprykina, 1976 and 1981; Staritsky *et al.*, 1981.

---

15 KURGINSKII
67°58′N; 35°10′E
Fig. 16

Kurginskii is located within the northeastern tectonic zone of the Kola aulocogen within Archaean gneisses of

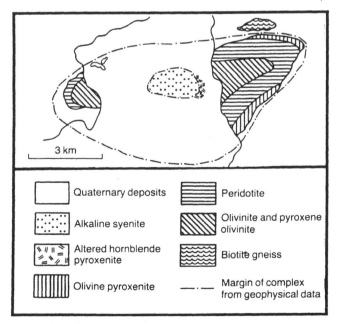

☐ Quaternary deposits	☰ Peridotite
⋮⋮⋮ Alkaline syenite	⧅ Olivinite and pyroxene olivinite
⧄ Altered hornblende pyroxenite	〰 Biotite gneiss
⫿⫿ Olivine pyroxenite	—·— Margin of complex from geophysical data

Fig. 16. Kurginskii (after Proskuryakov and Zak, 1966, Fig. 1).

the Kola series. It has an elliptical shape, is 9 km in length and has an area of about 30 km². The greater part of the intrusion lies beneath the waters of Lovozero Lake. It is a heavily faulted, steeply plunging intrusion of ultramafic rocks which, in the central part, is cut by a stock of alkaline and nepheline syenites (Kukharenko *et al.*, 1971). An alternative interpretation is that it is a downward-focusing, zoned intrusion (Proskuryakov and Zak, 1966). The ultrabasic complex consists, from the centre to the margin, of olivinites, wherlites and clinopyroxenites. The youngest rocks are nepheline syenites, malignites, tinguaites and cross-cutting veins of ultramafic rocks and nepheline syenite.

Age The ultramafic rocks have been dated, using pyroxene, at about 2440 Ma whereas K–Ar on biotite from alkaline syenites gave $405 - 430 \pm 40$ Ma (Kukharenko *et al.*, 1971). These data indicate that Kurginskii was formed in two quite different temporal and compositional episodes (Kukharenko *et al.*, 1971). However, Proskuryakov and Zak (1966) consider this occurrence to be part of the Afrikanda–Kovdor complex.

References Kukharenko *et al.*, 1971; Proskuryakov and Zak, 1966.

---

16 LOVOZERO
67°47′N; 34°45′E
Fig. 17

The Lovozero alkaline complex was first described by Ramsay (1890), his preliminary work laying the foundations for an expedition which between 1890 and 1899 made a detailed study of this complex and the adjacent Khibina intrusion (locality 17). Ramsay thought that the two intrusions were parts of a single complex of which Lovozero comprised the upper portion. He described for the first time the layered structures and was aware of the similar structures in occurrences in Greenland. He considered that the layering was generated by successive intrusion. Work then ceased until resumed by Fersman under whom teams of petrologists and mineralogists in the 1920s undertook systematic research. This and subsequent research, up to 1966, is briefly detailed in the Introduction to the book by Vlasov *et al.* (1966). Lovozero is a notably rich source of minerals and many species have been recorded for the first time from this complex.

The Lovozero complex is located in the central part of the Kola Peninsula. It consists of a mountainous upland with steep precipitous slopes and flat plateau-like summits, elevated almost a kilometre above the surrounding hilly plain. It is located between the two large lakes of Umbozero and Lovozero with Lake Seydozero nestling in the central part of the complex. The massif is rectangular in plan and has an area of 650 km². Lovozero lies in a northwesterly striking tectonic zone within which a sunken, east–west-trending belt of Palaeozoic rocks has been preserved. Within the complex are included areas of steeply dipping granite–gneisses and rare patches of highly aluminous rocks and ferruginous quartzite of Proterozoic age, all of which have a northwesterly strike. The roof of the intrusion is younger than these formations and consists of the Lovozero effusive-sedimentary formation of Middle Palaeozoic age. The rocks of the Lovozero formation occur as isolated xenoliths within the nepheline syenites of the massif.

The Lovozero extrusive rocks comprise picrite porphyries, augite porphyries, essexite porphyries and a suite of alkaline rhomb porphyries, porphyritic phonolites and pseudoleucite porphyries with the more alkaline varieties becoming more predominant with time. The age of these formations is similar to that of the

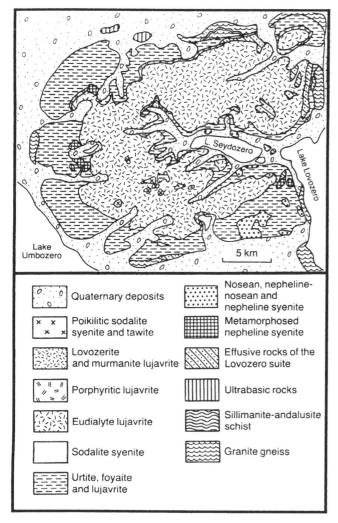

Lake Umbozero

L. Seydozero

Lake Lovozero

5 km

	Quaternary deposits
	Poikilitic sodalite syenite and tawite
	Lovozerite and murmanite lujavrite
	Porphyritic lujavrite
	Eudialyte lujavrite
	Sodalite syenite
	Urtite, foyaite and lujavrite
	Nosean, nepheline-nosean and nepheline syenite
	Metamorphosed nepheline syenite
	Effusive rocks of the Lovozero suite
	Ultrabasic rocks
	Sillimanite-andalusite schist
	Granite gneiss

Fig. 17. Lovozero (after Gerasimovsky et al., 1966, Fig. 1).

intrusive rocks, which is confirmed by the presence of a late Devonian flora in sedimentary rocks interbedded with the volcanics.

The Lovozero intrusive complex is emplaced in the Archaean granite gneiss sequence and has the form of a laccolith with a broad base. According to the geophysical work of Shablinsky (1963) the alkaline rocks can be traced to a depth of more than 7 km, but their lower limit could not be determined. Geophysical investigations have also shown that the intrusion consists of two structural units. The upper part comprises a layered intrusion about 2 km thick and covering 20 × 30 km. The lower stock-like part has been displaced towards the east relative to the centre of the layered body, measures 12 × 16 km in plan and extends to a depth of more than 5 km. The dip of the contacts of the intrusion with the surrounding rocks is steep and close to vertical. The lower part of the intrusion, according to the same geophysical investigations, has a concentrically zoned structure. The density of the rocks of successive zones increases from the centre towards the periphery, but the density of the rocks of the lower stock-like part of the complex remains close to the values known for the main varieties of nepheline syenites of the intrusion. The upper layered part of the alkaline intrusion has in large measure been exposed by erosion and comprises a complex single-phase formation. The geological structure of the complex and the sequence of formation of the rocks have been interpreted in different ways by different investigators. Table 2 summarizes the principal ideas on the general igneous relationships within the complex. The ideas developed by us on the structure of the massif, and presented here, are closest to the system of Eliseev and Fedorov (1953).

The rocks of the Lovozero complex comprise four assemblages, formed in four intrusive phases. The rocks of phase I occupy only about 5% of the total volume, but it is possible that the role of these rocks increases with depth. The rocks of phase II comprise the main area (77%) and those of phase III a lesser amount (18%). Rare dykes of phase IV are volumetrically insignificant (0.01%). The oldest intrusive formations (Phase I) are even-grained nepheline syenites, nepheline–nosean

Table 2 Interpretations of the igneous sequences in the Lovozero complex

Eliseev and Fedorov (1953)	Bussen and Sakharov (1972)	Vlasov et al. (1959)	Gerasimovsky et al. (1966); Kogarko (1977)
4. Complex of young veined rocks	5. Complex of veined alkaline rocks		4. Complex of alkaline lamprophyres
	4. Complex of eudialyte and porphyroid lujavrites	3. Veined rocks	3b. Porphyroid lujavrites 3a. Complex of eudialyte lujavrite and syngenetic poikilitic sodalite syenites and tawites
3. Complex of eudialyte lujavrites	3. Differentiated complex of urtite–foyaite–lujavrite	2. Complex of poikilitic syenites	
2. Differentiated complex of urtite–foyaite–lujavrite	2. Complex of porphyroid and poikilitic nepheline syenites		
1. Complex of alkaline and nepheline syenites, poikilitic nepheline and sodalite syenite, urtite, juvite and foyaite	1. Complex of metamorphosed alkaline rocks – nepheline syenite porphyries and rhomb porphyries	1b. Differentiated complex of urtite – foyaite and lujavrite 1a. Complex of eudialyte lujavrites	1. Nepheline syenite, nepheline–hydrosodalite and poikilitic hydrosodalite syenites and metamorphosed nepheline syenite

Table 3 Composition of the main rock types of the Lovozero complex

Rocks	Phase	Main rock-forming minerals	Characteristic accessory minerals
Nepheline and nosean syenite	I	K-Na feldspar, nepheline, nosean, aegirine-diopside ($Ac_{35}Hd_{15}Di_{50}$), magnesioriebeckite	Lavenite, ilmenite, titanite, apatite
Urtite	II	Nepheline, microcline, sodalite	Apatite, titanates
Foyaite	II	Microcline, nepheline, aegirine ($Ac_{85}Hd_6Di_9$), arfvedsonite, sodalite, analcime	Eudialyte, murmanite, lamprophyllite, villiaumite
Aegirine lujavrite	II	Microcline, nepheline, aegirine ($Ac_{75}Hd_{11}Di_{14}$)	Eudialyte, titanates, lamprophyllite, rinkolite
Amphibole lujavrite	II	Microcline, nepheline, arfvedsonite	Titanite, apatite
Eudialyte lujavrite	III	Microcline, eudialyte, aegirine ($Ac_{82}Hd_7Di_{11}$), arfvedsonite, nepheline	Murmanite, lovozerite, lamprophyllite
Poikilitic sodalite syenite	II and III	Microcline, nepheline, sodalite, aegirine	Eudialyte, villiaumite, chinglusuite

syenites, poikilitic nosean syenites and metamorphosed nepheline syenites. Amongst this assemblage of rocks there are suites of hypabyssal character (nepheline–syenite porphyries). The rocks of Phase I, occurring in their original position, are located in the marginal parts of the complex and exposed at low levels at the foot of the mountains. Outcrops in a small sector in the northwestern part of the massif indicate that the contact of the rocks of Phase I with Phase II dips towards the centre of the complex at an angle of 20–30°. It is probable that the rocks of the first phase underlie the rocks of intrusive Phase II, but at what depth and to what extent is unknown. Within the nepheline–nosean syenites cross-cutting dykes, including pegmatites, of intrusive Phase II have been identified. Abundant xenoliths of rocks of intrusive Phase I are found throughout the complex in rocks of later phases. In individual xenoliths the varieties of nepheline syenite of intrusive Phase I have gradual transitions between them but the *in situ* rocks of Phase I are mainly nepheline–nosean and poikilitic nosean syenites.

The second intrusive phase comprises a strongly differentiated complex of urtite, foyaite and lujavrite. Outcrops of these rocks are found on the slopes of the mountains. The geometry of this suite of rocks has not been finally established, although it has been studied, as a result of drilling, to a depth of 2 km, and probably has a circular form. Phase II consists of a layered sequence with, as seen in vertical sections, a regular alternation of layers of urtite, juvite, foyaite and aegirine and amphibole lujavrite which range in thickness from a few centimetres to hundreds of metres. The stratigraphical order of the rock types is the same in the various parts of the massif, and the inward dips of the layers are at low angles and vary little between the margins and the centre of the complex. In the marginal parts, bordered by older formations, the layering as a rule disappears and the facies of the rocks close to the outer contact consists of mesocratic, coarse-grained, sometimes pegmatitic foyaites. Among the lujavrites there are circular bodies of poikilitic sodalite syenite. In the lower part of the urtite–foyaite–lujavrite suite an eruptive breccia is sometimes developed which contains xenoliths of rocks of

intrusive Phase I. The individual layers of nepheline–syenite in this suite are enriched in xenoliths of Devonian extrusive rocks and of intrusive Phase I rocks for distances of several kilometres. A well-known eruptive breccia of altered tuffaceous rocks occurs in a thin horizon of apatite urtites, which can be traced around the entire perimeter of the suite.

The rocks of intrusive Phase III comprise a suite of eudialyte lujavrites which cut, and overlie, the upper part of the rocks of Phase II. The plane of contact between rocks of Phases II and III dips towards the centre of the complex with the angle of dip increasing from the margins towards the centre. In general the suite of eudialyte lujavrites has the form of a complex funnel-shaped body (ethmolith). The rocks of Phase III form the summits of the mountains of the Lovozero Massif, and the thickness of this suite reaches 450 m, but, because of erosion, decreases from northwest to southeast. Near the contact with the rocks of the second intrusive phase the eudialyte lujavrites contain an eruptive breccia of these rocks with xenoliths measuring from a few metres up to 2 km across. Moreover, in the eudialyte lujavrites there are also xenoliths of Phase I rocks with porphyritic textures. The rocks of intrusive Phase III include leucocratic, mesocratic and melanocratic eudialyte lujavrites, eudialyte foyaite and urtite and a coarser layering is developed than in the rocks of the second intrusive phase. On the boundary with the rocks of Phase II there are ubiquitously layered bodies of porphyritic lujavrite, which are late derivatives of Phase III. Individual dykes of porphyritic lujavrite, up to several kilometres long, cut the rocks of Phases I and II. Poikilitic sodalite syenite and tawite (sodalite with some aegirine and minor nepheline, alkali feldspar and eudialyte) are found in the form of equidimensional, sharply defined bodies amongst the eudialyte lujavrites of Phase III. The rocks of intrusive Phase IV consist of rare dykes of alkaline lamprophyres (monchiquite, fourchite, tinguaite, etc.) which cut all the older alkaline rocks, and also the surrounding granite gneisses. The principal rock-forming and characteristic accessory minerals of the Lovozero rocks are given in Table 3.

A general account of the complex, which concentrates

on the pegmatites, mineralogy and geochemistry of individual elements, is that of Vlasov *et al.* (1959), of which there has been an English translation (Vlasov *et al.*, 1966). Gerasimovsky *et. al.* (1966) have also described the geology, petrology and geochemistry of Lovozero and this is also available in English (Gerasimovsky *et al.*, 1968). Khomyakov (1987) has de scribed unstable and water soluble minerals.

Economic There is a eudialyte, apatite and loparite mineralization (Zr, Nb, REE, Sr, Ba, P) (Vlasov *et al.*, 1959; Gerasimovskky *et al.*, 1966; Eliseev and Fedorov, 1953; Osokin, 1980). Loparite has been produced for 40 years, as an ore of niobium, and there are now two underground mines, Umbozero on the northwest and Karnasurt on the western side of the complex; both are located in rocks of the second intrusive phase. The loparite is beneficiated locally, the production being 30,000 tons a year of a 95% concentrate. The concentrate is sent to Estonia and Kazakhstan for processing. The eudialyte in the eudialyte lujavrite complex is being assessed for possible exploitation as an ore of Zr and Y.

Age A Rb–Sr isochron gave an age of 362 ± 17 Ma (Kogarko *et al.*, 1983). K–Ar on nepheline and biotite from poikilitic nepheline syenite gave 418 ± 12 and 406 ± 12 Ma respectively and nepheline from the urtite-ijolite series varied from 386 ± 12 to 406 ± 12 Ma (Kononova and Shanin, 1971). Whole rock and mineral Rb–Sr isochrons gave ages of 371.6 ± 20.3 Ma for the first stage and 361.7 ± 1.1 Ma for the second stage of igneous activity (Kramm *et al.*, 1993).

References Bussen and Sakharov, 1972; Eliseev and Fedorov, 1953; Gerasimovsky *et al.*, 1966 and *1968; *Khomyakov, 1987; Kogarko, 1977; Kogarko *et al.*, 1983; Kononova and Shanin, 1971; *Kramm *et al.*, 1993; Osokin, 1980; Ramsay, 1890; Shablinsky, 1963; Vlasov *et al.*, 1959 and *1966.

---

17 KHIBINA
(Khibiny)

67°43′N; 33°47′E
Figs 18 and 19

The Khibina alkaline complex with an area of 1327 km² is second only to the Guli complex (Meimecha Kotui No. 2) in terms of size. It forms an isolated mountain massif between Lake Imandra in the west and Lake Umbozero to the east, the mountains rising above the ancient surrounding peneplain to a height of 700–1000 m.

The complex is emplaced into Archaean granite gneisses and Proterozoic volcanic-sedimentary rocks along steep outer contacts, which have been traced to a depth of 7 km by geophysical methods. Adjacent to the outer contacts albite–aegirine fenites and hornfelses are extensively developed. The intrusion has a concentric, zonal structure in which primary igneous layering is very well developed. Galakhov (1975) distinguished several zones in the complex, which correspond to distinct ring and conical intrusions formed as a result of successive phases of intrusion.

The earliest intrusions are alkaline and nepheline trachytes and rhomb and nepheline porphyries which form a steeply inclined body 0.5 km in thickness in the western part of the massif and xenoliths in the peripheral zones. The zones of the complex, from the periphery to the centre, are as follows (see Table 4): (1) alkaline syenite (umptekite) and nepheline syenite (0.3 km thick); (2) and (3) massive and trachytic khibinites (about 5.5 km thick); (4) rischorrite (biotite–nepheline syenite)-ijolite–urtite–apatite–nepheline rocks (2–3 km); (5) melteigite, ijolite and urtite; (6) and (7) heterogeneous nepheline syenite and foyaite (3.5–4 km); (8) carbonatite. The khibinites of zone (3) are trachytic and in the deepest parts of the exposed section are stratified in the form of alternating sequences of leucocratic nepheline syenite and melanocratic ijolitic rocks. The rischorrites of zone (4) comprise a complex ring-like intrusive body the rocks of which are characterized by poikilitic textures and the occurrence of dactylotypic and micropegmatitic intergrowths of alkali feldspar and nepheline. Metasomatic processes may have played an important role in the generation of these rocks. The melteigite–ijolite–urtite series that comprises the fifth zone develops a striking layered complex within which is the well-known apatite–nepheline ore deposit, described below.

In the northwestern part of the complex within the khibinites are a high density of xenoliths of peridotite, perovskite clinopyroxenite and ijolite which are similar in composition to the alkaline ultramafic rocks of the Afrikanda and Kovdor complexes. Throughout the intrusion there are alkaline pegmatites and numerous zones of albitization. In the eastern part, near to the bodies of rischorrite and urtite, there is a carbonatite stockwork with a diameter of about 800 m (Fig. 19), which lies at the focus of the multiple layered complex. Drilling has indicated that the carbonatite extends to a depth of at least 1.7 km. The carbonatites are considered to be younger than the principal layered units of the

Table 4 Composition of the principal rock types of the Khibina complex

Rocks	Phase	Rock-forming minerals	Characteristic accessories
Alkaline and nepheline syenites	1	Nepheline, K-feldspar, aegirine–augite	Titanite, apatite, biotite
Khibinite	2, 3	Microline, nepheline, aegirine, arfvedsonite	Eudialyte, titanite, lamprophyllite
Rischorrite	4	Microcline, orthoclase, nepheline, aegirine, aegirine–augite, arfvedsonite, biotite	Eudialyte, titanite astrophyllite, lamprophyllite apatite
Melteigite, ijolite and urtite	5	Nepheline, aegirine–augite	Apatite, titanite, titanomagnetite
Foyaite and aegirine–nepheline syenites	6 and 7	Microcline, nepheline, aegirine–augite, arfvedsonite, biotite	Eucolite, titanite, astrophyllite
Carbonatite	8	Calcite, aegirine, biotite	Apatite

complex and the dykes of tinguaite, alkaline trachyte and alkaline lamprophyres (monchiquite and damkjernite). Dykes are widespread in this part of the complex, where they cut the principal alkaline rocks with the formation of typical eruptive-explosive breccias which have lamprophyric, tinguaitic, orthoclasitic and chalcedonic cements. As well as the dykes in the eastern part of the complex, there are numerous pipes of eruptive breccia that have diameters of 50–100 m and contain fragments which include olivinite, clinopyroxenite, phoscorite, urtite and nepheline syenite set in a matrix which may be tinguaitic or of phlogopite picrite. The carbonatite stock has a complicated structure. Multistage carbonatite brec-cias are cut by a stockwork of carbonatites which extend through the central part of the stock. Vertical zoning is evident in the stock with in the upper part magnesian varieties of carbonatite and in the lower part aegirine–biotite–calcite carbonatites with apatite. Zones of carbonatization of the host-rocks, including foyaites and tinguaites, are related to the carbonatites. There are a number of formational stages of differing composition amongst the carbonatites including (1) biotite-, aegirine–biotite- and albite–calcite carbonatites, (2) manganiferous calcite–ankerite and siderite carbonatites and (3) manganiferous siderite and ankerite carbonatites with significant natrolite and dawsonite (Dudkin, 1991).

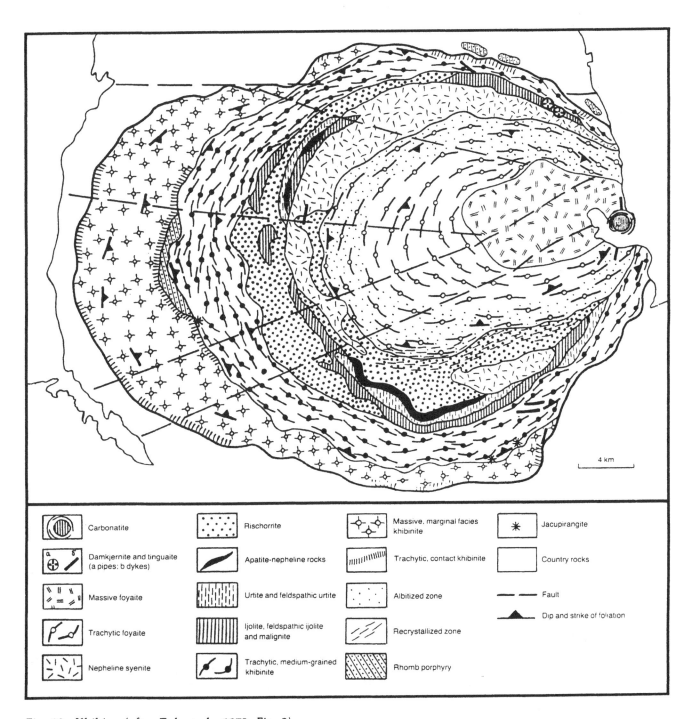

Fig. 18. Khibina (after Zak et al., 1972, Fig. 2).

Legend:

- Carbonatite
- Damkjernite and tinguaite (a pipes: b dykes)
- Massive foyaite
- Trachytic foyaite
- Nepheline syenite
- Rischorrite
- Apatite-nepheline rocks
- Urtite and feldspathic urtite
- Ijolite, feldspathic ijolite and malignite
- Trachytic, medium-grained khibinite
- Massive, marginal facies khibinite
- Trachytic, contact khibinite
- Albitized zone
- Recrystallized zone
- Rhomb porphyry
- Jacupirangite
- Country rocks
- Fault
- Dip and strike of foliation

Noteworthy chemical features of the carbonatites are a predominance of Na over K and high contents of Sr, Ba, Mn, F, S and rare earth elements. More than 80 minerals have been identified in the carbonatites, and closely associated rocks, including dawsonite, nahcolite, fluorite, cryolite, synchisite, parisite, burbankite and edingtonite (Dudkin, 1991). The carbonatites have been described in detail by Dudkin et al., (1984) and Dudkin (1991). The mineralogy of Khibina is described in the two volume monograph by Kostyleva-Labuntsova et al., (1978) and a range of unstable and water soluble minerals from the complex is described by Khomyakov (1987). The geological structure of the massif is outlined in the maps and papers of Eliseev et al., (1939) and Zak et al., (1972) while Galakhov (1975) has described the petrology of the complex in detail. There is a brief account in Gerasimovsky et al., (1974).

Economic The largest igneous apatite deposit in the world is located in the ijolite–urtite part (zone 5) of the complex (Ivanova, 1963). Eight major apatite ore bodies have been delineated along an arcuate zone some 75 km in length. The apatite-rich rocks have been classified into three groups which are referred to as I 'pre-ore', II 'ore' and III 'post- ore'. The rocks of the first group consist of an ijolite series which is interlayered with subordinate melteigite, urtite, juvite and malignite, the whole having a total thickness of less than 700 m. The second group consists of massive feldspathic urtite, ijolite–urtite and apatite ore with a total thickness of 200–700 m. The units of group III are from 10 to 1400 m thick and include urtite, ijolite, melteigite, juvite, malignite and lujavrite. The principal phosphate-ore deposits are found in group II where the apatite-rich rocks occur in the hanging wall of an ijolite–urtite intrusion. The deposit is both petrographically and geochemically zoned. The upper zone (apatite-rich ore) has been called patchy and patchy-banded ore and consists of 60%–90% euhedral apatite crystals ranging up to several tenths of a millimetre across. Clinopyroxene, titanite, feldspars, titanomagnetite and nepheline occur as intergranular minerals. Monomineralic nepheline layers sometimes alternate with monomineralic apatite layers. The lower zone (apatite-poor ore) is composed of lenticular- banded, net-like and block ore. The lenticular-banded ore consists of fine-grained ijolite separated by layers of apatite and fine-grained urtite, whereas the net-like ore is texturally and structurally similar to lenticular-banded ore differing from it only in the scarcity of urtite and apatite bands. The block ore appears to be pegmatitic. Occasional large crystals of nepheline (up to 15 cm across) occur in the nepheline–apatite rock and in the apatite segregations. The lower zone grades down into massive urtite which consists of 75%–90% large euhedral nepheline crystals with intergranular acmitic clinopyroxene, feldspar, titanomagnetite and aenigmatite. Small grains of euhedral apatite are also found in the mesostasis. An apparent late-stage eruptive breccia, containing xenoliths from the lower zone of the apatite ore in an ijolite–urtite matrix, also occurs in the zone. A crystal fractionation model for the apatite deposit is proposed by Khapayev and Kogarko (1987), based on observations of the rock-forming minerals; they give numerous analyses of nepheline, pyroxene, apatite and titanite. Frenkel and Khapayev (1990) consider a convective accumulation model for production of the apatite-rich rocks.

The resources have been reported to total 4000 million tonnes averaging 15% P_2O_5 (Ilyin, 1989) and this is by far the largest source of igneous phosphate in the world. Production began in 1929 and at present there are five mines, with both underground and surface working, but the central open pit accounts for about half of produc-

tion; there are three beneficiation plants with an estimated total capacity of 20 million tonnes annually. The flotation concentrate produced is maintained at 39.5% P_2O_5 with production in 1987 at 20.8 million tonnes (Ilyin, 1989). Nepheline from urtite, ijolite and the apatite ores is also produced and used for the production of aluminium. The carbonatites contain up to 9% REE, 6.5% Sr and 3% Ba (A. Zaitsev and M.J. Le Bas, unpublished data).

Age Rb–Sr and Sm–Nd isochrons on rock-forming minerals and whole rocks gave 365 ± 13 Ma (Kogarko et al., 1981 and 1986). K–Ar on biotite gave 400 ± 12 Ma (Shanin et al., 1967) and on nepheline and lepidomelane from rischorrite 392 ± 16 and 400 Ma respectively, while

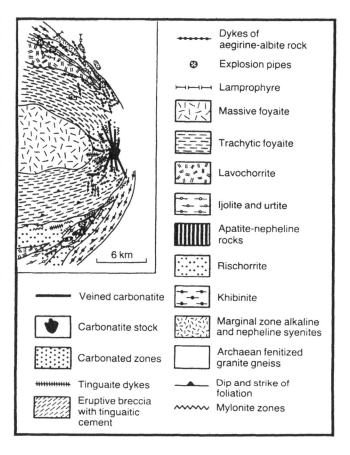

Legend:
- Dykes of aegirine-albite rock
- Explosion pipes
- Lamprophyre
- Massive foyaite
- Trachytic foyaite
- Lavochorrite
- Ijolite and urtite
- Apatite-nepheline rocks
- Rischorrite
- Khibinite
- Marginal zone alkaline and nepheline syenites
- Archaean fenitized granite gneiss
- Dip and strike of foliation
- Mylonite zones
- Veined carbonatite
- Carbonatite stock
- Carbonated zones
- Tinguaite dykes
- Eruptive breccia with tinguaitic cement

Fig. 19. Carbonatite centre at eastern end of Khibina complex (after Dudkin et al., 1984, Fig. 2).

nepheline from ijolite gave 386 ± 12 to 410 ± 12 Ma, from foyaite 412 ± 12 and from chibinite 392 ± 12 Ma (Kononova and Shanin, 1971). Five whole rock and mineral Rb–Sr isochrons ranged from 362.4 ± 4.5 to 377.3 ± 3.9 Ma (Kramm et al., 1993).

References *Dudkin, 1991; Dudkin et al., 1984; Eliseev et al., 1939; *Frenkel and Kapayev, 1989; Galakhov, 1975; *Gerasimovsky et al., 1974; *Ilyin, 1989; Ivanova, 1963; *Khapayev and Kogarko, 1987; *Khomyakov, 1987; Kogarko et al., 1981 and 1986; Kononova and Shanin, 1971; Kostyleva-Labunsova et al., 1978; *Kramm et al., 1993; Kukharenko et al., 1971; O'insky, 1935; Shanin et al., 1967. Zak et al., 1972.

18 SOUSTOVA
67°31′36′N; 33°36′E
Fig. 20

This occurrence is located 5 km south of the Khibina pluton and follows an east–west-trending pattern of fractures. It is lens-shaped and has a length of 20 km and width from 0.5 to 2.2 km. The contacts with volcanogenic country rocks are tectonized and complicated by mylonite zones and injected by veins of alkaline rocks. The intrusion consists of peralkaline, analcime and nepheline–analcime syenites. There are also dykes of syenite porphyry, syenite pegmatite and nordmarkite as well as albitites, which develop in the zones of syenite mylonitization.

Age U–Pb dating of syenites gave 2000 ± 50 Ma (Vinogradov *et al.*, 1985).

References Kukharenko *et al.*, 1971; Vinogradov *et al.*, 1985.

19 OZERNAYA VARAKA
67°26′N; 32°57′E
Fig. 21

Ozernaya Varaka lies to the east of the Afrikanda complex and is situated on an east–west-trending ridge of uplands. The massif intrudes and fenitizes gneisses of the Belomorskaya Archaean series. The total area of the complex is only 0.8 km² and it has a concentrically zoned structure. The outer zone, 250 to 300 m wide, is composed of ijolite with bands of melteigite. Towards the centre this is replaced by an area of melteigites, which are patchily enriched in melanite. The central part of the massif is composed of nepheline pyroxenites. Dyke rocks

are, in order of decreasing age, ijolite and urtite, biotite glimmerites, which contain pyroxene, amphibole and apatite, ijolite pegmatites, nepheline and cancrinite syenites and carbonatites; there are also younger dykes of ijolite porphyry, tinguaite and monchiquite. Carbonatites and silicate–carbonate rocks are encountered both in the inner parts of the massif and in the contact aureole. They occur as veins and lens-shaped bodies, which are considered to be metasomatic in origin, and are developed among the ijolites, melteigites, nepheline pyroxenites and fenites; occasionally they replace glimmerites. Less often, the carbonatites have sharp and cross-cutting contacts. The thickness of the carbonatite veins varies from several centime tres to 2.5 or even 3 m. The following types of carbonate-bearing rocks have been distinguished: aegirine–calcite carbonatites (ringites) with accessory zircon, pyrochlore, aegirine, nepheline and biotite; calcite, calcite–albite, cancrinite–calcite–albite and aegirine–calcite–albite rocks.

Age U–Th–Pb dating of schorlomite from ijolite pegmatite gave 338 Ma (Kukharenko *et al.*, 1965). Four whole rock and biotite Rb–Sr isochrons ranged from 376 ± 2.9 to 369 ± 5.3 Ma (Kramm *et al.*, 1993).

References *Kramm *et al.*, 1993; Kukharenko *et al.*, 1965.

20 AFRIKANDA
67°25′N; 32°48′E
Fig. 22

This complex has an isometric shape, an area of some 7 km², and lies on an east–west-trending fracture system. The country rocks are granite gneisses. Four zones can be distinguished from the periphery to the centre. The outermost zone, with a width of 500 m, contains nephe-

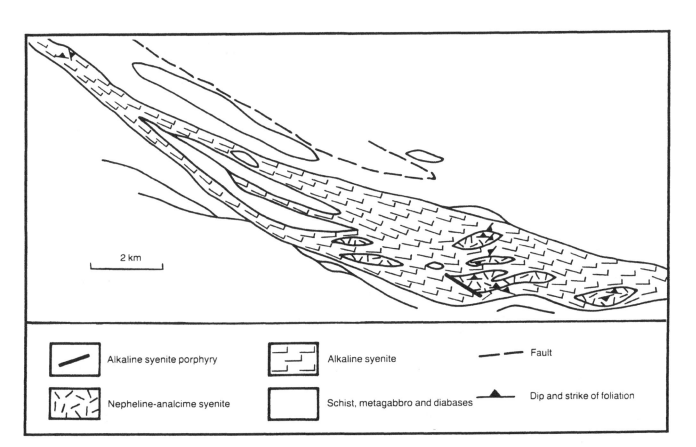

Alkaline syenite porphyry	Alkaline syenite	— — — Fault
Nepheline-analcime syenite	Schist, metagabbro and diabases	Dip and strike of foliation

Fig. 20. Soustova (after Kukharenko et al., 1971, Fig. 15).

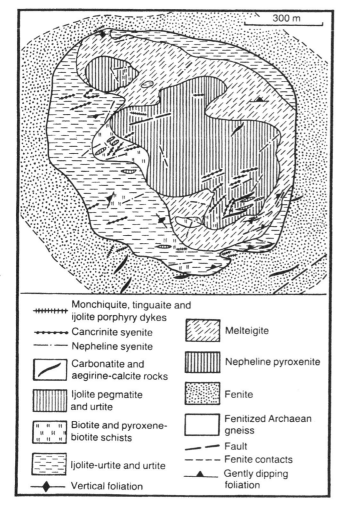

Monchiquite, tinguaite and ijolite porphyry dykes

Cancrinite syenite

Nepheline syenite

Carbonatite and aegirine-calcite rocks

Ijolite pegmatite and urtite

Biotite and pyroxene-biotite schists

Ijolite-urtite and urtite

Vertical foliation

Melteigite

Nepheline pyroxenite

Fenite

Fenitized Archaean gneiss

Fault

Fenite contacts

Gently dipping foliation

Fig. 21. Ozernaya Varaka (after Kukharenko et al., 1965, Fig. 51).

line and feldspar-bearing pyroxenites which grade into melteigites. The rocks in this zone have a layered structure and comprise nepheline, aegirine-augite and microcline with lesser amounts of melanite, amphibole, biotite, cancrinite, apatite, titanite, titanomagnetite, perovskite, sodalite, calcite and sulphides. The second zone includes fine-grained nepheline-bearing pyroxenites, enriched in apatite, mica and amphibole. The fine-grained pyroxenites are gradually replaced by medium- and coarse-grained nepheline pyroxenites towards the centre of the massif and these constitute the third zone. Some pyroxenites are of exceptionally coarse grain size and these are enriched in titanomagnetite. The main minerals of these pyroxenites are diopside–augite, olivine and titanomagnetite with subordinate serpentine, phlogopite, pargasite, titanite, chlorite and calcite. The rocks of the third zone are cut by phlogopite–pyroxene rocks. The fourth zone, occupying the central part of the complex, has the most complicated structure and comprises coarse-grained pyroxenites, olivinites, sometimes containing melilite, calcite–amphibole–pyroxene rocks with nests of perovskite, and alkaline pegmatites, of a range of compositions, which cut the olivinites and pyroxenites extensively. The mineralogy of the olivinites is olivine (12–14% Fa), magnetite and perovskite with lesser diop-

side, augite, phlogopite, melilite and monticellite. The calcite–amphibole–pyroxene rocks, as well as these minerals, also include perovskite, titanite, phlogopite, actinolite, apatite, nepheline, zirconolite and pyrochlore. Other veins include ilmenite–titanomagnetite-rich rocks, ijolite pegmatites with schorlomite, calcite carbonatites and calcite–zeolite veins with prehnite.

Economic Titanomagnetite–perovskite ores form an oval pipe-like body which has been drilled to 400 m. The principal ore types are ore olivinites, ore pyroxenites and ore-amphibole pyroxenites. The perovskite content of the ores varies from 18 to 48% and titanomagnetite from 20 to 76% (Gorbunov *et al.*, 1981).

Age K–Ar on biotite gave ages from 344 to 426 Ma (Kukharenko *et al.*, 1965) and a Rb–Sr mineral isochron 364 ± 3.1 Ma (Kramm *et al.*, 1993).

References Gorbunov *et al.*, 1981; *Kramm *et al.*, 1993; Kukharenko *et al.*, 1965.

---

21 LESNAYA VARAKA
67°23′N; 33°04′E
Fig. 23

This complex lies along the western margin of Seid lake and is situated in an upland part of an east–west-trending ridge. It is oval in outline and has an area of about 9 km². The intrusion is composed largely of fine-, medium- and coarse-grained olivinites, which occupy 85–90% of the area of the complex, that are enriched in titanomagnetite and display a well-pronounced banding. Coarse-grained olivinites which do not contain ore minerals form an arcuate body of about 2 km² on the eastern side of the complex. They are younger than the ore-bearing olivinites, as demonstrated by cross-cutting relationships with banding defined by ore-rich layers in the fine- and medium-grained olivinites. On the western, southwestern and southern margins of the complex the olivinites are rimmed by clinopyroxenites, into which they grade gradually through a zone of olivine clinopyroxenites (wherlites). In the immediate vicinity of the outer contact the pyroxenites contain orthoclase. In the central part of the complex dykes are widespread and include ijolite pegmatite, tinguaite and alkaline and cancrinite syenite. Postmagmatic veins include tremolite, clinohumite–tremolite, phlogopite, phlogopite–vermiculite, serpentine–hornblende, serpentine–tremolite-dolomite rocks, carbonatites and serpentine veins. Among the carbonatites, apatite–dolomite types, with which rare earth mineralization is associated, and monomineralic dolomite carbonatites are distinguished. Apart from the predominant dolomite and fluorapatite the carbonatites also contain tremolite, hydro-phlogopite, biotite and, as secondary minerals, calcite, quartz, magnetite, hydrogoethite, pyrrhotite, pyrite, titanite and accessory pyrochlore, natroniobite, zircon, dysanalyte, lueshite and phosphates, carbonates and fluocarbonates of rare earths. Three mineral associations, corresponding to three stages of carbonatite evolution, have been distinguished by Kukharenko *et al.* (1965). These are (a) zircon–apatite–dolomite, (b) pyrochlore–dolomite and (c) quartz–limonite with secondary phosphates and carbonates of rare earth elements. The gneisses into which the complex is emplaced are fenitized adjacent to the contacts.

Economic The olivinites are enriched in titanomagnetite (Kukharenko *et al.*, 1965).

Reference Kukharenko *et al.*, 1965.

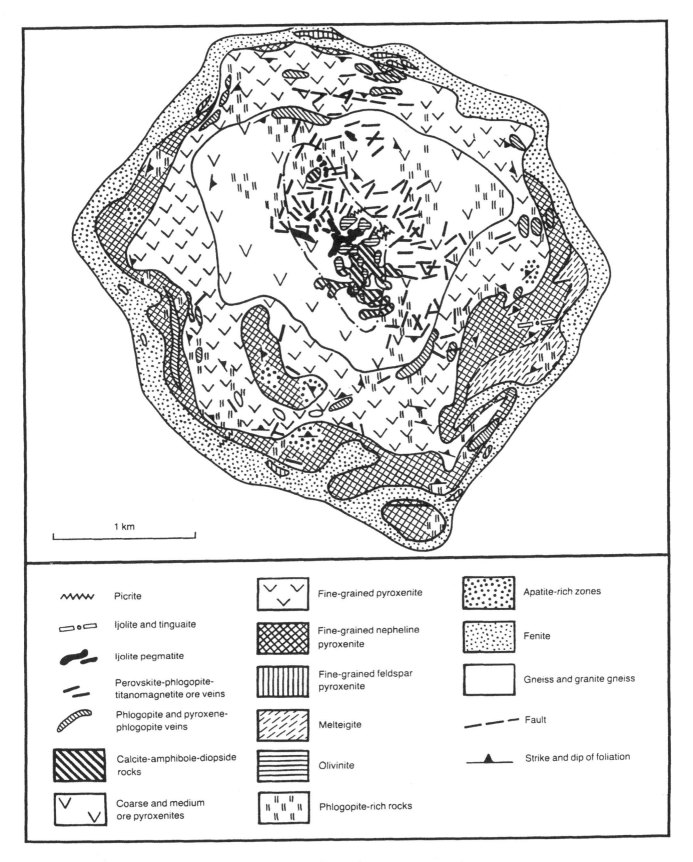

1 km

∿∿∿	Picrite	⌄ ⌄ ⌄	Fine-grained pyroxenite

Fig. 22. Afrikanda (after Kukharenko et al., 1965, Fig. 71).

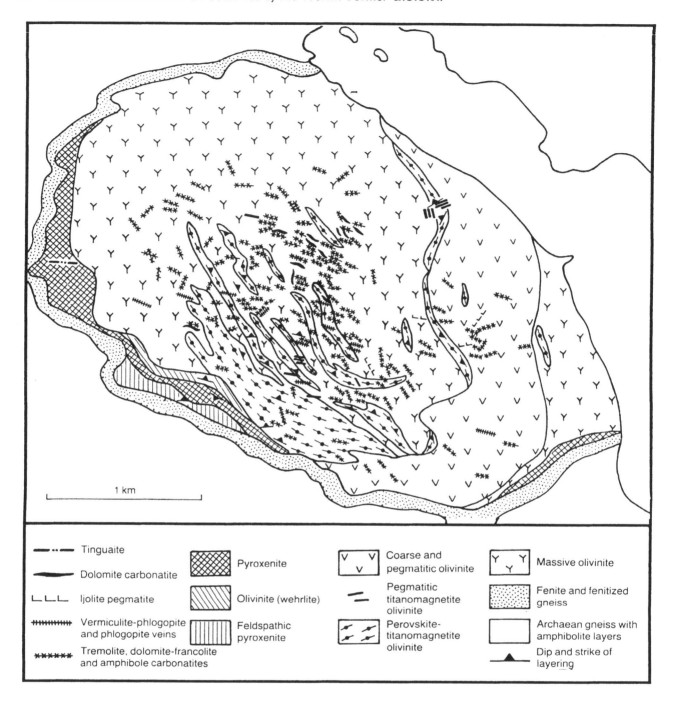

Fig. 23. Lesnaya Varaka (after Kukharenko et al., 1965, Fig. 27).

22 SALMAGORSKII
67°18′N; 33°30′E
Fig. 24

Together with the Afrikanda, Ozernaya Varaka and Lesnaya Varaka complexes, Salmagorskii constitutes one of the Khabozerskaya group of intrusions, the location of which is controlled by a single fracture zone. These complexes seem to be comagmatic. Salmagorskii is almost circular in shape and has an area of about 20 km². It has a partially zoned structure with an unusual, for the Kola–Karelia group of alkaline occur-

rences, combination of ultramafic and alkaline rocks. Olivinites and subordinate olivine–pyroxene rocks (wehrlites) and clinopyroxenites form steeply-dipping lens-shaped bodies in the peripheral parts of the massif and occupy about 15% of the area. These rocks are cross-cut and injected by an alkaline intrusion of melteigite and urtite which comprises the principal body of the complex. The melteigites are localized in the central area while ijolite and ijolite–urtite tend to be concentrated at the periphery. The ultramafic rocks of the southeastern part of the complex are surrounded by an aureole of essentially melilitic rocks, which include melilitolites,

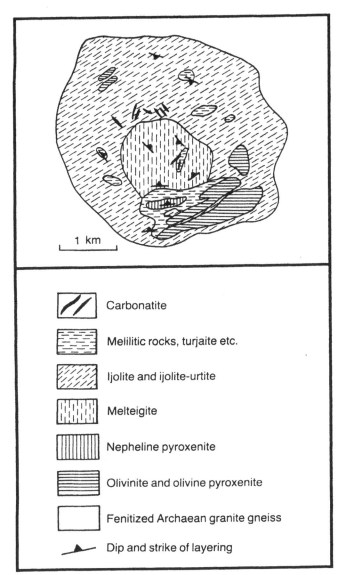

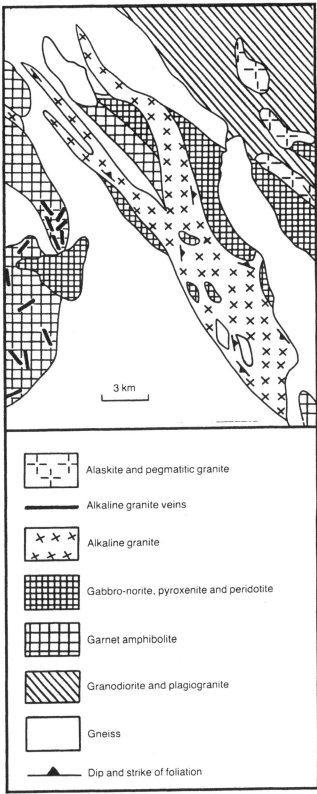

Fig. 24. Salmagorskii (after Kukharenko et al., 1965, Fig. 13).

monticellite rocks, monticellite–melilite turjaites and turjaites, and these represent high-temperature, contact reaction rocks. Within them an association of calcium silicates and micas is found including monticellite, andradite–grossular, diopside, cancrinite and phlogopite. A dyke facies is represented by ijolite porphyries and monchiquites; carbonatites occur as rare, relatively thin veins composed of calcite and ankerite. In calcite carbonatites aegirine–diopside, wollastonite, phlogopite, chlorite and zeolites are present in variable, but subordinate, amounts. The accessory minerals include apatite, hematite, titanite and zircon. In the calcite–ankerite carbonatites, calcite and ankerite account for 30 to 40% of the rock by volume and aegirine–diopside, chlorite and zeolites are also present with accessory apatite, magnetite, hematite, anatase, titanite, zircon and pyrochlore. At the contacts between the carbonatites and the alkaline, melilite and ultramafic rocks, zones of phlogopite–carbonate rocks are developed.

Age U–Pb dating on schorlomite from ijolite pegmatite gave an estimated age of 375 Ma (Kukharenko *et al.*, 1965) and K–Ar on nepheline from melteigite 480 ± 15,

Fig. 25. Kanozerskii (after Batieva, 1976, Fig. 3).

ijolite 540 ± 16 and urtite 494 ± 15 Ma (Kononova and Shanin, 1971).

References Kukharenko *et al.*, 1965; Kononova and Shanin, 1971; Orlova, 1959.

23 OGORODNYI   67°03′N; 33°10′

Associated with the Imandra–Varzugskii fault, the Ogorodnyi peralkaline granite occurrence is a northwest-trending, elongate body of 5 km². The rocks are characterized by gneissose textures which are particularly well developed at the margins. The principal minerals are microcline, albite, quartz, ferrohastingsite and variable amounts of aegirine–augite and biotite; accessory are magnetite, titanite, orthite, epidote, apatite, fluorite, garnet, hematite and sulphides. The massif is cut by amazonite-bearing pegmatites and quartz veins.

Age K–Ar on amazonite from pegmatite gave 1330 Ma.

Reference Batieva, 1976.

---

24 KANOZERSKII   .   67°05′N; 33°58′E
Fig. 25

Kanozerskii is controlled by a northwesterly-trending deep fault. It is an elongate body bifurcating to the northwest and has an area of 150 km². Dykes, veins and small intrusions of alkaline granites occur to the east of the massif which consists essentially of peralkaline granite containing lepidomelane and ferrohastingsite. The principal minerals are quartz, plagioclase (albite–oligoclase), microcline, ferrohastingsite, lepidomelane and aegirine–augite. Accessory minerals are titanite, zircon, orthite, apatite, fluorite, pyrite and magnetite.

Age The Pb method on zircon gave 2330–2365 Ma (Pushkarev, 1990).

References Batieva, 1985 and 1976; Pushkarev, 1990.

---

25 TURIY PENINSULA   66°35′N; 34°27′E
Fig. 26

The alkaline complex of the Turiy Peninsula consists of several intrusions of subvolcanic type and associated dykes and veins. The largest massifs of the complex – Central with an area of 28 km², Southern (13 km²) and Kuznavolokskii (15 km²) – are separate intrusive bodies of complex form. Their shapes were determined by the block structures and complicated tectonics of the enclosing rocks at the time of intrusion. The intrusions are composed, from the oldest to the youngest rocks, of (1) ultramafic rocks (olivinite and clinopyroxenite), (2) feldspathoidal mafic rocks (melteigite, ijolite and ijolite–urtite), (3) melilite-bearing rocks (okaite, uncompahgrite, melilitolite and turjaite), (4) eruptive breccia of ijolite porphyry, (5) phoscorite and carbonatite and (6) dyke rocks. The intrusions have complicated zonal structures which indicate that initial single intrusive bodies were emplaced which were subsequently transformed into confocal intrusions with central stocks of carbonatite and phoscorite. With the emplacement of each successive phase there was associated intense alteration of the earlier rocks – nephelinization, melilitization, growth of garnet and mica and carbonatization with multistage fenitization of the enclosing granodiorites and Tersk sandstones. The principal minerals of the olivinites and pyroxenites are olivine, diopside-augite and titanomagnetite with subordinate nepheline, melanite, perovskite, titanite, apatite, amphibole, wollastonite, phlogopite, cancrinite and calcite. The melteigites, ijolites and urtites comprise nepheline, aegirine–augite, titanomagnetite and melanite with lesser amounts of perovskite, titanite, apatite, phlogopite,

amphibole, sodalite, cancrinite and calcite. The predominant melilite-bearing rocks are turjaites and uncompahgrites, with melilitolites common but okaites rare. These rocks consist of melilite, nepheline and phlogopite with subordinate titanomagnetite, perovskite, apatite, diopside, olivine, schorlomite, amphibole, vesuvianite, sodalite and calcite. The petrogenesis of the turjaites is considered by Ronenson *et al.* (1981) who give whole rock and mineral chemical data. Among the carbonatites are calcitic, calcite–dolomitic and dolomitic types with ore phoscorites less abundant. The rock-forming minerals of the calcite carbonatites are calcite, forsterite and monticellite with accessory phlogopite, diopside, apatite, magnetite, serpentine, clinohumite, ilmenite, pyrite and baddeleyite. The calcite–dolomite and dolomite carbonatites consist of calcite and dolomite with subordinate pyrite, galena, sphalerite, tetraferriphlogopite, fluorite, magnetite and amphibole. The phoscorites of the Turiy Peninsula are much enriched either in apatite (5–15%) or in magnetite (up to 50–70%) with calcite (10–50%); minor minerals include forsterite, diopside, aegirine, hastingsite, actinolite, phlogopite and tetraferriphlogopite. Bulakh and Ivannikov (1984) emphasize the principal phoscorite associations of phlogopite–apatite–forsterite, forsterite–magnetite–calcite and apatite–magnetite–forsterite.

The Central and Southern massifs are intersected by dykes of alkaline lamprophyres and in the northwestern part of the Southern massif there is a neck of olivine melteigite porphyry. Among the dyke rocks there are three groups. Group 1 includes altered picrites, monchiquites, alnoites, melilitites and carbonatites with evidence for explosive emplacement and also intense fenitization. They are older than the central intrusions. Group 2 consists of rocks which are comagmatic with the large Turiy intrusive complexes. They include micromelteigite, olivine melteigite porphyry, turjaite, nephelinite and noseanite. They are associated with monomineralic veins in fenites and with carbonate rock veins, which form complicated dyke and vein systems. The third group, emplaced after consolidation of the principal alkaline intrusions, comprises, in order of decreasing age, monchiquite, olivine and olivine–melilite melanephelinite, melanephelinite, augitite, melilitite, nepheline melilitite and melilite nephelinite and calcite carbonatite. Among the carbonatites calcitic varieties with phlogopite, diopside and aegirine are the most prevalent; less common are melilite-bearing varieties, monomineralic calcite and dolomite types and dolomite–calcite carbonatites. Among hydrothermal veins there are apatite–aegirine, pyrhottite–aegirine-apatite, aegirine–anorthoclase with labuntsovite and strontianite, anorthoclase–wollastonite with eudialyte, quartz–albite–anorthoclase with arfvedsonite, narsarsukite and labuntsovite and quartz–calcite–fluorite with anatase. A detailed discussion of the carbonatites is given by Samoilov and Afanas'yev (1980) who give numerous analyses of rocks and many rock-forming minerals as well as estimates of temperatures of formation of carbonatites and associated quartz-bearing rocks.

Age K–Ar ages of 294–380 Ma have been obtained on turjaite, ijolite and nepheline pyroxenite and for a nephelinite dyke 355 Ma; a dyke from one of the main complexes gave 270 Ma (Shurkin, 1959). A whole rock and mineral isochron gave 373.1 ± 5.6 Ma (Kramm *et al.*, 1993).

References Bulakh and Ivannikov, 1984; *Kramm *et al.*, 1993; Kukharenko *et al.*, 1965; *Ronenson *et al.*, 1981; *Samoilov and Afanas'yev, 1980; Shurkin, 1959.

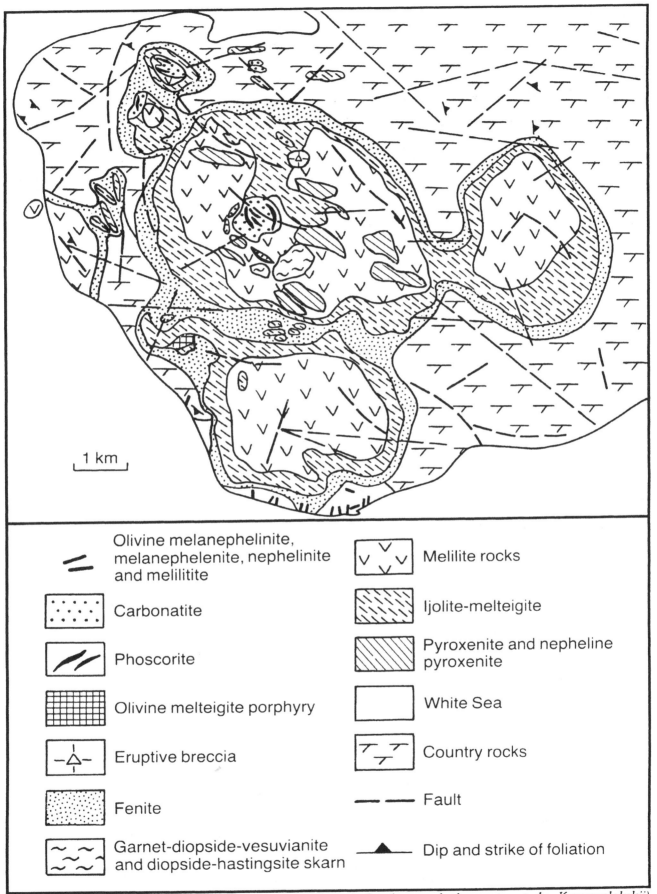

Olivine melanephelinite, melanephelenite, nephelinite and melilitite

Carbonatite

Phoscorite

Olivine melteigite porphyry

Eruptive breccia

Fenite

Garnet-diopside-vesuvianite and diopside-hastingsite skarn

Melilite rocks

Ijolite-melteigite

Pyroxenite and nepheline pyroxenite

White Sea

Country rocks

Fault

Dip and strike of foliation

1 km

Fig. 26. Alkaline complexes of the Turiy Peninsula (Central, Southern and, the most easterly, Kuznavolokskii) (after Bulakh and Ivannikov, 1984, Fig. 4).

26 KANDALAKSHSKII DYKES AND PIPES
66°35′–67°10′N;
32°22′–34°41′E
Fig. 27

Dykes and explosion pipes of alkaline ultramafic and
alkaline lamprophyric rocks are widespread on the coast
and islands of Kandalaksha Bay at the northwestern end
of the White Sea. The dykes form swarms with a general
north–south trend, a direction normal to the axis of the
Kandalaksha graben (or palaeorift). The most intensive
dyke swarms are the Kandalaksha and Tur'insk groups,
which are also the most intensively studied. In the
vicinity of the town of Kandalaksha and on the coast
at the head of Kandalaksha Bay three systems of dykes
can be distinguished which trend approximately north
(0–30°), northeast (50–60°) and east (80°). The dykes
have a vertical or very steep dip and two age groups have
been recognized. The first group is older than the dykes
associated with the Turiy intrusion (No. 25). They
include lamprophyric carbonatites, alkaline picrites,
monchiquites, olivine melilitites and carbonatites. From
dyke intersections the picrites prove to be older than the
monchiquites. The carbonatites are younger than the
alkaline lamprophyres and picrites. All the dykes are
choked with xenoliths derived from the enclosing rocks.
Dykes of the second age group are represented, in order
of decreasing age, by alkaline picrites, monchiquites,
limburgites, camptonites, fourchites, melanephelinites,
nephelinites, microshonkinites and alkaline syenite porphyries. They correspond to the third group of dykes
on the Turiy peninsula, but differ from those by the
presence of plagioclase-bearing rocks, notably camptonites; melilite-bearing varieties are not represented and
neither are carbonatites. In an alluvial layer derived
from the Romanenko pipe, concentrations are found of
transparent crystals of olivine, perovskite, chromite,
chrome-diopside, orange almandine, pyrope and purple
dichroic pyrope. Dichroic pyropes were also found in
heavy concentrates derived from picrites. On Telyachiy
Island, which is also situated in the most northerly part
of Kandalaksha Bay, a dyke of augite–biotite monchiquite has been located that contains a breccia with a
conglomerate-like structure, that is, in places, intensely
carbonated. It is similar in composition and morphology
to the 'boulder' dykes of the Turiy peninsula. According
to Dr Hillary Downes (pers. comm., 1993) it is a carbonatite dyke with xenoliths of carbonatite and amphibole–
phlogopite–pyroxene rocks.

Age Monchiquite and fourchite on the islands of
Ovechy, Borshchevets and Telachiy have given ages of
470–480 Ma by K–Ar; the dykes of the second age group
are equivalent in age to some of the dykes of Turiy, i.e.
270 Ma (Bulakh and Ivannikov, 1984).

References Bulakh and Ivannikov, 1984; Kurileva and
Nosikov, 1959; Putintsiva and Uvadiev, 1987; Shurkin,
1960.

27 ELOVYI ISLAND   67°05′N; 32°18′E
(Elovaya)

In the northern part of Kandalaksha Bay, on Elovyi
Island, a pipe (Elovaya) has been found which is
18 × 0.5 m and oval in section with the longest axis
trending northeast. A sketch map will be found in
Bindeman *et al.* (1990, Fig. 1). The contact with the
enclosing granites and granite gneiss is sharp. The pipe
is filled with an eruptive breccia, which is intersected by
a dyke of picrite or monchiquite containing numerous

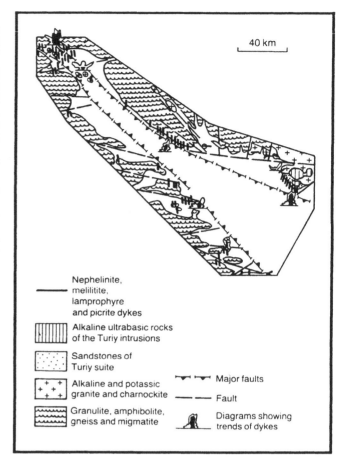

Fig. 27. *Dykes and pipes of the coast and islands of Kan-*
dalaksha Bay (after Bulakh and Ivannikov, 1984,
Fig. 3).

eclogite inclusions. The eruptive breccia consists of fragments of rocks derived from depth, including feldspar
and feldspar-free eclogites and rather fewer of modified
peridotites, as well as angular xenoliths of the country
rock Archaean granites and amphibolites and fragments
of carbonated ultramafic rocks. The groundmass constitutes 12 to 15% of the volume of the breccia, and consists of fine-grained carbonate. Biotite-bearing xenoliths
are described, with whole rock and mineral analyses, by
Bindeman *et al.* (1990). In the central and northern parts
of the island small (2 × 9 and 2 × 0.8 m) stocks of carbonatite occur. The carbonatites intersect a dyke of
picrite.

References *Bindeman *et al.*, 1990; Bulakh and Ivannikov, 1984.

28 KANDAGUBSKII   67°06′N; 32°16′E
Fig. 28

This occurrence is situated on the southern coast of the
small bay of Kanda-guba on the White sea. The intrusion, with an area of about 3.6 km², is partly covered
by Quaternary sediments. According to geophysical
data it has a structure which is close to vertical and it is
considered to be the uppermost part of a larger intrusion.
It has a zonal structure with small areas of apatite-
bearing ijolite in the centre, which change outwards
to malignites and carbonatites. The periphery of the
complex is composed of alkaline syenites. Fenitized

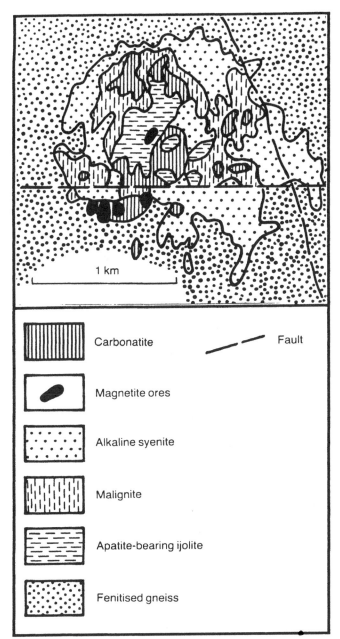

Carbonatite

Fault

Magnetite ores

Alkaline syenite

Malignite

Apatite-bearing ijolite

Fenitised gneiss

Fig. 28. Kandagubskii (after Chuvardinsky and Karaev, 1988, Fig. 2).

amphibole–biotite gneisses of the Belomorsk Archaean series are widely developed around the intrusion. The presence of lens-shaped bodies of magnetite ores has been predicted, by geophysical means, in the southern contact zone and towards the centre of the massif. The complex is characterized by a high content of rare elements of a carbonatite association type and by a high content of apatite in the ijolite, carbonatites and malignite.

Economic There are some magnetite lenses.

Reference Chuvardinsky and Karaev, 1988.

29 MAVRAGUBSKII   67°37′N; 31°17′E

Mavragubskii is located on the northern bank of Lake Verhnyaya Pirenga, near the mouth of the Mavra River.

It is a small intrusion of 0.5 × 1.5 km consisting of melteigite and ijolite. The central part is unexposed but geophysical data indicate that it probably consists of melanocratic, highly magnetic rocks such as clinopyroxenite or other ultramafic types. The melteigites and ijolites are texturally and structurally highly variable, grading from massive to banded and through coarse-, medium-, fine-grained and porphyritic varieties. The rocks comprise pyroxene, varying from diopside to aegirine (35–60%), nepheline (30–50%), titanite (2–8%), phlogopite (3–8%), apatite (0.5–5%) and olivine (1–3%) with accessory magnetite, sulphides, K–feldspar, cancrinite, calcite, sodalite and liebenerite. The country rock granite gneisses are fenitized, the width of the aureole varying between 1 and 1.5 km. Malignites have been formed in the inner aureole.

Age Probably Palaeozoic, from similarity of the rocks to those of complexes such as Kovdor and Turiy.

References Kirnarsky and Kozireva, 1978; Sudovikov, 1936.

30 KOVDOR   67°34′N; 30°29′E
Figs 29 and 30

The Kovdor complex has a drop-like shape in plan and its area is about 55 km². It is related to an east–west fracture system (Kukharenko *et al.*, 1971) and intrudes granite gneisses of Archaean age. It is a composite, multiphase intrusion with a concentric zonal structure. The following rock suites have been mapped from oldest to youngest: (1) ultrabasic rocks, including olivinites and pyroxenites, (2) alkaline rocks of an ijolite–melteigite series, (3) melilite rocks, (4) a complex of apatite phoscorite rocks and magnetite ores, (5) carbonatites and (6) nepheline syenites. The olivinites are located in the central part of the complex and they have been detected gravimetrically down to a depth of 7–10 km. The olivinites contain inclusions of titanomagnetite, and include apatite and calcite. Almost everywhere the olivines have been subjected to a process of phlogopitization. Pyroxenites envelope the olivinite body on the southern and western sides, forming a discontinuous zone 1–1.3 km in width. The pyroxenites are variable in grain size and have aggregates of titanomagnetite, and they are completely phlogopitized and amphibolized, particularly where in contact with the alkaline rocks. Pyroxenites are often observed in the olivinites as veins, and sometimes pyroxene–olivine rocks, which are essentially phlogopitized, are developed. The phlogopite which is mined for industrial purposes comes from these rocks, and individual crystals may reach several metres in diameter.

The alkaline rocks of the Kovdor complex are represented by ijolites, melteigites and nepheline pyroxenites with a subordinate part played by urtites and ijolite–urtites; ijolites, malignites and nepheline syenites are relatively insignificant. The rocks of the ijolite–melteigite series (Volotovskaya, 1960) form a conical intrusion which is developed as a discontinuous ring around the central massif of ultrabasic rocks. At the present level of exposure the width of the circle of alkaline rocks ranges from 100 to 800 m. Melilitic and monticellitic rocks form a semicircular zone up to 1 km wide between the ultrabasic and alkaline rocks as well as several rather small bodies amongst the ijolites and pyroxenites. These rocks have very mixed mineral compositions and are heterogeneous in texture and structure. They include turjaites, pyroxene turjaites and melilitites. Less commonly found are monticellite, melilite–pyroxene and melilite rocks with olivine. The 'ore complex' of

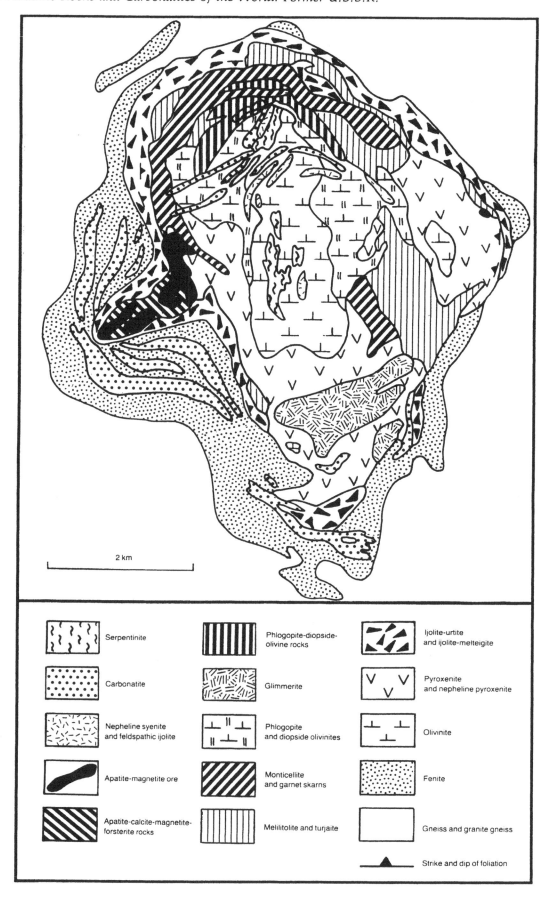

Fig. 29. Kovdor (after Kukharenko et al., 1965, Fig. 113).

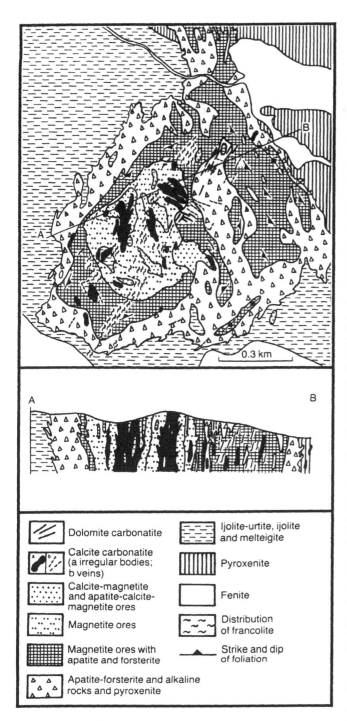

Fig. 30. Map and section of Kovdor magnetite deposit (after Kukharenko et al., 1965, Fig. 117). Line of section A–B indicated on map.

Legend:

- Dolomite carbonatite
- Calcite carbonatite (a irregular bodies; b veins)
- Calcite-magnetite and apatite-calcite-magnetite ores
- Magnetite ores
- Magnetite ores with apatite and forsterite
- Apatite-forsterite and alkaline rocks and pyroxenite
- Ijolite-urtite, ijolite and melteigite
- Pyroxenite
- Fenite
- Distribution of francolite
- Strike and dip of foliation

the Kovdor massif (Fig. 30) is situated in the south-western part and consists of apatite–forsterite rocks and magnetite-rich ores which form an elongate body of 0.8 × 3 km. The ore complex comprises the following series (1) forsterite, apatite–forsterite and apatite–phlogopite–forsterite rocks, (2) rocks containing a high proportion of magnetite and including apatite, forsterite and phlogopite and (3) ores consisting essentially of apatite and including magnetite–apatite and calcite–apatite rocks. The most characteristic accessory mineral

of the magnetite ores is baddeleyite.

Carbonatites of the Kovdor massif are represented by two varieties: calcite carbonatites, comprising bodies 100 to 500 m in diameter, and dolomite carbonatites, which form the greatest number of veins, that vary from 0.7 to 1 m wide. The main minerals of the calcite carbonatites are calcite, apatite, magnetite, phlogopite and pyrrhotite with subordinate forsterite, actinolite, tremolite, hastingsite, diopside, aegirine–diopside, andradite, wollastonite, ankerite, chalcopyrite, pyrochlore, baddeleyite, zircon, zirconolite and dysanalyte. The dolomite carbonatites comprise 90–98% dolomite with calcite, apatite, strontianite, vivianite, zircon, baddeleyite, pyrochlore, dysanalyte, lueshite, magnetite, phlogopite and vermiculite. Kononova and Yashina (1984) have determined $\delta^{18}O‰$ and $\delta^{13}C‰$ values for calcite from calcite carbonatite of +8.7 and −3.6 and for calcite–dolomite carbonatite of +7.6 and −3.4 respectively. $S‰$ for sulphides from calcite carbonatite vary from −2.4 to −4.5 and from dolomite carbonatite gave a value of −6.4 (Grinenko et al., 1970). Details of dolomite–calcite textures, which are utilized for geothermometry, are described by Zaitsev and Polezhaeva (1994). Nepheline and cancrinite syenites form a series of small veins which cut all the other rock types, including the carbonatites. Some authors (Kukharenko et al., 1965) consider that the melilitites and rocks of the ore complex are of metasomatic origin. Analyses of melteigites, ijolites and urtites and their rock-forming minerals will be found in Kononova (1971) and Kononova et al. (1965 and 1975). Major and trace element data on carbonates, and their thermoluminescence characteristics, are given and discussed by Sokolov (1985), and the same author (Sokolov, 1973) using the Mg contents of coexisting clinopyroxene and mica, estimated the temperature of formation of carbonatites and a range of silicate rocks. Epshteyn and Danil'chenko (1988) review the stages of emplacement, rock-forming minerals and principal rock types and their spatial distribution and conclude that replacement rather than magmatic injection was an important process in genesis, particularly for the apatite–magnetite ores. Kharlamov et al. (1981) discuss the origin of the carbonatites based on examination of primary melt inclusions.

Economic Apatite–magnetite, phlogopite, vermiculite and baddeleyite deposits (Kukharenko et al., 1965: Kampel and Shaposhnikov, 1989; Gorbunov et al., 1981) are mined. The phlogopite and vermiculite deposit was discovered in 1960 and in 1962 the production of iron ore from an open pit commenced. In 1964 an experimental plant for the extraction of apatite and baddeleyite came on stream. In that year 100 million tonnes of iron ore were extracted and in 1989 260 million tonnes. The apatite occurs principally as apatite–forsterite and apatite–forsterite–magnetite rocks. The apatite is produced as a by-product of magnetite production with 700,000 tonnes of apatite concentrate per annum, according to Ilyin (1989). In the upper zone of the mine there is strong weathering which has generated francolite and vermiculite. A brief review of the magnetite deposits is given by Smirnov (1989).

Age K–Ar on biotite gave 706 ± 12 Ma (Shanin et al., 1967). However, Ivanenko and Karpenko (1988) from investigation of two samples of nepheline, which gave maximum K–Ar ages of about 700 Ma, showed that they contain some 30% excess ^{40}Ar. After grinding, which destroyed the inclusions containing the excess Ar, the age was calculated to about 500 Ma. Bayanova et al. (1991) dated baddeleyite from tetraferriphlogopite carbonatite by U–Pb and obtained a date of 380 ± 4 Ma.

References Bayanova *et al.*, 1991; Epstein and Anikeeva, 1963; *Epshteyn and Danil'chenko, 1988; Gorbunov *et al.*, 1981; Grinenko *et al.*, 1970; *Ilyin, 1989; *Ivanenko and Karpenko, 1988; Kampel and Shaposhnikov, 1989; *Kharlamov *et al.*, 1981; Kononova, 1971; Kononova and Shanin, 1971; Kononova and Yashina, 1984; Kononova *et al.*, 1965 and 1975; Kukharenko *et al.*, 1965 and 1971; *Smirnov, 1989; *Sokolov, 1973 and 1985; *Zaitsev and Polezhaeva, 1994.

31 SALLANLATVI　　　　66°57′N; 29°10′E
(Sallanlatvinskii)　　　　　　　　Fig. 31

Sallanlatvi is the most westerly member of an east–west-trending group that includes the occurrences of Vuoriyarvi and Kovdozero. The complex is a concentrically zoned intrusion with an area of about 9 km². It comprises three major zones which from the periphery towards the centre decrease in age and are composed respectively of melteigite, urtite–ijolite and carbonatite. The width of the peripheral zone varies from 100–200 m in the north to 600–800 m in the west. The melteigites are fine-grained and characterized by a fluidal layered structure, which dips towards the centre of the complex. These rocks are gradually replaced by banded ijolites with nests and vein-like bodies of urtite, with in the western part of the inner zone the urtites becoming predominant. In the same part of the complex numerous arcuate bodies of ijolite pegmatite enriched in apatite and schorlomite occur, which are traceable along the

strike for up to 200 m, the thickness varying from 0.5 to 2 m. Carbonatites make up the central part of the massif, which also has a zonal structure. The outer ring, which is rather narrow, is composed of a series of closely spaced, steeply-dipping veins and irregular, possibly lens-shaped, bodies of calcite carbonatite with relicts of silicate rocks which have been much replaced by mica. The central part of the carbonatite zone is composed of dolomite–ankerite or dolomite–siderite carbonatites. Calcite carbonatites also form independent veins in the ijolite–urtite and ijolite–melteigite zone of the massif. They are similar in composition to the carbonatites of the ring. There are blocks (0.5–1.5 m diameter) of pyroxenites, much replaced by mica, among the carbonatites of the central core; the pyroxenites are injected by ijolite. All rocks, including the carbonatites, are intersected by thin dykes of monchiquite. The calcite carbonatites of the ring-shaped zone and veins contain 75–80% calcite together with natrolite, cancrinite, phlogopite, aegirine-diopside and secondary chlorite. The accessories include apatite, titanite, magnetite, zircon, pyrochlore, hydroxyapatite, ancylite and parisite. The dolomite–ankerite carbonatites also contain dolomite, rhodochrosite and baryte in the crust of weathering; they are also enriched in hematite, psilomelane, pyrolusite, aragonite and strontianite. Accesories are represented by magnetite, ilmenite, ancylite, parisite, pyrochlore, and an abundance of sulphides of which pyrite, sphalerite and arsenopyrite are typical.

Economic Apatite, baryte and rare earth minerals are concentrated in the weathered crust. Apatite mineralization in the ijolites and urtites reaches 10–15% and there are concentrations (up to 9%) of titanomagnetite in the pyroxenites (Kukharenko *et al.*, 1965).

References Kukharenko *et al.*, 1965; Orlova, 1963.

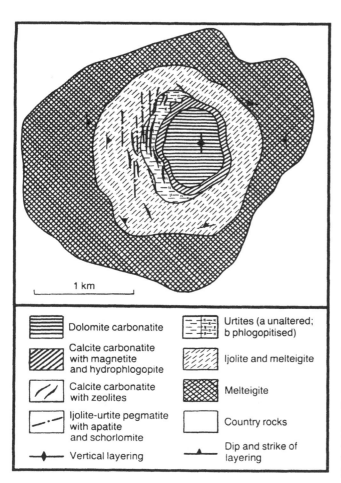

▦ Dolomite carbonatite	Urtites (a unaltered; b phlogopitised)
▨ Calcite carbonatite with magnetite and hydrophlogopite	Ijolite and melteigite
⟋ Calcite carbonatite with zeolites	▥ Melteigite
Ijolite-urtite pegmatite with apatite and schorlomite	Country rocks
◆ Vertical layering	▲ Dip and strike of layering

Fig. 31. Sallanlatvi (after Kukharenko et al., 1965, Fig. 165).

32 VUORIYARVI　　　　66°48′N; 30°07′E
　　　　　　　　　　　　　　　Fig. 32

This complex occurs within the gneisses and migmatites of the Belomorskaya Archaean series where the latter is associated with a Proterozoic sedimentary and volcanogenic sequence. The intrusion, the total area of which is about 20 km², has an elliptical shape and a concentrically zoned structure, which is described in detail by Kapustin (1976). The central and largest part (12 km²) is composed of clinopyroxenites containing individual blocks and lenses of olivinite and olivine–pyroxene rocks. The peripheral zone is made up by melteigites and ijolites with schorlomite, and occasionally with K–feldspar. In the eastern part of the massif is situated a stockwork of phoscorites, which contain calcite and dolomite, that have been proved by drilling to extend to a depth of several hundred metres. In the eastern part of the outer contact zone of the complex are outcrops of breccias, which are similar to those of the Kovdozerskii complex, and probably produced by explosive eruption. The enclosing rocks are intensely fenitized. Four to six kilometres east of the complex in a zone of approximately east–west-trending fractures are outcrops of actinolite–calcite, dolomite–calcite, ankerite–calcite and siderite–dolomite carbonatites and quartz–carbonate rocks. Carbonatites are widely distributed through the complex, mainly as sheets and veins, but also as a stock-like body towards the eastern margin. They are spatially and genetically associated with magnetite–forsterite–apatite rocks (phoscorites), which form a series of thick bodies elongated in an east–west direction. Early and late carbonatites have been distinguished (Kapustin, 1976), the former comprising about 90% of the total.

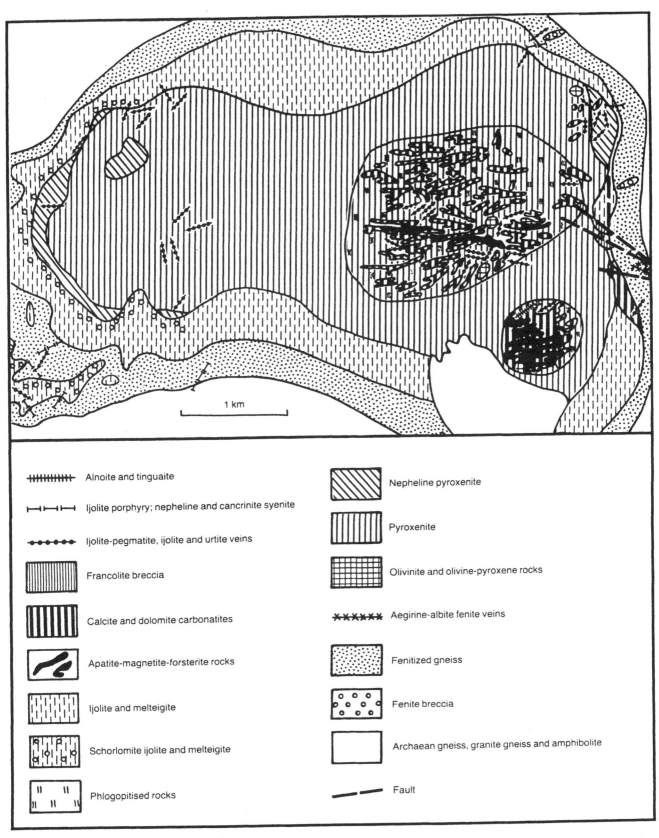

Fig. 32. Vuoriyarvi (after Kukharenko et al., 1965, Fig. 150).

Carbonatite dykes of what is called an 'explosive facies' are described and illustrated with photomicrographs and numerous analyses by Lapin (1981). Veins and dykes are represented within the central complex by ijolite pegmatite, ijolite porphyry, monchiquite, nephe line and cancrinite syenites and tinguaite. Phoscorites are usually characterized by massive, spotted, lens-shaped banded structures and rarely by orbicular (liquation) structures with magnetite ovoids 0.5 cm in diameter, surrounded by a mesostasis of calcite–forsterite with apatite and phlogopite (Fizhenko, 1970; Landa, 1979). Some of these are illustrated by Lapin (1978) who considers the association of carbonatite and phoscoritic rocks in terms of immiscibility. Both apatite phoscorites and carbonatites contain a suite of rare earth and other accesssory minerals including baddeleyite, dysanalyte, pyrochlore, zirconolite, zircon, titanite and sulphides such as pyrrhotite and chalcopyrite. δS^0/oo values for sulphides from melteigite vary from -1.0 to -2.2 and for calcite carbonatite from -3.8 to -4.3 (Grinenko *et al.*, 1970).

Economic High concentrations of titanomagnetite, apatite, phlogopite and vermiculite occur in the complex. There is an apatite–titanomagnetite deposit and forsterite–magnetite, apatite–magnetite and calcite–magnetite ores are developed in which subordinate minerals include baddeleyite, tetraferriphlogopite, pyrrhotite, melilite and diopside (Kukharenko *et al.*, 1965; Orlova and Dyadkina, 1978; Gorbunov *et al.*, 1981).

Age Phlogopite and biotite from clinopyroxenites gave 380–402 Ma by K–Ar (Kukharenko *et al.*, 1965).

References Fizhenko, 1970; Gorbunov *et al.*, 1981; Grinen ko *et al.*, 1970; Kapustin, 1976; Kukharenko *et al.*, 1965; Landa, 1979; *Lapin, 1978 and 1981; Orlova and Dyadkina, 1978.

---

33 KOVDOZERSKII   66°45′N; 31°55′E
(Kovdozero)   Fig. 33

This occurrence is located on a number of small islands in Kovdozero Lake. Melteigite and ijolite form a body extending for nearly 0.5 km across several islands, and these rocks are sometimes distinguished by the presence of orbicular structures. About 75 dykes have been located up to several kilometres to the south. Sudovikov (1946) has distinguished two groups of dykes: (1) melteigite, which is represented by fenitized porphyries and (2) syn- and post-melteigite dykes of nephelinite, monchiquite and foyaite. The country rocks are granite gneisses and amphibolites of the Belomorsk Archaean series and these have been intensively fenitized.

Economic There are high concentrations of titanomagnetite (Kukharenko *et al.*, 1965).

References Kukharenko *et al.*, 1965; Shurkin, 1959; Sudovikov, 1946.

---

34   TIKSHOZERSKII   66°17′N; 31°40′E
                                           Fig. 34

Tikshozerskii is emplaced in Archaean granite gneisses which are altered to form unusual scapolite–microcline fenites. The layered intrusion is formed of two parts: a semi-circular body 5 km in diameter and a northeasterly-trending lens-like body about 3 km long. The complex is composed of olivinite, theralite, gabbro, titanomagnetite clinopyroxenite, nepheline pyroxenite, melteigite–ijolite and numerous dykes of micromelteigite and microijolite. In the northern part of the massif

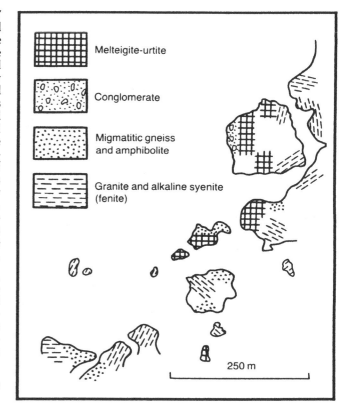

Fig. 33. *Kovdozerskii (after Sudovikov, 1946).*

a small area of gabbroic rocks is intersected by veins of trachytic nepheline syenite and dykes of olivine-bearing lamprophyre. In the outer part of the intrusion are developed cancrinite (20–23%), calcite (25–35%), amphibole rocks and carbonatites which form veins and eruptive breccias. The carbonatites are composed of calcite, Mn–calcite, ankerite and dolomite with accessory magnetite, ilmenite, rutile, corundum, sulphides, ancylite and rare-earth phosphates. These rocks are enriched in La, Ce, Ba (up to 0.5%), Sr (up to 0.6%), Mo and V.

Age From its geological position, petrographic composition and geochemistry the complex is considered to be part of the Elet'ozerskii complex and thus 1800–1900 Ma. However, it is suggested that the complex is possibly polyformational and polychronous (Bogachev *et al.*, 1987).

References Bogachev *et al.*, 1987; Safronova, 1982; Safronova and Gavrilova, 1982.

---

35   ELET'OZERSKII   66°04′N; 31°57′E
(Elet'ozero)   Fig. 35

A layered intrusion, with an area of about 100 km², Elet'ozerskii cuts through Archaean granite gneisses which have been altered to skarns and melted. The complex has a concentric, zoned structure, the peripheral part being composed of a layered gabbro series; the central area is occupied by nepheline syenites. Between these two rock groups is an arcuate body of clino pyroxenites. The gabbros are intersected by dykes of dolerite, spessartite, alkaline syenite pegmatite, syenite porphyry, bostonite and carbonatite. The gabbro series is represented by plagioclase-bearing wherlites which grade upwards into orthoclase gabbros and orthoclase

In the legend of Fig. 33:
Melteigite-urtite
Conglomerate
Migmatitic gneiss and amphibolite
Granite and alkaline syenite (fenite)
250 m

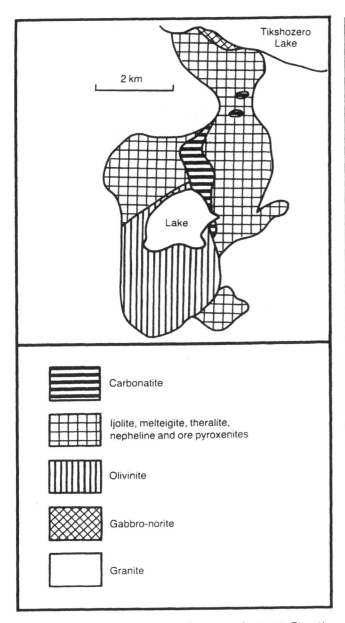

Fig. 34. Tikshozerskii (after Klunin et al., 1987, Fig. 1).

plagioclasites. Rhythmic layering involves alternation of leucocratic layers composed essentially of plagioclase and melanocratic layers of pyroxene, olivine and titanomagnetite. The nepheline syenites, the central area of which covers about 10 km^2, also form dykes and veins which intersect the layered gabbroid series and also the accompanying dyke rocks (dolerite and spessartite). The nepheline syenites comprise nepheline, microcline and aegirine with lesser arfvedsonite, biotite, apatite, zircon, titanite, ilmenite, titanomagnetite and spessartine. The pegmatites comprise microcline, sodic plagioclase and nepheline with accessory pyrochlore, zircon, titanite, columbite, fersmite, thorite and orthite. Carbonatite constitutes the matrix of an "explosion" breccia in the northern part of the complex. It is a calcite carbonatite with biotite, titanite and rare-earth minerals.

Economic The gabbroic rocks are enriched in ilmenite and titanomagnetite and apatite ores are present (Kukharenko *et al.*, 1969 and 1971; Bogachev *et al.*, 1963.)

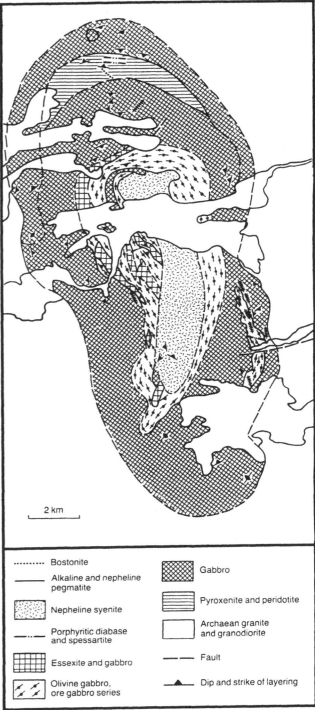

Fig. 35. Elet'ozerskii (after Kukharenko et al., 1969, Fig. 2).

Age K–Ar on biotite from nepheline–syenites gave 1910 Ma and biotite and muscovite from nepheline syenites 1800–1830 Ma, while whole rock determinations on nepheline syenite gave 1705–1620 Ma. Pb–U–Th determinations on orthite, pyrochlore and zircon gave 1860 and 1740 Ma (Pushkarev, 1990). A Sm–Nd isochron age of 2080 ± 180 Ma has been obtained by Kogarko and Karpenko (in press).

References Abakumova, 1966; Bogachev et al., 1963; Kogarko and Karpenko, in press; Kukharenko et al., 1969 and 1971.

References Gorkovets et al., 1981; Orlova and Shadenkov, 1992; *Proskuryakov et al., 1990.

36   KOSTOMUKSHSKII DYKE COMPLEX
64°37′N; 30°36′E
Fig. 36

Dykes of micaceous picrites are confined to north-south-trending fault zones in the Archaean Kontokskaya and Gimolskaya series. The dykes are uniform with thicknesses varying from 0.5 to 3.5 m. The rocks are characterized by porphyritic textures with olivine phenocrysts constituting 10–20%; they are generally replaced by serpentine and less often by talc and carbonate. The groundmass is composed of phlogopite (20–40%), serpentine after olivine (2–35%) and diopside replaced by actinolite (5–15%), with subordinate amounts of talc, calcite and rarely hypersthene. Less frequently the groundmass is made up of pseudoleucite (up to 30%), phlogopite (25%), olivine (15%) kaersutitic amphibole (3–5%) and calcite. The accessory minerals are apatite, magnetite, pyrite, baryte, chrome picotite, titanite, leucoxene and anatase, with more rarely columbite and rutile. In the Russian literature the rock chemistry and mineral compositions are taken to indicate that these rocks are lamproites.

Age K–Ar on phlogopite gave 1120–1300 Ma (Gorkovets et al., 1981).

37   ELISENVAARSKII
(Kaivimyaki and Raivimyaki)

61°25′N; 29°55′E
Fig. 37

Two areas of outcrop of alkaline gabbroic rocks and syenites, called Kaivimyaki and Raivimyaki, which are surrounded by fenites, appear to represent parts of a single, larger occurrence with an area of about 30 km². They are located within the approximately north–south-trending deep-seated fracture zone of Western Preladoga. Here in an area 20 km long and about 5 km wide occur outcrops of alkaline gabbroids and syenites, as well as related fenites, the last consisting of albite, microcline, phlogopite, hastingsite and apatite. The alkaline gabbroids are the most abundant rocks. Peralkaline syenites and melasyenites consist of orthoclase-perthite, aegirine–diopside and hastingsite with enrichment of apatite in places. Among the peralkaline syenites Khazov (1982) has described melanocratic varieties containing 1–20% of aegirine–diopside, 1–50% of hornblende, 5–45% biotite, along with orthoclas-perthite (20–70%) and apatite (2–8 to 15–20%). The apatite-rich varieties of these rocks have been called nevoite and ladogalite by Khazov (1982 and 1983). The rocks described above are intersected by dykes of syenite pegmatite, syenite aplite and alkaline lamprophyres, the last forming eruptive

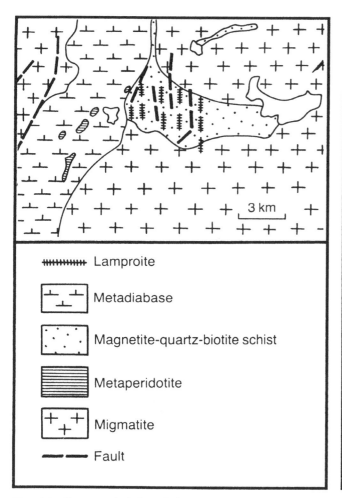

Lamproite

Metadiabase

Magnetite-quartz-biotite schist

Metaperidotite

Migmatite

Fault

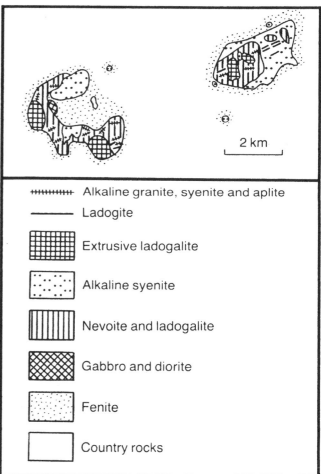

2 km

Alkaline granite, syenite and aplite

Ladogite

Extrusive ladogalite

Alkaline syenite

Nevoite and ladogalite

Gabbro and diorite

Fenite

Country rocks

Fig. 36. Kostomukshskii dyke complex (after Orlova and Shadenkov, 1992, Fig. 1).

Fig. 37. Elisenvaarskii (Raivimyaki and Kaivimyaki) (after Khazov et al., 1987, Fig. 2).

breccias. There is some apatite mineralization associated with these massifs.

Age K–Ar determinations on five biotite samples gave 1600–1900 Ma (Khazov, 1982; Khazov et al., 1987).

References Ivashenko et al., 1985; Khazov, 1982 and 1983; Khazov and Ivashinko, 1980; Khazov et al., 1987.

Kola and Karelia references

ABAKUMOVA, N.B. 1966. Alkaline pegmatites of the Elet'ozerskii massif gabbroid and alkaline rocks (N. Karelia). *Soviet Geology*, 5: 131–40.

BATIEVA, I.D. 1976. *The petrology of alkaline granitoids of the Kola peninsula*. Nauka, Leningrad. 223 pp.

BATIEVA, I.D. 1985. Alkaline granites of the Kanozero-Kolvitsky lake. 1958. In *Alkaline granites of the Kola peninsula*. Izdatelstvo Akademii Nauk SSSR, Moscow and Leningrad. 146–70.

BATIEVA, I.D. and BEL'KOV, I.V. 1984. *The rocks and minerals of the Sakhar'oksky alkaline massif*. Akademii Nauk SSSR, Apatity. 133 pp.

BATIEVA, I.D. et al. 1985. *Magmatic formations of the north-eastern part of the Precambrian Baltic Shield*. Nauka, Leningrad. 175 pp.

BAYANOVA, T.B., KIVNARSKYI, Yu.M., GANNIBAL, A.F., KOSHEEV, O.A. and BALASHOV, Yu.A. 1991. U–Pb dating of baddeleyite from the Kovdor carbonatite massif. In Abstracts of All-Union School Seminar 'Methods of isotope geology'. 21–25 October, 1991. Zvenigovod, St Petersburg. 31.

BEL'KOV, I.V. (ed.) 1985. *Magmatic Precambrian formations of the northeastern part of the Baltic shield*. Nauka, Leningrad. 87 pp.

BEL'KOV, I.V., KRAVCHENKO, E.V. and PUSHKAREV, Yu.D. 1975. Rb–Sr and K–Ar ages of the Rautingsky and Iyokan'sky massifs (Kola Peninsula). *Geochronology of the west-east platform and junction of the Caucasus-Karpat system*. (Abstract) Akademii Nauk SSSR, Moscow. 31–2.

BEL'KOV, I.V., BATIEVA, U.D., VINOGRADOVA, G.V. and VINOGRADOV, A.I. 1988. *Mineralization and fluid regime of the contact zones of the alkaline granite intrusions*. Akademii Nauk SSSR, Apatity.

BILIBIN, Yu.A. 1941. *Post Jurassic intrusions of the Aldan region*. Akademii Nauk SSSR, Moscow and Leningrad. 160 pp.

BILIBIN, Yu.A. 1947. *Petrology of the Yulimakhsky intrusive*. Gosgeolitizdat, Moscow and Leningrad. 240 pp.

*BINDEMAN, I.I., SHARKOV, Ye. V. and IONOV, D.A. 1990. Xenoliths of biotite–garnet–orthopyroxene rocks from a dike-like diatreme on Yelovy Island, White Sea. *International Geology Review*, 32: 905–15.

BLYANKIN, D.S. and VLODAVETS, V.I. 1932. The alkaline complex of Turij Mys. *Works of the Petrography Institute, Akademii Nauk SSSR*, 2: 45–71.

BOGACHEV, A.I., VLODAVETS V.I., KLUNIN, S.F. and KIRILLOV, A.I. 1987. Metallogeny of the Tiksheozersky-Elet'ozersky ultrabasic gabbro-alkaline complex. *Magmatism, metamorphism and geochronology of the west-east Precambrian platform*. 102–4. Akademii Nauk SSSR, Kola Branch, Petrozavodsk.

BOGACHEV, A.I., ZAK, S.I., SAFRONOVA, G.P. and IRINA, K.A. 1963. *Geology and petrology of the Elet'ozero massif gabbroid rocks (geology, petrography, petrology, metallogeny)*. Akademii Nauk SSSR, Moscow and Leningrad. 158 pp.

BORODIN, L.S. 1974. *The main provinces and formations of the alkaline rocks*. Nauka, Moscow. 376 pp.

*BROWN, D.A. 1970. The geological-petrographic features of the massifs of alkaline-ultrabasic rocks of the Kola Peninsula and Northern Karelia. Translation of Kukharenko et al., 1965. *The Caledonian complex of ultrabasic and alkaline rocks and carbonatites of the Kola Peninsula and Northern Karelia*. 1–288. Australian National University, Canberra.

*BROWN, D.A. 1980. The petrology of the alkaline-ultramafic massifs. Translation of Kukharenko et al., 1965. *The Caledonian complex of ultrabasic and alkaline rocks and carbonatites of the Kola Peninsula and Northern Karelia*. 645–754. Australian National University, Canberra.

BULAKH, A.G. and IVANNIKOV, V.V. 1984. *Problems of mineralogy and carbonatite petrology*. Leningradskii Gosudarstvennyi Universitet, Leningrad. 242 pp.

BUSSEN, I.V. and SAKHAROV, A.S. 1967. *Geology of the Lovozero tundra*. Nauka, Leningrad. 125 pp.

BUSSEN, I.V. and SAKHAROV A.S. 1972. *Petrology of the Lovozero alkaline massif*. Nauka, Leningrad. 296 pp.

BUTAKOVA, E.L. and EGOROV, L.S. 1962. The Maimecha-Kotui complex of alkaline and ultrabasic formations. *Petrography of Eastern Siberia*, 1: 417–589. Izdatelstvo Akademii Nauk SSSR, Moscow.

CHUVARDINSKI, V.G. and KARAEV, S.S. 1988. *New data in the investigations of mineral raw materials in the Murmansk region*. 12–13. Akademii Nauk SSSR, Kola Branch, Apatity.

*DUDKIN, O.B. 1991. Carbonatite and the sequence of formation of the Khibiny pluton. *International Geology Review*, 33: 375–84.

DUDKIN, O.B. (ed.) 1987. *Geology and mineralogy of apatite deposits of the Kola peninsula, USSR*. Akademii Nauk SSSR, Kola Branch, Geological Institute, Apatity. 69 pp.

DUDKIN, O.B., POLYAKOV, K.I. and KULAKOV, A.N. 1978. The conditions of formation, complexity and quality of the Kola Peninsula nepheline ore deposits. *Nepheline raw material*, 26–32. Nauka, Leningrad.

DUDKIN, O.B., MINAKOV, F.V., KRAVCHENKO, M.P., KYLAKOV, A.N., POLEZHAIVA, L.E., PRIPACHKIN, V.A., PUSHKAREV, Yu.D. and RUNGENEN, G.E. 1984. *Carbonatites of the Khibina*. Akademii Nauk SSSR, Kola Branch, Apatity. 98 pp.

ELISEEV, N.A. 1953. *Structural petrology*. Leningrad State University, Leningrad. 310 pp.

ELISEEV, N.A. 1957. On the classification of nepheline rocks. *Transactions of the Mineralogical Society*, 5: 48–53.

ELISEEV, N.A. 1958a. Alkaline granites of the Chagvedaiv massif. In *Alkaline granites of the Kola peninsula*. 208–12. Izdatelstvo Akademii Nauk SSSR, Moscow and Leningrad.

ELISEEV, N.A. 1958b. Petrology and metallogeny of ultrabasic and basic nickel-bearing intrusions of the Kola peninsula. *Transactions of the II All-Union petrographic meeting, Tashkent*, 353–7.

ELISEEV, N.A. and FEDOROV, E.E. 1953. The Lovozero pluton and its deposits. *Materiali Laboratorii Geologii Dokembraya*, 1.

ELISEEV, N.A., OZSHINSKY, N.S. and VOLODIN, E.N. 1939. *Geological map of the Khibina tundras*. Geolupravleniye, Leningrad, 19: 68 pp.

ELISEEV, N.A., VANIDOVSKAYA, A.V., POKROVSKY, S.D., SAKHAROV, A.S. and UNKSOV, V.A. 1937. About palaeosols in the central part of the Kola peninsula. *Problems of Soviet Geology*, 7: 283–94.

*EPSHTEYN, Y.M. and DANIL'CHENKO, N.A. 1988. A spatial-genetic model of the Kovdor apatite-magnetite deposit, a carbonatite complex of the ultramafic, ijolite and carbonatite rock association. *International Geology Review*, 30: 981–93.

EPSTEIN, E.M. and ANIKEEVA, L.I. 1963. Some aspects on geology and petrology of the ultrabasic rocks complex. Physical-chemical problems of the formation of ore. *Doklady Akademii Nauk SSSR*, 11: 182–96.

FERSMAN, A.E. 1955. *Geochemistry*. III. Leningrad Himteorisdat. 283 pp.

FERSMAN, A.E. 1958. *Selected works*. III and IV. Akademii Nauk SSSR, Moscow. 798 and 555 pp.

FIZHENKO, V.V. 1970. On the genesis of orbicular texture in apatite–calcite–forsterite–magnetite ores of the Vuorijarvi and Gornozero massifs. In *Petrological and structural analysis of crystalline formations*. 167–73. Nauka, Leningrad.

FRAKTSESSON, E.V. 1968. *Petrology of kimberlites*. Nedra, Moscow. 199 pp.

*FRENKEL, M.Y. and KHAPAYEV, V.V. 1989. A convective cumulation model of crystallization differentiation of the melt and formation of the apatite deposits in the Khibiny ijolite–urtite intrusion. *Geochemistry International*, 27(4): 101–12.

GALAKHOV, A.V. 1961. A geological-petrographical sketch of the Khibina alkaline massif and problems on further scientific investigations. *Problems in geology, mineralogy and petrography of Khibina tundras*. Akademii Nauk SSSR, Moscow. 35 pp.

GALAKHOV, A.V. 1962. Peculiarities of the content of the ore-forming nepheline of the Khibina alkaline massif. *Materials of mineralogy of the Kola peninsula*. 35–42. Akademii Nauk SSSR, Apatity.

GALAKHOV, A.V. 1966. Manifestations of alkaline-ultrabasic magmatism in the Khibina tundras (Kola peninsula). *Doklady Akademii Nauk SSSR, Moscow*, 170: 657–60.

GALAKHOV, A.V. 1975. *The petrology of the Khibina alkaline massif*. Izdatelstvo Akademii Nauk SSSR, Kola Branch, Leningrad. 256 pp.

GALAKHOV, A.V., KOSIREVA, L.V. and ZUBAREV, A.I. 1978. Titano-magnetite deposits connected with alkali-ultrabasic massifs of the Kola peninsula. 99–124. *Alkaline rocks of the Kola peninsula and related apatite mineralization*. Akademii Nauk SSSR, Apatity.

*GERASIMOVSKY, V.I., VOLKOV, V.P., KOGARKO, L.N. and POLYAKOV, A.I. 1974. Kola Peninsula. In H. Sorensen (ed.), *The alkaline rocks*. 206–21. John Wiley, London.

GERASIMOVSKY, V.K., VOLKOV, V.P., KOGARKO, L.N., POLYAKOV, A.I., SAPRYKINA, T.V. and BALASHOV, Yu.A. 1966. *Geochemistry of the Lovozero alkaline massif*. Nauka, Moscow. 395 pp.

*GERASIMOVSKY, V.K., VOLKOV, V.P., KOGARKO, L.N., POLYAKOV, A.I., SAPRYKINA, T.V. and BALASHOV, Yu.A. 1968. *The geochemistry of the Lovozero alkaline massif*. (trans D.A. Brown), 1: Geology and petrology 1–224; 2: Geochemistry 1–369. Australian National University Press, Canberra.

GERLING, E.K. 1942. On the age of the pyroxenitic intrusions of Africanda and Ozernaya Varaka on the Kola peninsula. *Izvestiya Akademii Nauk SSSR*, 35: 168–9.

GINSBURG, A.N. and FELDMAN, Z.G. 1978. Tantalum and niobium deposits. *USSR ore deposits*, 3: 341–2. Nedra, Moscow.

GLEVASSKY, E.B. and KRIVDIK, S.G. 1981. *The Precambrian carbonatitic complex of the Azov region*. Naukova Dumka, Kiev. 227 pp.

GORBUNOV, G.I., BEL'KOV, I.V., MAKIEVSKY, S.I., GORYAINOV, P.M., SAKHAROV, A.S., YUDIN, B.A., ONOKHIN, F.M., GONCHAROV, Yu.V., ANTONYUK, E.S. and VESELOVSKY, N.N. 1981. *Mineral deposits of the Kola Peninsula*. Nauka, Leningrad. 272 pp.

GORKOVETZ, B.Ya., RAYEVSKAYA, M.B, BELOUSOV, E. and IRINA, K.A.. 1981. *Geology and metallogeny of the Kostomuksha ore deposit region*. 107–9. Petrozavodsk, Karelia. 143 pp.

GRIB, B.G., ORLOVA, M.P., POLUNINA, L.Y. SELIVANOVSKAYA, T.V., UDALOVA, A.A. and YANOVA, E.N. 1981. *Evolution and metallogeny of the sedimentary cover of the Russian Platform*. (ed. U.G. Staritsky). Nedra, Leningrad. 227 pp.

GRINENKO, L.N., KONONOVA, V.A. and GRINENKO, V.A. 1970. Isotopic composition of sulphur of sulphides from carbonatites. *Geokhimiya*, 1: 66–75. *Geochemistry International*, 645–53.

*ILYIN, A.V. 1989. Apatite deposits in the Khibiny and Kovdor alkaline igneous complexes, Kola Peninsula, north-western USSR. In A.J.G. Notholt, R.P. Sheldon and D.F. Davidson (eds), *Phosphate deposits of the world, 2: Phosphate rock resources*. 485–93. Cambridge University Press, Cambridge.

*IVANENKO, V.V. and KARPENKO, M.I. 1988. ^{39}Ar–^{40}Ar data on excess argon–40 in nepheline from the Kovdor massif, Kola Peninsula. *Geochemistry International*, 25(1): 77–82.

IVANOVA, T.N. 1963. *Apatite deposits of the Khibina tundras*. Gosgeoltekhizdat, Moscow. 288 pp.

IVANOVA, T.N. 1968. Apatite deposits and ores in nepheline syenites and other alkaline rocks. *Apatite*. 59–85. Nauka, Moscow.

IVANOVA, T.N, ARZAMASCHEV, A.A. and KONDRATOVICH, U.U. 1978. *Alkaline rocks of the Kola peninsula and their apatite-bearing rocks*. 16–32. Akademii Nauk SSSR, Kola Branch, Apatity.

IVASHENKO, V.I., OVCHINNIKOVA, L.V. and VORONOVSKY, S.N. 1985. Genesis and age of apatite-bearing rocks of the Elisenvaarsky massif. *Doklady Akademii Nauk SSSR*, 280: 973–6.

*KAPUSTIN, Yu.L. 1976. Structure of the Vuoriyarvi carbonatite complex. *International Geology Review*, 18: 1296–304.

*KHAPAYEV, V.V. and KOGARKO, L.N. 1987. Composition of rock-forming minerals in the Khibiny apatite-bearing intrusion and the origin of the apatite deposits. *Geochemistry International*, 24(12): 21–32.

*KHARLAMOV, Ye.S., KUDRYAVISEVA, G.P., GARANIN, V.K., KORENNOVA, N.G., MOSKALYUK, A.A., SANDOMIRSKAYA, S.M. and SHUGUROVA, N.A. 1981. Origin of carbonatites of the Kovdor deposit. *International Geology Review*, 23: 865–80.

KHAZOV, R.A. 1982. *Metalogeny of the Ladozsh-Botnichesky geoblock of the Baltic shield*. Nauka, Leningrad, 192 pp.

KHAZOV, R.A. 1983. Ladogalite – a new apatite-bearing alkaline ultrabasic rock. *Doklady Akademii Nauk SSSR, Leningrad*. 268: 1199–203.

KHAZOV, R.A. and IVASHINKO, V.I. 1980. New manifestations of alkaline magmatism and apatite ores on the Baltic sheild. *Doklady Akademii Nauk SSSR*, 252: 944–7.

KHAZOV, R.A., POPOV, M.G. and BISKE, N.S. 1987. K–alkaline rocks of the Ladoga region and their ores. *Materials on Karelia metallogeny*. 78–88. KOLFAN, Petrozavodsk.

*KHOMYAKOV, A.P. 1987. Salt minerals in ultra-agpaitic rocks and the ore potential of alkaline massifs. *International Geology Review*, **29**: 1446–56.

KIRICHENKO, L.A. 1970. Kontozero series of the coal rocks of the Kola peninsula. *Materials on Geology and Mineral Deposits of north-west Russia*, Nedra, Leningrad, **9**: 110.

KIRNARSKY, Yu.M. 1962. Apatite and sphene in the rocks of the massif 'Kholodnoye'. *Mineralogy of the Kola Peninsula*. 188–99. Nauka, Moscow and Leningrad.

KIRNARSKY, Yu.M. and KOZYEVA, L.V. 1978. New data on the Mavragubskii alkaline massif. *Geology and ore deposits of the Kola Peninsula*. 109–16. Doklady Akademii Nauk SSSR, Kola Branch, Apatity.

KLUNIN, S.F., SAFRONOVA, G.P. and BELO-BORODOV, V.I. 1987. New apatite deposits in North Karelia. *Geology and mineralogy of apatite deposits*. 37–43. Akademii Nauk SSSR, Kola Branch, Apatity.

KOGARKO, L.N. 1977. *Genetic problems of agpaitic magmas*. Nauka, Moscow. 294 pp.

*KOGARKO, L.N. 1987. Alkaline rocks of the eastern part of the Baltic Shield (Kola Peninsula). In J.G. Fitton and B.G.J. Upton (eds), *Alkaline Igneous Rocks*, Geological Society of London, Special Publication **30**: 531–44.

KOGARKO, L.N. and KARPENKO, S.F. (in press). New data on the genesis and sources of the alkaline massifs Gremyakha-Vyrmes and Elet'ozero. *Geochemistry Journal*.

*KOGARKO, L.N., KRAMM, U., DUDKIN, O.B. and MINAKOV, F.V. 1986. Age and genesis of carbonatites of the Khibiny alkalic pluton, as inferred from rubidium-strontium isotope data. *Transactions (Doklady) of the USSR Academy of Sciences, Earth Science Sections*, **289**: 196–8.

KOGARKO, L.N., KRAMM, U., BLAXLAND, A., GRAUERT, B. and PETROVA, E.N. 1981. Age and origin of the alkaline rocks of the Khibina massif (Rb and Sr isotopes). *Doklady Akademii Nauk SSSR, Geochemistry*, **260**: 1001–4.

KONONOVA, V.A. 1971. On the role of the magmatic and metasomatic processes for the origin of the melteigite-urtite series of rocks. In A.I. Tugarinov (ed.) *Geochemistry, petrography and mineralogy of alkaline rocks*. 35–52. Nauka, Moscow.

KONONOVA, V.A. 1976. *The jacupirangite-urtite series of alkaline rocks*. Nauka, Moscow. 214 pp.

*KONONOVA, V.A. and SHANIN, L.L. 1971. On the possible application of nepheline for alkaline rock dating. *Bulletin Volcanologique*, **35**: 1–14.

*KONONOVA, V.A. and YASHINA, R.M. 1984. Geochemical criteria for differentiating between rare-metallic carbonatites and barren carbonatite-like rocks. *Indian Mineralogist*, 136–50.

KONONOVA, V.A., ORGANOVA, N.I. and LOME'KO, E.I. 1965. On the composition and temperature of the nepheline crystallization from the ijolite-melteigite series of rocks. *Izvestiya Akademii Nauk SSSR, Seriya Geologiya*, **7**: 65–73.

KONONOVA, V.A., LAPUTINA, I.P., LOME'KO, E.I. and TIMOFEEVA, I.A. 1975. Typomorphic features of the rock-forming minerals from the jacu-pirangite-urtite series of rocks. *Izvestiya Akademii Nauk SSSR. Seriya Geologiya*, **7**: 45–58.

KOSTYLEVA-LABUNTSOVA, E.E., BORUTSKY, B.E., SOKOLOVA, M.N., SHLUKOVA, Z.V., DORFMAN, M.D., DUDKIN, O.B., KOZYREVA, L.V. and IKORSKY, S.V. 1978. *Mineralogy of the Khibina massif*. Nauka, Moscow. 2 volumes.

*KRAMM, U., KOGARKO, L.N., KONONOVA, V.A. and VARTIAINEN, H. 1993. The Kola alkaline pro-vince of the CIS and Finland: precise Rb–Sr ages define 380–360 Ma age range for all magmatism. *Lithos*, **30**: 33–44.

KUKHARENKO, A.A., ORLOVA, M.P. and BAGDA-SAROV, E.A. 1969. Alkaline gabbroids of Karelia. Leningrad State University Press, Leningrad. 184 pp.

KUKHARENKO, A.A., BULAKH, A.G., IL'INSKY G.A., SHINKAREV, N.F. and ORLOVA, M.P. 1971. *Metallogenic peculiarities of alkaline formations of the eastern part of the Baltic shield*. Trudy Leningrad-skogo Obshchestva Estestvoispytatelei, **122** (2): 278 pp.

KUKHARENKO, A.A., ORLOVA, M.P., BULAKH, A.G., BAGDASAROV, E.A., RIMSKAYA-KORSA-KOVA, O.M., NEPHEDOV, E.I., IL'INSKII, G.A., ERGEEV, A.S. and ABAKUMOVA, N.B. 1965. *The Caledonian complex of ultrabasic alkaline rocks and carbonatites of the Kola peninsula and north Karelia*. Nedra, Moscow. 772 pp.

KURBATOVA, G.S. 1978. The distribution of phos-phorus in the rocks of Vuorijarvi. *Alkaline rocks of the Kola Peninsula and their apatite*, 64–70. Akademii Nauk SSSR, Apatity.

KURILEVA, N.A. and NOSIKOV, V.V. 1959. Volcanic eruptive pipes on the Kola Peninsula. *Razvedka i ohrana nedra*, **3**: 5–8.

LANDA, E.A. 1979. On orbicular picrite of the Vuoriyarvi massif and genesis of similar rocks. *Doklady Akademii Nauk SSSR*, **245**: 700–2.

LANDA, E.A. and EGOROV, L.S. 1974. *Apatite deposits of carbonatitic complexes*. Nedra, Leningrad. 145 pp.

*LAPIN, A.V. 1978. Geologic examples of limited miscibility in ore–silicate–carbonate melts. *Doklady Earth Science Sections. American Geological Insti-tute*, **231**: 163–6.

*LAPIN, A.V. 1979. Mineral parageneses of apatite ores and carbonatites of the Sebl'yavr massif. *International Geology Review*, **21**: 1043–52.

*LAPIN, A.V. 1981. Carbonatites and the explosive and dike facies (systematics of the geological facies of carbonatites). *International Geology Review*, **23**: 761–75.

MILASHEV, V.A. 1974. *Kimberlitic provinces*. Nedra, Leningrad. 238 pp.

*MITCHELL, R.H., PLATT, R.G. and LOWNAY, M. 1987. Petrology of lamproites from Smoky Butte, Montana. *Journal of Petrology*, **28**: 645–78.

O'INSKY, I.S. 1935. Lovchorrite-rinkolite deposits of the Khibina external belt. *Proceedings of the All-Union Mineralogical Society*. Ser 2, **64**: 355–415.

ORLOVA, M.P. 1959. On the genesis of turjaite of the Salmagorsky massif on the Kola peninsula. *Materials on the mineralogy of ore deposits*. All-Union Scien-tific Research Geological Institute, New Series, **26**: 10–8.

ORLOVA, M.P. 1963. Some questions of petrochem-istry and petrology of the Caledonian complex of alkaline ultrabasic rocks of the Kola Peninsula. *All-Union Scientific Research Geological Institute, New Series*, **96**: 18–41.

ORLOVA, M.P. and DYADKINA, I.Yu. 1978. *Regional and local varieties of the distribution of phlogopite deposits*. Nedra, Leningrad. 85 pp.

ORLOVA, M.P. and SHADENKOV, E.M. 1992. Lamproites of Kostomuksha. *Zapiski Vsesoyuznogo Mineralogicheskogo Obshchestva*, **6**: 33–43.

OSOKIN, E.D. 1980. Rare-earth mineralization and genetic aspects of the Lovozero massif formation. *Ore geochemistry and geology of magmatogenic deposits*. 168–78. Nauka, Moscow.

*OZHOGIN, V.A. 1969a. Statistical processing of

silicate analyses for the upper Ponoy alkalic granite and its host rocks (Kola peninsula). *Doklady Akademii Nauk SSSR*, 182: 156–9.

*OZHOGIN, V.A. 1969b. Zircon of the upper Ponoy alkalic granite, Kola peninsula, as an indicator of its origin. *Doklady Earth Science Sections. American Geological Institute*, 182: 159–62.

POLKANOV, A.A. 1938. The pluton of Chagve-Uaiv alkaline rocks. *Izvestiya Akademii Nauk SSSR, Seriya Geologiya*, 2: 771–801.

POLKANOV, A.A. and ELISEEV, N.A. 1941. *The petrology of the Gremyakha-Vyrmes pluton (Kola peninsula)*. Gosudarstvennyi Universitet, Leningrad. 112 pp.

POLKANOV, A.A., ELISEEV, N.A., ELISEEV, E.N. and KAVARDIN, G.I. 1976. *The Gremyakha-Vyrmes massif on the Kola peninsula*. Nauka, Moscow. 51 pp.

POPOV, A.S. 1967. On the Palaeozoic age of Kola peninsula volcanism. *Doklady Akademii Nauk SSSR*, 174: 173–6.

PROSKURYAKOV, V.V. and ZAK, S.I. 1966. The Kurginsky massif of ultrabasic and alkaline rocks on the Kola Peninsula. *Alkaline rocks of the Kola Peninsula*. 44–54. Nauka, Moscow and Leningrad.

PUSHKAREV, Yu.D. 1990. *Megacycles in the evolution of the crust-mantle system. Leningrad*. Nauka, Leningrad. 300 pp.

PUSHKAREV, Yu.D., OSOKIN, A.C., KRAVCHENKO, M.P. and RYUNGENEN, G.I. 1987. Isotope composition of lead and strontium in apatites from the Gremyakha-Vyrmes alkaline massif as a basis for the determination of the age and source of the primary magmatic melt. *In* Abstracts of All-Union School Seminar 'Methods of isotope geology'. 1–13 December 1987. Moscow. 174–5.

PUTINTSIVA, E.V. and UVADIEV, L.I. 1987. On the evolution of dyke magmatism of the Kandalaksha shield (Kola peninsula). *Magmatism, metamorphism and geochronology of the west--east Precambrian platform*. 91–3. Akademii Nauk SSSR, Kola Branch, Petrozavodsk.

*PYATENKO, I.K. and OSOKIN, Ye.D. 1988. Geochemical features of the carbonatite paleovolcano at Kontozero, Kola Peninsula. *Geochemistry International*, 25 (12): 101–13.

*PYATENKO, I.K. and SAPRYKINA, L.G. 1976. Carbonatite lavas and pyroclastics in the Paleozoic sedimentary volcanic sequence of the Kontozero district, Kola Peninsula. *Doklady Earth Science Sections. American Geological Institute*, 229: 185–7.

PYATENKO, I.K. and SAPRYKINA, L.G. 1981. Petrological peculiarities of alkaline basaltoids and volcanic carbonatites of the Russian platform. *Petrology and petrochemistry of magmatic formations*. 233–55. Nauka, Moscow.

RAMSAY, W. 1890. Geologische Beobachtungen auf der Halbinsel Kola. *Fennia. Bulletin de la Societe de Geographie de Finlande. Helsingfor*, 3 (7): 1–52.

RAMSAY, W. 1899. Das Nephelinsyenitgebiet auf der Halbinsel Kola, II. *Fennia. Bulletin de la Societe de Geographie de Finlande. Helsingfor*, 15 (2): 1–27.

RAMSAY, W. and HACKMAN, V. 1894. Das Nephelinsyenitgebiet auf der Halbinsel Kola, I. *Fennia. Bulletin de la Societe de Geographie de Finlande. Helsingfors*, 11 (2): 1–225.

RONENSON, B.M. 1966. Origin of miaskites and their connection with rare-earths. *Geology of the rare-earths deposits*. Nedra, Moscow. 270 pp.

*RONENSON, B.M., AFANAS'YEV, B.V. and LEVIN, V.Ya. 1981. Turjaite paragenesis of Tur'ya Peninsula. *International Geology Review* 23: 535–43.

RUDENKO, S.A. 1964. On the genesis of apatite deposits of the Khibina massif. *Zapiski Leningradskogo Krasnogo Znameni Gornogo Instituta im G.V. Plekhanova*, 47: 49–70.

RUSANOV, M.S., ARZAMASTSEV, A.A. and SHEVCHENKO, S.A. 1989. *Ivanovsky volcanic-intrusive complex-new appearance of alkaline magmatism on the Kola Peninsula*. Akademii Nauk SSSR, Kola Branch, Apatity. 29 pp.

SAFRONOVA, G.P. 1982. Perspectives of the Tikshozero massif metallogeny in connection with new types of alkaline and carbonatitic rocks. *Metallogeny of Karelia*. 143–60. Nauka, Petrozavodsk.

SAFRONOVA, G.P. and GAVRILOVA, L.M. 1982. On carbonatites of the Tikshozersky massif (data on oxygen isotopes in carbonate). *Metallogeny of Karelia*. 161–7. Publishing House of the Geological Institute, Petrozavodsk.

SAMOILOV, V.S. 1977. *Carbonatites (facies and conditions of formation)*. Nauka, Moscow. 271 pp.

*SAMOILOV, V.S. and AFANAS'YEV, B.V. 1980. New data on the carbonatites of the alkaline-ultrabasic complex of Turiy Peninsula. *International Geology Review*, 22: 39–50.

SANINA, M.Ya. 1975. *Dolomitic and ankeritic carbonatites of eastern Siberia*. Nedra, Moscow. 176 pp.

SEMENOV, E.I. 1963. *Mineralogy of the rare-earths. (Mineralogy, genetic types of mineralization and main features of rare-earth elements geochemistry)*. Editorial House of the USSR Academy of Sciences, Moscow. 412 pp.

SHABLINSKY, G.N. 1963. On the minor structure of the Khibina and Lovozero plutons. *Trudy Leningradskobo Obshchestva Estestvoispytatelei*, 74: 41–3.

SHANIN, L.L., KONONOVA, V.A. and IVANOV, I.B. 1967. On the application of nepheline in K–Ar geochronometry. *Izvestiya Akademii Nauk SSSR, Seriya Geologiya*, 5: 19–30.

SHURKIN, K.A. 1959. On Palaeozoic pseudoconglomerates of North Karelia and the Kola peninsula. *Doklady Akademii Nauk SSSR*, 125: 1329–32.

SHURKIN, K.A. 1960. On Kandalaksha conglomerates of Turij Mys. *Works of the Precambrian Laboratory of the USSR*, 9: 398–411.

*SMIRNOV, V.I. 1989. European part of the U.S.S.R. *In* F.W. Dunning, P. Garrard, H.W. Haslam and R.A. Ixer (eds). *Mineral deposits of Europe. 4/5: Southwest and eastern Europe, with Iceland*. 279–407. The Institution of Mining and Metallurgy and The Mineralogical Society, London.

*SOKOLOV, S.V. 1973. A geothermometric study of carbonatite complexes. *Geochemistry International*, 10: 1110–16.

*SOKOLOV, S.V. 1985. Carbonates in ultramafite, alkali rock, and carbonatite intrusions. *Geochemistry International*, 22(4): 150–66.

STARITSKY, Yu.G. 1981. History of development and mineralogy of the Russian platform. *Works of the All-Union Scientific Research Geological Institute. Nedra, Leningrad*, 308: 226.

STARITSKY, Yu. G., ORLOVA, M.P., POLYNINA, L.A. *et al*. 1981. Magmatism of the Russian Platform. *In Magmatism of the platform regions of the USSR*. 112–99, Nedra, Leningrad.

SUBBOTIN, V.V. and MICHAELIS, S.A. 1986. Genetic types of apatite ores of the Seblyavr complex deposit. *Deposits of nonmetallic raw materials of the Kola Peninsula*. 27–35. Akademii Nauk SSSR, Kola Branch, Apatity.

SUDOVIKOV, N.G. 1946. Petrology of the Kovdozersky complex of alkaline rocks. *Uchenye Zapiski Leningradskogo Gosudarstvennyi Universitet. Leningrad. Seriya Geologiya*, 93: 275–349.

TERNOVA, V.I. 1977. *Carbonatite massifs and their mineral deposits.* Leningradska Gosudarstvennyi Universitet, Leningrad. 166 pp.

TERNOVA, V.I., AFANAS'EV, B.V. and SULIMOV, B.I. 1969. *Geology and prospecting of the Kovdor vermiculite-phlogopite deposit.* Nedra, Leningrad. 28 pp.

TIKHONENKOV, I.P. 1963. *Nepheline syenites and pegmatites of the northern part of the Khibina massif and the role of postmagmatic occurrences in their formation.* Editorial House of the USSR, Moscow. 247 pp.

TIKHONENKOV, I.P. and SEMENOV, Ye.I. 1963. Arsenides of cobalt, nickel and iron in alkaline pegmatites. *Doklady Akademii Nauk SSSR,* 150: 888–9.

VASILIEV, Yu.R. and ZOLOTUKHIN, V.V. 1975. *Petrology of the ultrabasites from the north of the Siberian platform and some other problems of their genesis.* Nauka, Novosibirsk. 271 pp.

VETRIN, B.P. 1970. On the age correlation of leucocratic granites–alaskites and alkaline granites–syenites. *Materials on geology and metallogeny of the Kola peninsula.* 1: 143–7.

VINOGRADOV, A.N., BATIEVA, I.D., BELKOV, I.V. and VINOGRADOV, G.V. 1985. Petrochemical types and succession of formation of intrusive series in the polyformational Gremyakha-Vyrmes massif. *Petrology and mineralogy of alkaline, alkaline-ultrabasic and carbonatite complexes of the Karelia-Kola region.* 61–9. Akademii Nauk SSSR. Kola Branch, Apatity.

VLASOV, K.A., KUZ'MENKO, M.Z. and ES'KOVA, E.M. 1959. *The Lovozero alkaline massif.* Akademii Nauk SSSR, Moscow. 623 pp.

*VLASOV, K.A., KUZ'MENKO, M.Z. and ES'KOVA, E.M. 1966. *The Lovozero alkali massif* (trans D.G. Fry and K. Syers). Oliver and Boyd, Edinburgh and London. 627 pp.

VOLOTOVSKAYA, N.A. 1960. The Karelia-Kola petrographic province of ultrabasic, alkaline and carbonatitic rocks. *Magmatism and its connection with mineral deposits.* Nedra, Moscow. 451–4.

VOLOTOVSKAYA, N.A. (ed.) 1972. *The Khibina alkaline massif.* Nedra, Leningrad. 215 pp.

VOROB'OVA, O.A. 1969. The main peculiarities of distribution and formation of alkaline rocks. *Problems of the geology of mineral deposits, petrology and mineralogy.* Nauka, Moscow. 2: 62–81.

YASHINA, R.M. 1982. Alkaline magmatism of the folded blocky regions (southern margins of the Siberian Platform). *Alkaline magmatism of folded-block areas.* Nauka, Moscow. 274 pp.

*ZAITSEV, A. and POLEZHAEVA, L. 1994. Dolomite-calcite textures in early carbonatites of the Kovdor ore deposit, Kola peninsula, Russia: their genesis and application for calcite–dolomite geo-thermometry. *Contributions to Mineralogy and Petrology,* 115: 339–44.

ZAK, S.I., KAMENEV, E.A. and MINAKOV, F.V. 1972. *The Khibina alkaline massif.* Nedra, Moscow. 175 pp.

ZAVARITSKY, A.N. and ZAVARITSKY, V.A. 1973. *Petrography of the Itimsky alkaline massif,* Nauka, Moscow. 184 pp.

*In English

KANIN-TIMAN

Two periods of alkaline magmatism are distinguished in the Kanin-Timan Province: one of Upper Proterozoic and one of Cambrian and Upper Devonian age. The occurrences of alkaline rocks are located on the Kanin Peninsula (Fig. 38), close to the north coast and to the south of Timan. It was formerly thought that the Kanin-Timan basement was of Upper Proterozoic age and separated from the Russian platform by the West Timan marginal fault (Bashilov and Kaminsky, 1975). However, according to new drilling results the Timan basement is composed of Archaean and Lower Proterozoic crystalline rocks (Tisova, 1984). The Kanin-Timan Precambrian block structure is apparently very similar to that of the East European platform, the magmatism having many features that are typical for such platforms.

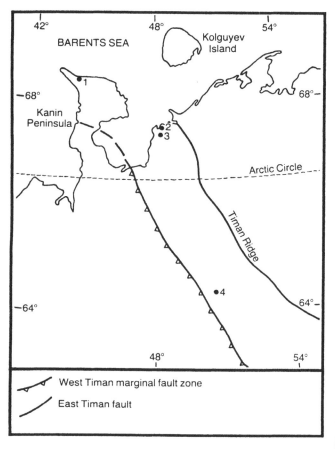

Fig. 38. *Distribution of alkaline rocks in the Kanin-Timan area (after Ivensen, 1964, Fig.1 and Bashilov and Kaminsky, 1975).*

1 Kanin Nos	3 Malyi Kameshek
2 Rumyanichnyi	4 Chetlasskii

1 KANIN NOS   68°34'N; 45°04'E

This small intrusive complex includes syenite-monzonite with lenses of nepheline syenite, which are highly altered. The nepheline syenites are composed of large crystals of albitized microcline and nepheline replaced by muscovite, biotite and aegirine.

Age Devonian, according to geological data.

Reference Ivensen, 1964.

2 RUMYANICHNYI   67°42'N; 47°48'E
Fig. 39

The peralkaline and nepheline syenites of this occurrence have a migmatitic structure and extend over an area of about 20 km². They are cut by lamprophyric dykes and veins of varying composition including carbonatites. The nepheline syenites are composed of alkali feldspar ($Or_{59}Ab_{41}$ – $Or_{46}Ab_{54}$; 35–60%), nepheline (up to 28%), plagioclase with variable composition from An_{1-9} to An_{20-36} (up to 23%), biotite (up to 2.5%), hastingsite (2–17%), and sometimes clinopyroxene (augite, aegirine). Secondary carbonates, fluorite and other accessory minerals are widespread. Lamprophyres are of variable composition, some of them containing nepheline. They include picrite porphyry, alkaline monticellite basalt, alnöite and fourchite, all of them very much altered. Minor bodies of fourchite 6 m thick, consist of clinopyroxene phenocrysts in a groundmass of aegirine-bearing clinopyroxene, zeolites and apatite. Among the youngest veins are nepheline–feldspar rocks with biotite or alkaline amphibole. One alkali picrite porphyry dyke, 0.2–1 m wide and cutting syenitic migmatite, with SiO_2 36–38%, Na_2O about 1% and K_2O 5% has been found along the sea coast. Veins of carbonatite 10–15 cm wide cut this dyke. Picrite porphyries are altered to glimmerites with relics of olivine, augite

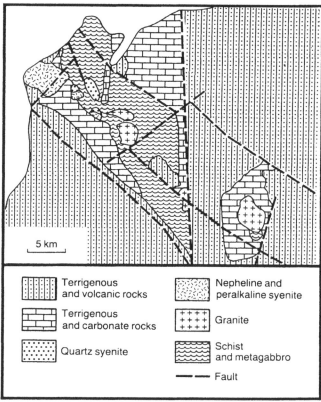

Fig. 39. *Distribution of alkaline rocks in the northern part of the Kanin-Timan province (after Smirnov et al., 1980, Fig. 1). Rumyanichnyi is the large intrusion in the northwest of the province and Malyi Kameshek lies 6 km southeast from it.*

and diopside and pseudomorphs after melilite, amongst others.

Age K–Ar on lamprophyres gave 505–546 Ma (Tisova, 1984) and on phlogopite from alkaline picrite porphyry 510 ± 18 Ma (Shchukin and Smirnov, 1987).

References Cherny *et al.*, 1977; Ivensen, 1964; Shchukin and Smirnov, 1987; Smirnov *et al.*, 1980; Tisova, 1984.

---

3 MALYI KAMESHEK   67°38′N; 48°00′E
Figs 39 and 40

This massif, which has an area of 1.5 km², consists mainly of quartz syenite and peralkaline syenite but with small areas of nepheline syenite. The nepheline syenites are composed of alkali feldspar $Or_{50-52}Ab_{48-50}$ (66–86%), nepheline (up to 10%), plagioclase An_{33-38} (up to 3.5%), biotite (7–9%) and amphibole.

Age Devonian – according to geological relationships.

References Ivensen, 1964; Smirnov *et al.*, 1980.

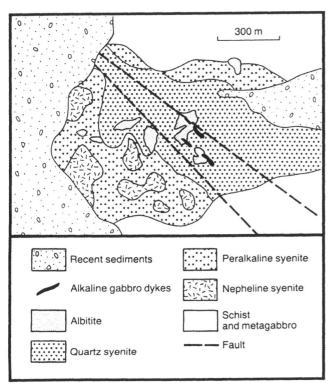

Legend:
- Recent sediments
- Alkaline gabbro dykes
- Albitite
- Quartz syenite
- Peralkaline syenite
- Nepheline syenite
- Schist and metagabbro
- —— Fault

Fig. 40. Malyi Kameshek (after Smirnov et al., 1980, Fig. 2a).

---

4 CHETLASSKII   64°13′N; 50°30′E

This occurrence is composed mainly of peralkaline syenites. They include alkali feldspar (25–30%), oligoclase (3–6%), amphibole (30%), which is variable in composition (barkevikite, hastingsite, arfvedsonite, riebeckite), and aegirine (15%). Accessory minerals are apatite, titanite, rutile and monazite. Dykes and veins of lamprophyre and carbonatite cut the peralkaline syenites.

The lamprophyres are much altered and divided into two groups according to their age. The earlier group is represented by small porphyry dykes and veins the phenocrysts including intensively hydrated biotite or phlogopite and aggregates of serpentine and talc, which apparently replaced olivine. The groundmass is brown and intensely turbid and only small plates of much altered mica can be identified. Accessories include apatite, rutile, chrome spinels, epidote and corundum. The lamprophyres of the later group are ultrabasic rocks (SiO_2 about 40%) with rather high contents of alkalis (K_2O and Na_2O about 2–3%) and they form dykes 3–8 m thick and 0.8–6.2 km long and also stock-like bodies. Again these rocks are highly altered but fresh samples of a green colour can be found occasionally. Lamprophyres of this group have a brecciated structure, with xenoliths of the country rocks, and variable texture (fine- and medium-grained, porphyritic and poikilitic). Poikilitic amphiboles are up to 0.5 cm diameter and enclose diopside–augite and low-Fe biotite crystals, while aggregates of talc, serpentine and calcite pseudomorph olivine. The groundmass is composed of clinopyroxene, biotite, epidote, serpentine and chlorite. Accessories comprise chrome spinel, magnetite, rutile and anatase. Many veins of carbonatite of variable composition are known and include phlogopite–calcite and calcite–feldspar types with alkaline amphibole and aegirine.

Age K–Ar on biotite from lamprophyre gave 540–685 Ma (Stepanenko, 1975)

References Cherny *et al.*, 1977; Ivensen, 1964; Stepanenko, 1975.

Kanin–Timan references

BASHILOV, V.I. and KAMINSKY, F.V. 1975. On the tectonic features and magmatism of Timan. *Sovetskaya Geologiya*, 6: 127–33.

CHERNY, V.G., VASSERMANN, B.Y., CHERNAY, I.P., SHAFRAN, E.B. and LESHCHENKO, S.E. 1977. Magmatism and formation of the folded margin of the north-east part of the Russian platform. In A.K. Krems (ed.) *Geology and oil-gas indications of the north-east European part of the USSR*, 4: 39–45.

IVENSEN, U.P. 1964. Magmatism of Timan and the Kanin Peninsula. Nauka, Moscow and Leningrad. 126 pp.

SHCHUKIN, V.S and SMIRNOV, M.Yu. 1987. Alkali-ultrabasic rocks in the northern Timan. *Doklady Akademii Nauk SSSR*, 294: 195–8.

SMIRNOV, M.U., DOMNIA, M.I., DONSKIH, A.V. and SHINKAREV, N.F. 1980. On the genetic relationship of the metabasites and granitoids with alkaline rocks of the North Timan (according to geological and experimental data). *Zapiski Vserossiiskogo Mineralogicheskogo Obshchestva. Moskva*, 109: 412–23.

STEPANENKO, V.I. 1975. Dykes of the alkaline-ultrabasic association from middle Timan. In M.V. Fishman (ed.) *Geology and ore deposits of the north-east European part of the USSR*. 99–105. Syktyvkar.

TISOVA, N.L. 1984. New data on the platform magmatism of Timan in the Baikalian (Upper Proterozoic) tectono-magmatic epoch. *Izvestia Vysshikh Uchebnykh Zavedenii. Geologiya i Razvedka. Moskva*. 11: 110–11.

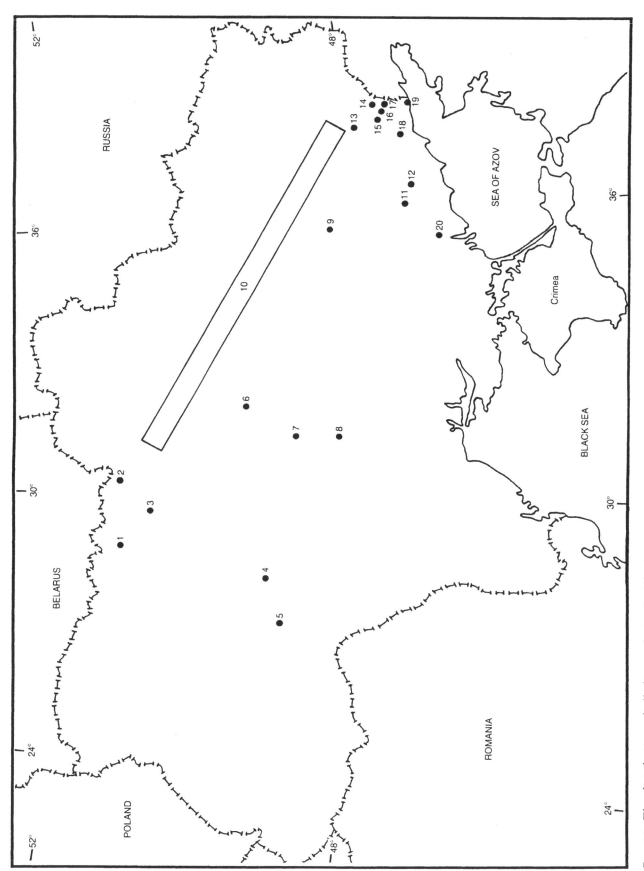

Fig. 41. The distribution of alkaline rocks in the Ukraine.

UKRAINE

At present about 20 complexes and other occurrences of alkaline rocks and carbonatites are known within the boundaries of the Ukrainian shield. Twelve of them are central intrusions with diameters over a kilometre, while others include thin veins, dykes, stockworks and bodies of which the dimensions, shape, structure and mode of occurrence are at present uncertain. The alkaline rocks, according to Krivdik and Tkachuk (1990), belong to two series having different ages: an alkali-ultrabasic (carbonatite) series and a gabbro-syenite series. Most of the occurrences of alkaline rocks within the shield, particularly the central intrusions, are located either in the Azov Sea coastal area or in the northwestern region (Fig. 41).

The alkali-ultrabasic series includes the complexes of Chernigovskii and Proskurov and a belt of meta-jacupirangite dykes adjacent to the Azov Sea coast. The gabbro-syenite series encompasses the majority of the central alkaline intrusions including Malotersyanskii, Pokrovo-Kireevo and the well-known Oktyabr'skii massif. Intrusions composed predominantly (South Kalchik, Kalmius and Elanchik) and completely (Yastrebets) of syenites are also included in this series.

The age of the intrusions of the alkali-ultrabasic series is about 2000 Ma and of the gabbro-syenite series, except for the Pokrovo-Kireevo massif, 1700 Ma. Phanerozoic alkaline rocks within the boundaries of the Ukrainian shield have only a limited development. Devonian sub-volcanic alkaline igneous rocks are widely developed in the Dnieper-Donetz and Pripiat' basins, which limit the Ukrainian shield on the northeastern and northern sides. The Pokrovo-Kireevo complex, which is Devonian in age, is located at the boundary of the shield with the Donetz Coal Basin, and the alkaline granites of the Kuznets-Michailovsk belt are of a similar age. The alkaline rocks of the Ukraine have been described by many authors, the main contributions being those of Tsarovsky (1954, 1972), Tsarovsky et al. (1980), Yeliseev et al. (1965), Krivdik and Tkachuk (1986, 1987, and 1990), Krivdik et al. (1988) and Sherback et al. (1978 and 1989). A general account of the carbonatites is given by Shramenko et al. (1992).

1 Yastrebets	11 Chernigovskii
2 Davidki	12 Azov Sea coast dykes
3 Korosten	13 Oktyabr'skii
4 Antonovskii	14 Pokrovo-Kireevo
5 Proskurovskii	15 Kalmius
6 South Cherkasskii sill	16 Elanchik
7 Korsun-Novomirgorod	17 Kuznetsko-Mikhailovskii
8 Korsun-Novomirgorod	belt
margin	18 South Kalchik
9 Malotersyanskii	19 Primorskii
10 Dnieper-Donetz basin	20 Melitopolskii

1 YASTREBETS
51°14′N; 28°51′E
Fig. 42

The Yastrebets massif is located in the northwest of the Ukrainian shield, a short distance northwest from the Korosten anorthosite–rapakivi granite pluton. The massif is confined structurally to the Sushano-Pershansk tectonic zone, where the Yastrebets, Pershansk and Plotnitsky faults intersect. The massif is nearly oval in shape, its area being about 4 km². It is considered to be a layered intrusion supposedly having the form of a lopolith (but see section – Fig. 42). By analogy with

other intrusions, the syenites have been divided into outer contact syenites, syenites of the main layered series, quartz syenites of the central core and syenites of the upper laminated group. The outer zone of the outer contact part of the massif consists of varieties of fine-grained syenite, which are a chilled facies, the thickness of which approaches 50–70 m. Towards the centre of the massif the grain size of the syenites increases. Syenites of the layered series and the central core are usually medium-grained, but in the main layered series coarse-grained and pegmatitic syenites are also encountered. Syenites of the main layered series are more than 1000 m thick, the upper laminated group 100–200 m, while the quartz syenites of the central core are 350–450 m in thickness. The syenites are essentially of mesoperthite: plagioclase is confined to the outer contact syenites, but only forms up to 5%. The dark-coloured minerals in the syenites are notable for their extremely high Fe:Mg

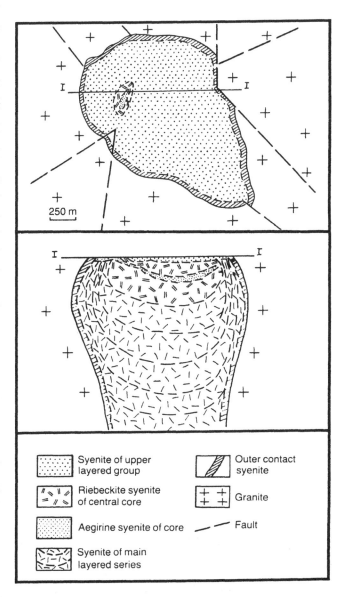

Fig. 42. Map and section of Yastrebets (after Tkachuk, 1987, Fig. 1). Line of section indicated on map.

49

ratios (90 and more) and include hedenbergite, aegirine-hedenbergite, aegirine, ferrohastingsite, ferroedenite, taramite, riebeckite and annite. Andradite is encountered occasionally. Accessory minerals include zircon, fluorite, orthite, ilmenite, bastnasite, parisite, britholite and, in the outer contact facies only, apatite.

· Rhythmic 'and cryptic layering occur throughout the intrusion. Rhythmic layering manifests itself in the alternation of leucocratic and mesocratic syenites, while melanocratic varieties or extremely melanocratic schlieren-like segregations composed only of dark-coloured minerals (clinopyroxene, amphibole, biotite), occur more rarely. Cryptic layering is expressed in the change in chemical composition of the rock-forming minerals towards the central core of the massif, that is in an upward direction in the main laminated group. In the same direction there is an increase in the alkalinity of the pyroxenes and amphiboles, from ferrohastingsite to riebeckite, the silica content of the biotite decreases, and the proportion of orthoclase component in the alkali feldspar increases. The same trend is also reflected by some of the accessories, for example, orthite disappears but britholite appears in the sequence. From the outer contacts towards the centre of the intrusion trends occur involving both changes in the chemical composition of the rock-forming minerals and in their paragenesis. Thus, small amounts of apatite and plagioclase (oligoclase) are encountered only in the outer contact syenites, which also show an elevated content of accessory ilmenite and orthite. However, alkaline varieties of the dark-coloured minerals are not observed in the outer contact syenites, the pyroxene being hedenbergite and amphibole ferrohastingsite.

Economic There is rare-earth and zirconium mineralization of the syenites (Tkachuk, 1987).

Age Lead determinations on zircon from the syenites gave ages of 1720–40 Ma (Tkachuk, 1987).

References Krivdik and Tkachuk, 1986; Tkachuk, 1987.

---

2 DAVIDKI   51°14′N; 30°30′E
                                                         Fig. 43

The Davidki intrusion is located on the northeastern outskirts of the Korosten pluton in the area of intersection between the east–west-trending Ovruch basin and the north–south-trending Vilcha basin. The massif has a rounded outline, a concentric zonal structure and occupies an area of about 30 km². The periphery is composed of gabbroic rocks of which a narrow strip, 150 to 300 m wide, seems to encircle the massif completely; there is a fine-grained chilled zone. Syenites occupy almost the whole central area of the massif and at the present level of erosion predominate. The thickness of the syenites in the centre, from drilling data, is over 360 m, decreasing to 30 m towards the periphery. Plagioclasites probably form a continuous ring around the syenites of the central part. The dips of the contacts of the intrusion are inwards, generally at angles of 40–50°. The Davidki intrusion is considered to be layered (Krivdik *et al.*, 1986) with syenites and plagioclasites at the centre forming the main layered series, while the gabbroic rocks of the periphery form a marginal layered group. The marginal layered group shows a distinct rhythmic layering which manifests itself in the alternation of leucocratic and melanocratic gabbros, troctolites and gabbro-diabase. The drilling also indicates the presence of cryptic layering which is represented by an upward increase in the Fe:Mg ratio of titanaugite (from 40 to 45), and a change in the plagioclase from An_{45-50} to An_{28-30}. In syenites the Fe:Mg ratio of clino-

pyroxene (titaniferous ferrohedenbergite and hedenbergite) increases upward in a borehole from 80 to 92; aegirine is sometimes present. Further, transverse-columnar rocks (crescumulates) similar to those of the Skaergaard intrusion have been found in gabbroic rocks and syenites (Krivdik *et al.*, 1986).

Economic Apatite and titanomagnetite–ilmenite mineralization is associated with the gabbroic rocks, in the form of apatite- and magnetite-rich layers. Accessory baddeleyite and zircon in both the gabbroic rocks and syenites is of considerable mineralogical and petrological interest.

Age The intrusion is assumed to have an age of 1,500 Ma. It is cut by volcanic rocks of the Vilenchansk depression and is probably younger than the Korosten granites (Skobelev *et al.*, 1980).

References Borisenko *et al.*, 1980; Krivdik and Tkachuk, 1986; Krivdik *et al.*, 1986; Proskurin *et al.*, 1977; Skobelev *et al.*, 1980.

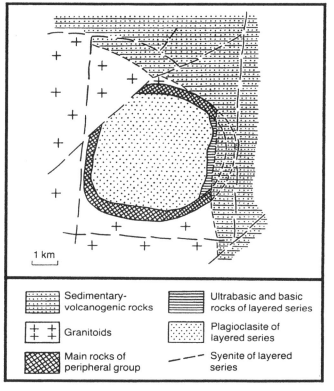

Sedimentary-volcanogenic rocks

Granitoids

Main rocks of peripheral group

Ultrabasic and basic rocks of layered series

Plagioclasite of layered series

Syenite of layered series

1 km

Fig. 43. Davidki (after Krivdik et al., 1986, Fig. 1).

---

3 KOROSTEN   50°52′N; 29°39′E

Finds of aegirine syenites in the Korosten and Korsun-Novomirgorod (No. 7) plutons were first reported by Sobolev (1940). Aegirine syenites are now known in many zones of the Korosten anorthosite–granite pluton, for instance, near the villages of Guta Potievka, Mikhailovka, Stavki and Stavishche. The aegirine syenites are among the last derivatives of these plutons since they are often observed as veins in granites of the rapakivi group and in monzodiorites or quartz diorites. The aegirine syenites are, for the most part, quartz-bearing and the feldspars are represented by mesoperthite, microcline-perthite, microcline, albite and, rather rarely, oligoclase. The predominant dark-coloured mineral is aegirine-augite or aegirine, which may form 5–10% of the rock. Subordinate minerals include alkali amphibole of a riebeckite type and andradite or a titaniferous andradite; accessory minerals include zircon, titanite, fluorite, apatite and ilmenite. The amounts of aegirine syenite within the anorthosite–rapakivi granite plutons are exceedingly small, generally forming thin veins, and only near the village of Guta Potievka does the aegirine syenite cover a significant area (2.5 × 0.4 km). The petrogenesis of these aegirine syenites is not clear, but Krivdik and Tkachuk (1988) consider them to be fenites associated with a hypothetical carbonatite complex. They describe and illustrate the gradual mineralogical changes of the fenitization process and give analyses of rocks, pyroxenes, amphiboles and micas.

Age A K–Ar age determination gave 170 ± 30 Ma (Krivdik and Tkachuk, 1990).

References Krivdik and Tkachuk, 1988 and 1990; Sobolev, 1940.

---

4 ANTONOVSKII   49°16'N; 28°13'E

The alkaline rocks of this massif, which were discovered by drilling, are situated one kilometre northeast of the town of Letichev. The complex consists of oligoclase and nepheline gabbroic rocks (essexites), jacupirangite, ijolite, sovite, nepheline syenite and fenite. The probable size of the massif is 2.2 × 1.2 km.

Age A U–Pb determination on zircon gave 2060 Ma (Sherback *et al.*, 1989).

References Sherback *et al.*, 1989; Tsarovsky *et al.*, 1980.

---

5 PROSKUROVSKII   49°01'N; 27°01'E
                                             Fig. 44

The Proskurovskii massif was discovered by drilling in 1978 (Tsarovsky *et al.*, 1980). It is located 43 km south of the town of Khmelnitsky and lies within Chudnovo-Berdichev garnet granitoids (with hypersthene and biotite). It has a symmetrical outline and is elongated to the northwest. Syenites and granosyenites are the most widely developed rocks of the complex, occupying 75–80% of the total area, and these prove to be fenites. Nepheline-rich rocks, including ijolite–melteigite and nepheline syenites (juvite, pulaskite), are considered to be intrusive. Malignites, which are a feldspathic facies of the ijolites and melteigites, alkaline pyroxenites, jacupirangites and tveitasites are occasionally encountered. It is thought that carbonatites are likely to be present. The first explorers of the massif (Tsarovsky *et al.*, 1980) thought that all the alkaline rocks were metasomatites which had been generated by nephelinization of the enclosing heterogeneous granitoids and pyroxenites. There is a certain regularity in the distribution of the nepheline rocks in the massif: thus, melanocratic rocks of ijolite–melteigite composition develop predominantly in the central area and form a core which seems to be bounded on the south and northeast by elongate, horseshoe-shaped bodies of feldspathic ijolite and nepheline syenite. Varieties of nepheline syenite include biotite–pyroxene, biotite–amphibole–pyroxene, amphibole–biotite and biotite types. Apart from the common nepheline–pyroxene rocks of the ijolite–melteigite series, types of pyroxenite with primary biotite also occur. Certain features of the Proskurov complex, notably the presence of jacupirangite–melteigite–ijolite and nepheline syenite, the abundance of calcite and apatite, and also the extensive fenitization of the enclosing granitoids, indicate that the massif is a representative of an alkali-ultrabasic association (in the sense of the Soviet literature). It remains to be considered whether some specific compositional properties, such as the low Ti content of the rocks of the jacupirangite-ijolite series, the presence of amphiboles of a composition transitional from hastingsite to katophorite, and aegirine–salitic pyroxenes in jacupirangites and ijolites are characteristic of the Proterozoic alkali-ultrabasic (carbonatite) complexes.

Economic High concentrations of apatite occur in the ijolites, melteigites and pyroxenites (Krivdik *et al.*, 1986).

Age Ages of 1245 and 1700 Ma were determined by K–Ar on nepheline and biotite (Tzarovsky *et al.*, 1980), and of 2100 ± 40 Ma by the thermoemission method on zircon from nepheline syenite (Krivdik *et al.*, 1986).

References Krivdik *et al.*, 1986; Krivdik and Bratlavsky, 1986; Tsarovsky *et al.*, 1980.

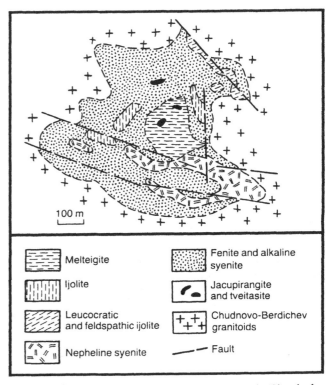

▭ Melteigite	▦ Fenite and alkaline syenite
▦ Ijolite	◖ Jacupirangite and tveitasite
▨ Leucocratic and feldspathic ijolite	+ + + Chudnovo-Berdichev granitoids
▨ Nepheline syenite	— Fault

Fig. 44. Proskurovskii (after Krivdik and Tkachuk, 1990, Fig. 11).

---

6 SOUTH CHERKASSKII SILL   49°20'N; 32°00'E

This sill of metamorphosed lamproite is located in the junction zone of the Ukrainian crystalline shield and the Dnieper-Donetz basin and is related to a deep, north-south-trending fault (lineament) and the peripheral ring fracture of the circular Kirovograd structure. The rock extends over some hundreds of square metres and is a microporphyritic phlogopite–amphibole–pseudoleucite lamproite. The mineralogy is microcline, K-arfvedsonitic and pargasitic amphibole and phlogopite with accessory rutile, Mn-ilmenite (5.3–10% MnO), apatite, titanite and magnetite.

Age A K–Ar determination gave an age of 1370 Ma (Geiko *et al.*, 1991).

Reference Geiko *et al.*, 1991.

7 KORSUN-NOVOMIRGOROD   48°49'N; 31°17'E

Aegirine syenite was first reported from the Korsun-Novomirgorod pluton by Sobolev (1940) and Tkachuk (1940). It occurs near the villages of Ternovka and Mala'a Smelianka where it forms veins cutting rapakivi granite, monzonite and quartz diorite. Petrographically it is similar to the aegirine syenites of the Korosten pluton (No. 3)

Age Krivdik and Tkachuk (1990) report a K–Ar age of 170 ± 30 Ma.

References Krivdik and Tkachuk, 1990; Sobolev, 1940; Tkachuk, 1940.

8 KORSUN-NOVOMIRGOROD MARGIN
48°14'N; 31°17'E
Fig. 45

In the southeast of the Korsun-Novomirgorod pluton rocks of the monzonite–syenite series, mostly quartz-bearing syenites, which are unusual in such plutons, have been identified (Krivdik et al., 1988). They show extreme iron enrichment of mafic minerals (fayalite, hedenbergite, ferrohastingsite) and a higher content, compared with gabbroids and rapakivi granites, of Zr (0.1–0.2%), light rare earths (up to 0.2%) and Nb (up to 0.02%). Accessory minerals typical of syenites include zircon, ilmenite, orthite, chevkinite and apatite. These syenites are considered to be co-magmatic rocks of the anorthosite–rapakivi granite pluton. Fayalite-hedenbergite syenites and monzosyenites occur as narrow bodies in the rapakivi granites or in the contact zone with enclosing granitoids. The largest massif, having dimensions of 4.5 × 2 km, is located in the southeast of the region. Drilling has detected intrusive contacts between monzonite–syenites and enclosing (new Ukrainian) granites. The relationships with the rapakivi granites of the pluton are uncertain. Predominantly medium-grained syenites become more fine-grained in the contact zone with enclosing granites. Veins of monzosyenite are fine-grained.

Economic There are concentrations of rare earths and zirconium in the syenites (Semenenko et al., 1977).

References Krivdik et al., 1988; Semenenko et al., 1977.

9 MALOTERSYANSKII   48°20'N; 35°39'E
Fig. 46

The Malotersyanskii intrusion of alkaline rocks is located on the northern slope of the Ukrainian shield adjacent to the Dnieper-Donetz basin (Timoshenko, 1975). Tectonically it is likely to be associated with the deep Orekhovo-Pavlograd fault. The massif is an assymetrical, sheared, laccolith-like intrusion which has a long axis of about 12 km and a width ranging from 2.5 to 4.5 km; the area is some 42 km². Alkaline rocks are also encountered to the north of the massif along the Orekhovo-Pavlograd fault. Malotersyanskii is composed of subalkaline (titanaugite) gabbro and gabbro-diabase in the outer zone and alkaline syenites and foyaites in the central zone. Dyke analogues, including pegmatites, of the alkaline rocks are encountered. Enclosing granitoids in the outer contact zone are fenitized. Fenites and, more rarely, intrusive rocks of the massif include aegirine, albite, mica and carbonate as metasomatic phases. Accessory minerals occurring in the metasomatic rocks include zircon, pyrochlore-hatchettolite, bastnäsite, orthite and thorite. In the opinion of Krivdik and Tkachuk (1990) the gabbroic rocks represent the earliest intrusive phase, while the alkali syenite–foyaite series of the complex was formed in the course of crystal differentiation of a single injection of magma, the composition of which was similar to that of the outer olivine-bearing alkaline syenites. Alkaline syenites and foyaites display features of cryptic layering: latent lamination has also been detected. The alkalinity and Fe:Mg values of rocks and dark coloured minerals

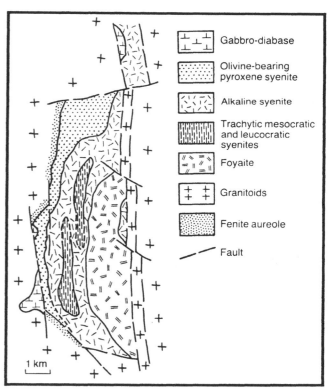

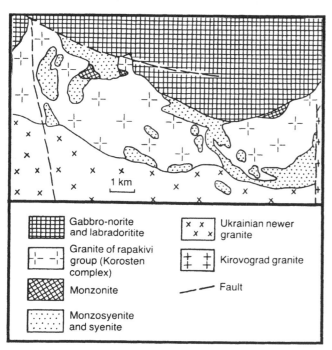

Fig. 45. Southern part of the Korsun-Novomirgorodskii pluton (after Krivdik et al., 1988, Fig. 1).

Fig. 46. Malotersyanskii (after Krivdik and Tkachuk, 1990, Fig. 14).

increase from the outer contacts towards the centre of the massif. Pyroxene evolves from aegirine-bearing ferrosalite (10% acmite) to aegirine ferrosalite and aegirine-hedenbergite (48% acmite). Alkali feldspar shows an increase in orthoclase content (26 to 63%). Late veins of alkaline rocks include typical agpaitic minerals such as aenigmatite and astrophyllite. Like Oktyabr'skii (No. 13), the Malotersyanskii complex has an agpaitic trend of evolution from gabbro through alkaline syenites and foyaites to nepheline syenites of agpaitic composition.

Economic Apatite and rare-earth mineralization, including pyrochlore, hatchetollite, bastnsite, orthite, thorite and zircon, are connected with albite metasomatites at the western, northern and southern margins of the complex. Nepheline syenites could be considered as potential raw materials for ceramics (Timoshenko, 1975).

Age Amphiboles from nepheline syenites were dated at 1740 Ma by the K-Ar method.

References Krivdik and Tkachuk, 1986 and 1990; Timoshenko, 1975; Tsarovsky, 1972.

---

10 DNIEPER-DONETZ BASIN 50°55′–48°00′N; 31°09′–37°47′E

During drilling in the search for oil at depths of 900–2300 m in the Dnieper-Donetz basin effusive and intrusive alkaline rocks have been found in a long narrow zone over an area of more than 1800 km². They can be divided into two series: (a) olivine basalt, analcime basalt, basalt-trachyandesite and trachyte-quartz porphyry and (b) alkaline picrite, meimechite, ankaramite, nephelinite and leucitite (Lyashkevich and Zavialova, 1974). Melteigite, ijolite, malignite (Voloshina, 1977), bergalite (Gladkikh, 1988), monchiquite, limburgite and augitite (Buturlinov *et al.*, 1973) have also been described. It is assumed that the alkaline magmatism of the Dnieper-Donetz basin is closely connected with the palaeo-rift system which extends westwards towards the Pripyat basin.

Age K-Ar on biotite gave 340–392 Ma (Semenenko *et al.*, 1977).

References Buturlinov *et al.*, 1973; Gladkikh, 1988; Lyashkevich and Zavialova, 1974; Semenenko *et al.*, 1977; Voloshina, 1977.

---

11 CHERNIGOVSKII 47°14′N; 36°15′E
(Novopoltavskii) Fig. 47

This massif was discovered over 20 years ago and has been described in a number of publications, including a comprehensive report by Glevassky and Krivdik (1981b) and a more recent one by Shramenko *et al.* (1992). The occurrence of alkaline rocks and carbonatites is confined to the so-called Chernigovka (after the village of Chernigovka, Zaporozhye region) fault zone of highly metamorphosed (granulite facies) Archaean rocks, which include gneiss, charnockite, migmatite, granite, metabasite and two-pyroxene schists. The alkaline igneous rocks and carbonatites occur intermittently in the fault zone over a length of about 20 km. The Chernigovskii massif has an elongated shape and consists of two lenticular parts, a northern larger part, Chernigovka proper, the so-called New Poltava, and a southern, or Begim-Chokrak part. The most widely developed rocks of the massif are alkaline syenites, tveitasites, a nordmarkite series and carbonatites; nepheline syenites (canadites) and alkaline pyroxenites are subordinate. Other rocks, including ijolite-melteigite, phlogopite

olivinite and peridotite, are found either as inclusions in carbonatites or in bodies of small size which occur sporadically (kimberlitic carbonatites, albite–barkevikite essexites). Enclosing granitoids and metabasites in the contact zone are variably fenitized. Carbonatites, alkaline pyroxenites and canadites occur as steeply dipping, up to vertical, dyke-like bodies which strike approximately north–south. The maximum thickness of rocks in the complex, including fenites, is 600–700 m. Four major types of carbonatites can be distiguished, namely sovites, alvikites, beforsites and kimberlitic carbonatites. Closely associated with the beforsites are phoscorites – rocks rich in apatite and/or magnetite. Carbonate–silicate rocks, which are of rare occurrence and transitional in composition from syenite–tveitasite to carbonatite, are classified as ringites. The sovites are calcite carbonatites containing apatite, aegirine–salite,

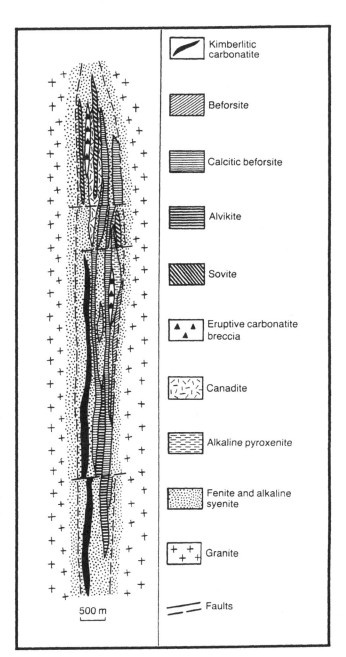

Fig. 47. Chernigovskii (after Krivdik and Tkachuk, 1990, Fig. 4).

biotite and amphibole. Accessory minerals include pyro-chlore–hatchettolite, rare columbite, zircon, titanite, monazite, orthite, magnetite and sulphides. In spite of their widespread occurrence the sovites are of sub-ordinate development, accounting for only one quarter of the total mass of carbonatites. The alvikites are cal-cite carbonatites containing apatite, olivine, diopside or aegirine–diopside, phlogopite or tetraferriphlogopite and magnetite. Typical accessories include zircon, ilme-nite (picro-ilmenite), sulphides and rarely anatase, col-umbite, pyrochlore, hatchettolite and monazite. The alvikites account for about 15% of the total mass of car-bonatites in the complex: they are best developed in the southern part of the massif (Fig. 47). Beforsites are calcite–dolomite, dolomite–calcite and dolomite-carbonatites containing apatite, olivine, phlogopite and magnetite. Accesories are represented by spinels, sul-phides, columbite, cerian fergusonite, pyrochlore–hat-chettolite, baddeleyite, zircon, monazite and ancylite. Beforsites are the most widely developed of the carbon-atites of the Chernigovka massif, accounting for about 60% of the carbonatites. With increase in the content of apatite, magnetite and phlogopite in the beforsites they grade into predominantly apatite, apatite–magnetite, olivine–apatite and phlogopite–apatite rocks, that is phoscorites. Kimberlitic carbonatites are composition-ally transitional between carbonatites and kimberlites. They are brecciated or massive phlogopite–olivine (serpentine)–calcite rocks containing a smaller amount of carbonate (25–50%) as compared with the carbona-tites described above. Kimberlitic carbonatites occur only in the southern part of the massif in the form of thin, rarely more than 10 m thick, vertical dykes. Car-bonatites of the Chernigovskii massif are also charac-terized by inclusions of the associated cumulative ultrabasic rocks, of which the silicate minerals (olivine, pyroxene, amphibole, biotite, phlogopite) are identical to those occurring in the carbonatites. Thus, olivine carbonatites, beforsites and alvikites generally contain inclusions of olivinite, olivine-bearing glimmerite and peridotite, while biotite–amphibole and pyroxene car-bonatites, which are usually sovites, include biotite-amphibole and amphibole–pyroxene–biotite ultrabasic rocks. Whole rock carbonatite analyses and extensive modal and other data on the minerals of the carbonatites are given by Vil'kovich and Pozharitskaya (1986), and data on REE in carbonatites by Shramenko and Kostyuchenko (1985). Vozynak *et al.* (1981) have described melt and mineral inclusions from baddeleyite in carbonatite. Although the Chernigovskii massif as a whole has many features characteristic of carbonatite complexes, it also has a number of peculiar structural, petrological and mineralogical features. The elongated form of the massif is distinct from that of the central type, which comprise the majority of carbonatite com-plexes. Olivines in alkaline ultrabasic rocks and car-bonatites of the massif are characterized by abnormally high iron content (up to Fa_{70}); the olivine is frequently associated with graphite. Nepheline syenites are repre-sented by biotite–albite canadites in which Ca–Na amphiboles are transitional between edenite–hastingsite and katophorite. Accessory cerian fergusonite (broce-nite) is only the second occurrence in the world.

Economic There is rare metal, apatite–magnetite and magnetite–ilmenite mineralization associated with the massif (Glevassky and Krivdik, 1981b; Zshukov *et al.*, 1973).

Age A minimum age of 1820 Ma was obtained by K–Ar on phlogopite from phlogopite olivinite, and a max-imum age of 2190 Ma was determined by the lead

method on zircon from carbonatite (Sherback *et al.*, 1978).

References Glevasskii and Krivdik, 1981b; Sherback *et al.*, 1978; *Shramenko and Kostyuchenko, 1985; Shra-menko *et al.*, *1991 and 1992; *Vil'kovich and Pozharit-skaya, 1986; *Vozynak *et al.*, 1981; Zshukov *et al.*, 1973.

---

12 AZOV SEA COAST DYKES   47°11'N; 36°34'E

A belt of alkaline ultrabasic dykes occurs along the western part of the Azov Sea Coast (Glevasskii and Kriv-dik, 1985). They have chemical compositions identical to jacupirangites, alkaline pyroxenites and gabbroids (of shonkinite type) of the Begim-Chokrak part of the Cher-nigovskii massif. The dyke rocks are enriched with trace elements typical of alkaline ultrabasic rocks, i.e. Ti, P, Nb and LREE. There are 25 dykes and they form a belt 110 km wide of northwesterly strike which can be traced for over 35 km. The dykes are probably connected with the Chernigovskii carbonatite complex.

Reference Glevasskii and Krivdik, 1985.

---

13 OKTYABR'SKII   47°50'N; 37°32'E
(Mariupol)   Fig. 48

The Oktyabr'skii massif, formerly called the Mariupol or Azov Sea coastal massif, is one of the best explored alkaline massifs of the Ukrainian shield (Morozewicz, 1930; Ainberg, 1933; Tsarovsky, 1954; Yeliseev *et al.*, 1965) and the most representative one of the gabbro-syenite series. The massif is widely known through the studies of Morozewicz, who described from here the rock type mariupolite, and it is the type locality for the minerals beckelite and taramite. The massif is oval in plan and emplaced in various granitoids including normal granites, granites of the rapakivi group and enderbite charnockites, and comprises subalkaline (titanaugite) basic and ultrabasic rocks (gabbro, pyrox-enites, peridotites) and alkaline and nepheline syenites (foyaites, mariupolites). Pegmatites, veins and dykes of alkaline rocks are also encountered. Agpaitic phonolites, containing eudialyte and astrophyllite, and dykes of nepheline syenite are the youngest intrusions of the com-plex (Krivdik and Tkachuk, 1989 and 1990). The com-plex has an indistinctly pronounced concentric, zonal structure. The outer zone is represented by a discon-tinuous ring of basic and ultrabasic rocks and syenites, while in the inner zone an arcuate body of foyaites is present. The succession of differentiates, namely sub-alkaline gabbro and pyroxenites–syenites–pulaskites–foyaites–mariupolites–agpaitic nepheline syenites, shows an increase in alkalinity and Fe:Mg ratios of rocks and mafic minerals. Pyroxene in gabbro and pyroxenites is represented by titanaugite, in outer zone syenites by aegirine-bearing ferrosalite and aegirine–ferrosalite, and in foyaites, mariupolites, agpaitic phonolite and nephe-line syenites by aegirine–hedenbergite and aegirine. Changes in the composition of Zr-bearing minerals are also regular: in gabbro and outer zone syenites they are represented by baddeleyite and zircon, in foyaites and mariupolites by zircon and in agpaitic phonolites by eudialyte. The last differentiates, mariupolites, agpaitic phonolites and foyaites, accumulated rare elements (Zr, Nb, RE) forming separate mineral phases including pyrochlore, britholite (beckelite), zircon and eudialyte. The content of Ba and Sr decreases correspondingly. The data presented above might suggest that the fractional crystallization of original alkali basaltic magma was the main petrogenic mechanism in the formation of the

Oktyabr'skii massif leading to the formation of alkaline rocks, including the agpaitic nepheline-bearing rocks.

Economic The complex is characterized by an Fe–Ti–V metallogenic association (Fomin, 1984). The nepheline syenites form potential ores for the production of alumina.

Age Zircon from nepheline syenite gave ages of 1730–1770 Ma by the U–Pb method (Vinogradov *et al.*, 1957).

References Ainberg, 1933; Fomin, 1984; *Krivdik and Tkachuk, 1989 and 1990; Morozewicz, 1930; Semenenko *et al.*, 1977; Tsarovsky, 1954; Vinogradov *et al.*, 1957; Yeliseev *et al.*, 1965.

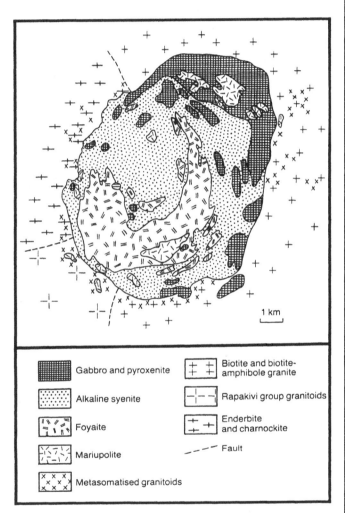

Fig. 48. Oktyabr'skii (after Krivdik and Tkachuk, 1990, Fig. 13).

14 POKROVO-KIREEVO   47°38′N; 38°16′E
Fig. 49

Pokrovo-Kireevo is a collective name for alkaline and subalkaline intrusive, dyke and effusive rocks encountered in the area of intersection of the Ukrainian shield along the Azov Sea coast with the Donetz coal basin near the village of Kumachovo (formerly Pokrovo-Kireevo). Intrusive rocks are represented by gabbro, pyroxenite, peridotite, malignite and nepheline and pseudoleucite syenites; effusive and dyke rocks by subalkaline and nepheline basalts, limburgite, augitite and bergalite (Buturlinov *et al.*, 1973). The relationship between the intrusive and effusive rocks is not clear. Quartz-bearing rocks of the normal and alkaline series, such as liparites and grorudites, also occur in the area but their geological status is uncertain. The best-known intrusion is oval in shape and lies within Precambrian granosyenites and Devonian basalts; it occupies an area of about 10 km² and has a zonal structure. The periphery is composed of pyroxenites and associated rocks rich in ore minerals,

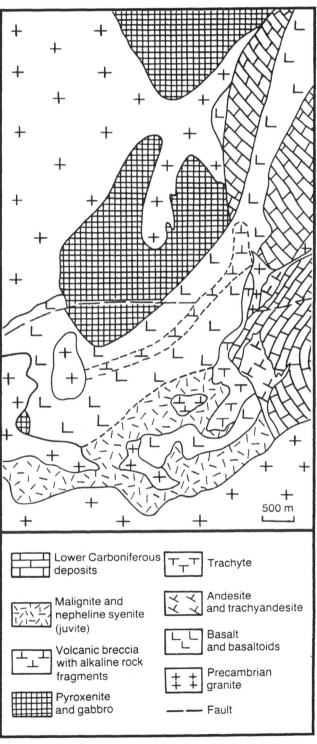

Fig. 49. Pokrovo-Kireevo (after Krivdik and Tkachuk, 1990, Fig. 15).

while gabbroic rocks of the central zone range from plagioclase pyroxenite to leucocratic gabbro. Malignites and juvites occupy a similar or smaller area and between these rocks a gradual transition is observed. Poikilitic textures are a salient feature of the juvites and malignites within which coarse segregations of K-feldspar contain numerous inclusions of fine-grained nepheline. Nepheline is also found included in pyroxene, biotite and calcic rinkolite. Large crystals of pyroxene in malignite are zoned from pinkish-brown titanaugite in the centre to green aegirine–ferrosalite at the periphery. Smaller grains of pyroxene are optically homogeneous and of aegirine–hedenbergite composition. Hence, the alkaline rocks of the Pokrovo-Kireevo massif, like those of Oktyabr'skii and Malotersyanskii, are spatially and apparently genetically linked with subalkaline gabbroic rocks and alkaline basalts, an association that seems to be supported by the pyroxenes. Pokrovo-Kireevo shows a similar agpaitic trend of development to the other complexes, from subalkaline gabbroids and basalts to hyperalkaline malignite–juvites in which agpaitic minerals such as calcic rinkolite occur. A description of a drilled specimen of bergalite is given by Gladkikh (1973).

Economic Titanomagnetite is the most abundant ore mineral of the Pokrovo-Kireevo massif, forming about 35% in the ore-rich pyroxenites. The amount of apatite is widely variable but in some rocks reaches 20–30%. High concentrations of Ta, Nb, Be, Pb, Zr, Hg and F are associated with the later magmatic phases (Buturlinov *et al.*, 1973).

Age According to geological data and absolute age determinations by the K–Ar method the complex was formed between 408 ± 20 and 326 ± 12 Ma (Buturlinov *et al.*, 1973).

References Buturlinov *et al.*, 1973; *Gladkikh, 1973.

---

15 KALMIUS   47°30′N; 37°50′E

This massif is considered to be the surface manifestation of a large batholith – the so-called Azov Sea coastal batholith, which represents a deeply eroded syenite-granosyenite analogue of the anorthosite–rapakivi granite plutons (Karamsin, 1979). The batholith has been traced for 100 km in an east–west direction and for 55 km north–south, the total outcrop being 1500 km². Alkali pyroxenes and amphiboles are found in some rocks. Veins of carbonate and fluorite are developed in the syenites and associated with faulting.

Age A K–Ar determination on amphibole from amphibole syenite gave 1700 Ma.

References Karamsin, 1979; Krivdik and Tkachuk, 1986; Semenenko *et al.*, 1977.

---

16 ELANCHIK   47°28′N; 38°12′E

Elanchik, and the two previously described occurences (Nos 14 and 15), have zonal structures, although all rock phases may not be involved in this zonality. Melanocratic rocks, including fayalite–hedenbergite and hedenbergite–ferrohastingsite syenites, tend to be concentrated at the margins, whereas leucocratic granosyenites and granites occur towards the centres of the massifs. The fayalite–hedenbergite syenites are similar, or identical, to those of the Korsun-Novomirgorod pluton (No. 7), which is taken to indicate a genetic relationship between the Azov Sea coastal complex and the anorthosite–rapakivi granite formation. High Fe:Mg

ratios of the dark-coloured minerals (fayalite, hedenbergite, ferrohastingsite) and high Zr values (0.14%) are characteristic features of the Azov Sea coastal syenites, which suggest that they are products of residual melts. Alkali pyroxenes and amphiboles are met with in a number of the rock types.

Age K–Ar on amphibole from granosyenite indicated a maximum age of 2105 Ma (Sherbak *et al.*, 1978).

References Karamsin, 1979; Krivdik and Tkachuk, 1986; Semenenko *et al.*, 1977; Sherbak *et al.*, 1978.

---

17 KUZNETSKO-MIKHAILOVSKII BELT
                                 47°24′N; 39°09′E

This occurrence consists of a system of parallel dykes within an area of 35×2 km. They are of peralkaline microgranites which contain phenocrysts of alkali feldspar and quartz in a groundmass of feldspar, quartz, riebeckite, aegirine and zeolites.

Age Ages of 340–326 Ma have been determined by the K–Ar method.

References Shatalov, 1986; Yeliseev *et al.*, 1985.

---

18 SOUTH KALCHIK   47°15′N; 37°30′E

Syenite, quartz syenite, granosyenite and associated rocks including quartz monzosyenite and amphibole-biotite granite, are components of the South Kalchik complex. The syenite–granosyenite series also includes small amounts of gabbro and peridotite rich in apatite and ilmenite, gabbro-norite, plagioclasite and granitoids of the rapakivi group containing fayalite and ferroaugite. Alkali pyroxenes and amphiboles are found in some rocks.

Age A maximum age of 2105 Ma was determined by K–Ar on amphibole from granosyenite (Sherbak *et al.*, 1978).

References Karamsin, 1979; Krivdik and Tkachuk, 1986; Semenenko *et al.*, 1977; Sherbak *et al.*, 1978.

---

19 PRIMORSKII   47°16′N; 38°13′E

Peralkaline syenites and granosyenites containing alkali amphibole have been discovered at a depth of 2890 m during drilling.

Age Middle Proterozoic (Kon'kov and Marchenko, 1980).

References Kon'kov and Marchenko, 1980.

---

20 MELITOPOLSKII   47°07′N; 35°25′E

An intrusion of amphibole syenite and granosyenites, with a diameter of 5–8 km, has been discovered by drilling. These rocks are peralkaline and contain alkali amphibole.

Age Middle Proterozoic (Kon'kov and Marchenko, 1980).

References Kon'kov and Marchenko, 1980.

Ukraine references

AINBERG, L.F. 1933. The Azov region alkaline massif. *Trudy sesoyuznogo Geologo-Razvedochnogo Ob'edineniya*, 196: 48–64.

BORISENKO, L.F., TARASENKO, B.S. and PROSKURIN, G.N. 1980. Ore-bearing gabbroids of the Korosten pluton. *Geologiya Rudnykh Mestorozhdenii*, 6: 27–37.

BUTURLINOV, N.V., GONSHAKOVA, V.I. and ZARITSKY, A.I. 1973. The Devonian alkaline-basic alkaline-basaltoid complex of the conjunction zone of the Donbass with the Preazov part of the Ukrainian shield. *Bazit-giperbazitovii magmatizm i mineralogiya yuga Vostochno-Evropeiskoi platformy*. 171–263. Nedra, Moscow.

FOMIN, A.B. 1984. *Geochemistry of hyperbasites of the Ukrainian Shield*. Naukova Dumka, Kiev. 232 pp.

GEIKO, Yu.V., ORLOVA, M.P. and FILONENKO, V.P. 1991. Pseudoleucitic lamproites of the Ukraine. *Transactions of the All-Union Mineralogical Society*, 5: 52–6.

*GLADKIKH, V.S. 1973. Bergalite of the Pokrovo-Kireyevskaya structure. *Doklady Earth Science Sections*. American Geological Institute, 203: 166–8.

GLADKIKH, V.S. 1988. Tholeiitic and subalkaline basaltic volcanism of the East European and Siberian platform. *Izvestiya Akademii Nauk SSSR. Seriya Geologicheskaya*, 9: 3–17.

GLEVASSKII, E.B. and KRIVDIK, S.G. 1981a. *Precambrian rocks of the Azov region*, Naukova Dumka, Kiev. 227 pp.

GLEVASSKII, E.B. and KRIVDIK, S.G. 1981b. Peculiarities of metallogeny of the Chernigov carbonatite massif. *Ore formation and metallogeny*. 72–6. Naukova Dumka, Kiev.

GLEVASSKII, E.B. and KRIVDIK, S.G. 1985. The belt of Precambrian dykes of alkaline meta-ultrabasics in the western part of the Azov area. *Geologicheskii Zhurnal*, 45(4): 58–64.

KARAMSIN, B.S. 1979. The Azov batholith and its structure. *Geologicheskii Zhurnal*, 4: 137–43.

KON'KOV, G.G. and MARCHENKO, Ye.Ya. 1980. Alkaline rocks and carbonatites in the surroundings of the Ukrainian shield. *Doklady Akademii Nauk SSSR*, 252: 937–41.

KORZUN, V.N. and MAKHNACH, A.S. 1977. *The Upper Devonian alkaline volcanogenic formation of the Pripyat basin*. Nauka i Tekhnika, Minsk. 162 pp.

KRIVDIK, S.G. and BRATSLAVSKY, P.F. 1986. Nepheline rocks of the Proskurov massif (Pridnestrov region) as indicators of its crystallization. *Mineralogicheskii Zhurnal*, 8: 74–9.

KRIVDIK, S.G. and TKACHUK, V.I. 1986. Formational classification of the alkaline rocks of the Ukrainian shield. *Tezisy Dokladov VII Vesesoyuznogo Petrograficheskogo Obshchestva*. Novosibirsk, 85–7.

KRIVDIK, S.G. and TKACHUK, V.I. 1987. Formational accessory minerals of alkaline metasomatites of the Beryozovaya Gat' (Zhitomir region, Ukrainian SSR). *Doklady Akademii Nauk Ukrainskoi SSR: Seriya B*. 1: 1–13.

KRIVDIK, S.G. and TKACHUK, V.I. 1988. Fenites of Berezov Gat. *Geologicheskii Zhurnal, Kiev*, 131–40.

*KRIVDIK, S.G. and TKACHUK, V.I. 1989. Eudialyte-bearing agpaitic phonolites and dike nepheline syenites in the October intrusion, Ukrainian shield. *Geochemistry International*, 26(3): 54–60.

KRIVDIK, S.G. and TKACHUK, V.I. 1990. *Petrology of the alkaline rocks of the Ukrainian Shield*. Naukova Dumka, Kiev. 407 pp.

KRIVDIK, S.G., ORSA, V.I. and BRYANSKY, V.P. 1988. Fayalite-hedenbergite syenites of the south-western part of the Korsun-Novomirgorod pluton. *Geologicheskii Zhurnal*, 6: 43–53.

KRIVDIK, S.G., TKACHUK, V.I., GLUKHOV, A.P.

and SHVAIBEROV, S.K. 1986. The Davidki gabbro-syenite massif, Ukrainian Shield. *Geologiya Mestorozhdenii*, 6: 58–71.

KUSMENKO, V.I. 1986. The Petrovo-Gnutovskoye deposit of parisite (Ukraine SSR). *Sovetskaya Geologiya*, 6: 58–71.

LYASHKEVICH, Z.I. and ZAVIALOVA, T.V. 1974. *Volcanism of the Dneiper-Donetz basin*. Naukova Dumka, Kiev, 178 pp.

MARCHENKO, E.A., KON'KOV, G.C. and VASENKO, V.I. 1980. About the carbonatite origin of the Petrovo-Gnutovo fluorite–carbonatite dyke of the Priazov region. *Doklady Academii Nauk Ukrainskoi SSR, Seriya B*. 1: 24–7.

MOROZEWICZ, J. 1930. Der mariupolit und seine Blutaverwandten. *Mineralogisch und Petrographisches*. *Mittelunden*, Nue Folge B.40, H, V-VI-S. 335–436.

PROSKURIN, G.P., PROSKURINA, V.F., FOMIN, A.B. and METALIDI, S.V. 1977. About ore troctolites of the Korosten pluton (Ukrainian Shield). *Doklady Akademii Nauk Ukrainskoi SSR, Seriya B*. 12: 1080–3.

SEMENENKO, N.P. SAVCHENKO, N.A. and BRITCHENKO, A.D. 1977. Cycles of volcanism of the Dnieper-Donetz basin and Donbass and metal-bearing problems of the northern slope of the Ukrainian Shield. *Volcanism and ore-formations of the Dneprovo-Donetz basin and Donbass*. AN Ukraine SSR, 156 pp.

SHATALOV, N.N. 1986. *Dykes of the Priazov region*. Naukova Dumka, AN Ukraine SSR, Institute of Geological Sciences. 192 pp.

SHERBACK, N.P., ARTEMENKO, G.V. and BARNITSKY, Ye.N. 1989. *Geochronoligical scale of the Precambrian of the Ukrainian Shield*. Naukova Dumka, Kiev, 144 pp.

SHERBACK, N.P., SKOBELEV, V.M. and VERCHOGLAD, V.M. 1987. Isotope age of the granites of the rapakivi of the Korosten and Korsun-Novomirgorod plutons of the Ukrainian Shield. *Doklady Akademii Nauk SSSR, Seriya Geologicheskaya*, 7: 39–41.

SHERBACK, N.P. (ed.), ZLOBENKO, V.G., ZSHUKOV, G.V. et al. 1978. *Catalogue of isotope data of the rocks of the Ukrainian Shield*. Naukova Dumka, Kiev. 233 pp.

*SHRAMENKO, I.F. and KOSTYUCHENKO, N.G. 1985. Rare-earth elements in Azov carbonatites. *Geochemistry International*, 22(8): 43–6.

SHRAMENKO, I.F., STADNIK, V.A. and OSADCHII, V.K. 1992. *The geochemistry of carbonatites from the Ukrainian shield*. Naukova Dumka, Kiev. 211 pp.

*SHRAMENKO, I.F., LEGKOVA, G.V., IVANITSKIY, V.P. and KOSTYUCHENKO, N.S. 1991. Mineralogical and geochemical studies of the petrogenesis of the Chernigov carbonatites. *International Geology Review*, 28(8): 102–9.

SKOBELEV, V.M., YELISEEVA, G.D., DEVKOVSKAYA, N. Yu. et al. 1980. Isotope dating of palaeovolcanites of the Zbrankovskaya group of the Ovruch series. *Doklady Akademii Nauk SSSR, Seriya B*. 12: 25–7.

SOBOLEV, V.S. 1940. Alkaline syenites of the complex Korosten pluton (Zhitomir region, Ukraine SSR). *Zapiski Vsesoyuznogo Mineralogicheskogo Obshchestva*. 69: 321–30.

TIMOSHENKO, O.D. 1975. Geology, composition and perspectives of ore-formation of the Malotersyan alkaline massif (central Azov region). Thesis abstract, Kharkov, 24 pp.

TKACHUK, L.G. 1940. Charnockites and their Precambrian accessories (in Ukrainian). *Geologicheskii Zhurnal*, 7: 153–99.

TKACHUK, L.G. 1987. Genesis of the Yastrebets syenite massif. *Geologicheskii Zhurnal*, 47: 106–11.

TSAROVSKY, I.D. 1954. Geological types of structures of alkaline rocks of the Ukraine SSR. *Izvestiya Akademii Nauk SSSR, Seriya Geologicheskaya*, 4: 101–12.

TSAROVSKY, I.D. 1972. Syenite complex. *Stratigraphy of the Ukraine SSR, Volume 1, Precambrian.* Naukova Dumka, Kiev, 287–98.

TSAROVSKY, T.D. and BRATSLAVSKY, N.F. 1980. *Nepheline rocks of the Dnestrov-Bursk region (geology, age and composition).* AN Ukraine SSR, Institute of Geochemistry and Physics of Minerals. Naukova Dumka. Kiev. 46 pp.

TSAROVSKY, I.D. BRATSLAVSKY, P.F., GEVORKYAN, S.B. and KUZNETZOV, T.V. 1980. Apatite of the Proskurov alkaline massif western slope of the Ukrainian Shield. *Doklady Akademii Nauk SSSR, Seriya B.* 12: 28–32.

*VIL'KOVICH, R.V. and POZHARITSKAYA, L.K. 1986. Compositional evolution of carbonatites in the Chernigov zone, Azov region. *Geochemistry International*, 23(7): 92–100.

VINOGRADOV, A.P., TUGARINOV, A.I., FEDOROVA, V.A. and ZYKOV, S.I. 1957. Age of the Precambrian rocks of the Ukraine. *Geokhimiya, Communication 3*, 7: 559–66.

VOLOSHINA, Z.G. 1977. Volcanogenic Devonian formations of the central part of the Dneprovo-Donetz basin. *Volcanism and ore formations of the Dneprovo-Donetz basin and Donbass.* 55–75. Naukova Dumka, Kiev.

*VOZYNAK, D.K., KVASNITSA, V.N. and KROCHUK, V.M. 1981. Solidified melt inclusions in baddeleyite from carbonatite of the Azov region. *Doklady Earth Science Sections. American Geological Institute*, 259: 167–70.

YELISEEV, N.A., KUSHEV, V.G. and VINOGRADOV, D.P. 1965. *Proterozoic intrusive complex of the eastern Priazov region.* Nauka, Moscow and Leningrad, 204 pp.

ZSHUKOV, G.V., BARKHOTOV, V.A. and SAKHATSKY, I.I. 1973. About the discovery of phosphate mineralization in the western Azov region. *Geologicheskii Zhurnal*, 33: 144–6.

* In English

CAUCASUS (ARMENIA, AZERBAI'AN, GEORGIA)

The Caucasus is an Alpine orogenic structure generated by collision between the Eurasian and Afro-Arabian plates, the area of contact being marked by the Lesser Caucasus ophiolite zone and intense volcanic activity. The Alpine alkaline volcanism in the Caucasus occurs in two distinct structural settings, the first of which is palaeoceanic and characterized by volcano-sedimentary sequences which are closely associated with the ophiolites. In the second setting the alkaline rocks are associated with an intra-arc extensional regime and were emplaced immediately after the main stage of continental collision. The alkaline magmatism is essentially Eocene to Miocene in age and is principally developed in the countries of Armenia, Azerbai'an and Georgia (Fig. 50). The greatest contributions to our knowledge of the Caucasian alkaline rocks were made by Adamyan (1955), Aslanyan (1958), Bagdasaryan (1966), Gevorkyan (1965), Kotlyar (1939), Meliksetyan (1963a and 1963b) and Azizbekov et al. (1979).

1 Adzharo-Trialetskaya	7 Aiotsdzor
2 Garnasar	8 Elpinskii
3 Bunduk	9 Bargushatskii
4 Sevano-Shirakskaya	10 Shvanidzorskii
5 Tezhsar	11 Pkhrutskii
6 Vedi-Azatskii	12 Talyshskii

1 ADZHARO-TRIALETSKAYA   41°58'N; 42°01'E

In this region Palaeogene volcanic rocks increase in alkalinity from east to west. Among the alkaline effusive rocks there are leucite tephrites, leucite basalts, limburgites and analcime trachytes. Fuller information seems not to be available.

Reference Lordkipanidze et al., 1979.

2 GARNASAR   40°49'N; 44°47'E

This complex occurs within the Prisevanskaya zone, in the central part of the Bazumsky range of the Caucasus Mountains. The Garnasar outcrops of alkaline rocks cover an area of 12 km² and represent a subvolcanic, laccolith-like body which is confined to a system of northeast-trending faults. The rocks are alkaline syenites and trachytes in which, as seen characteristically in vertical sections, the medium- or fine-grained intrusive syenites are consistently replaced by trachytes. The syenites are hypidiomorphic rocks with trachytic textures and vary little in composition; potassium prevails over sodium. The rocks of the effusive facies are predominantly alkaline trachytes which are commonly aphanitic but variable in composition; they have the same potassium to sodium ratio as the syenites. The

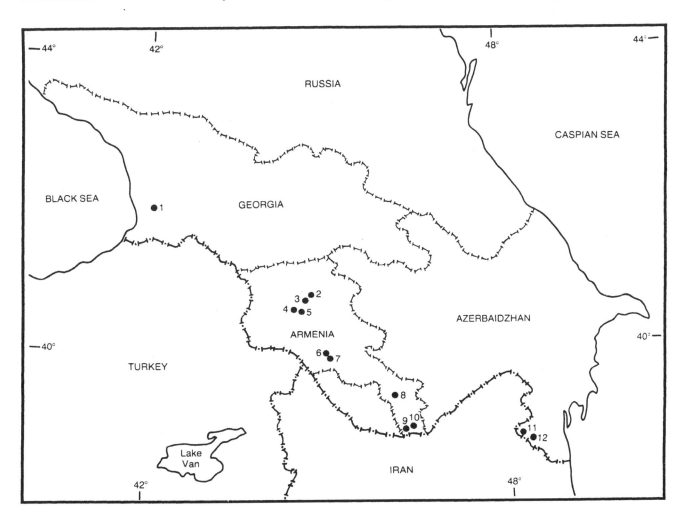

Fig. 50. Distribution of alkaline rocks in the Caucasus.

59

main rock-forming minerals of both the syenites and trachytes are orthoclase, albite–oligoclase, stilbite, natrolite, analcime and biotite. Various types of postvolcanic alteration are manifest including zeolitization, feldspathization, sericitization and the formation of clay minerals. The Garnasar complex is thought to be closely related to Tezhsar (No. 5).

Age K–Ar on whole rocks gave 34–35 Ma (Bagdasaryan and Meliksetyan, 1966)

References Adamyan and Mkrtchyan, 1959; Bagdasaryan, 1956; Bagdasaryan and Meliksetyan, 1966; Dzhsrbashyan and Meliksetyan, 1974; Kochinyan, 1975; Kotlyar, 1939; Meliksetyan, 1971.

---

3 BUNDUK   40°45′N; 44°45′E
Fig. 51

The Bunduk intrusion is situated several kilometres northeast of the Tezhsar complex on the southern slope of the Bazumsky mountain range. This is a dyke-like body expanding at its eastern end to a width of 2 km. There are also several smaller bodies in the vicinity with areas up to 1 km². The Bunduk intrusion is composed of various textural and mineralogical varieties of peralkaline syenites including leucosyenites, one variety of which is an oligoclase rock with a little arfvedsonite. The principal rock-forming minerals are perthite, oligoclase (An_{15-30}), diopside–augite, arfvedsonite, ferrohastingsite and biotite with accessory zircon, davidite, hellandite, orthite, baddeleyite, titanite, ilmenite and thorite. There is a little nepheline syenite. Various types of secondary alteration, including the development of biotite, amphibole and albite, are manifest.

Economic There is rare-earth mineralization in the oligoclasites.

Age K–Ar determinations on whole rocks gave 36 Ma (Bagdasaryan and Gukasyan, 1962).

References Abovyan *et al.*, 1981; Bagdasaryan and Gukasyan, 1962.

4 SEVANO-SHIRAKSKAYA   40°42′N; 44°32′E

Alkaline basalts, trachytes and phonolites with a thickness of 300–400 m are developed in the Palaeogene deposits of this region. Leucite tephrites and leucite phonolites are found in the succession.

Reference Ismail-zade *et al.*, 1990.

---

5 TEZHSAR   40°41′N; 44°39′E
Fig. 52

The Tezhsar alkaline complex has an area of about 80 km², is of central type and is developed in an elevated part of the Prisevanskaya zone of the Pambaksky range. A distinguishing feature of the complex is its ring-shaped structure and the combination of effusive and intrusive facies which it contains. The alkaline effusive series is 600 m thick and occurs as a broad, ring-shaped zone 12 to 15 km² in area towards the outer part of the complex. This series consists of three almost equally thick, pyroclastic formations which, from bottom to top, consist of (1) trachyandesites, (2) trachytes and (3) leucite phonolites or pseudoleucite porphyries. Also found in this series are alkaline potassium-rich basalts, leucite tephrites, shoshonites, sanidine and leucite trachytes, phonolites, leucitophyres and italites consisting almost solely of leucite. There are dykes of bostonite, essexite porphyry and alkaline lamprophyres, as well as an arc of dykes 1–5 m thick of pseudoleucite phonolite and syenite porphyry. There is replacement of the extrusive rocks by albite, muscovite and fluorite. The intrusive rocks of Tezhsar comprise (1) the alkaline syenites of the outer ring, (2) pseudoleucite syenites, (3) nepheline syenites, (4) nordmarkites and (5) peralkaline syenites and quartz syenites. These rocks define concentric zones, the pseudoleucite syenites (15 km²) and peralkaline syenites (16 km²) in particular forming well-defined arcs. To the marginal parts of the core are confined zones of nepheline-rich pseudoleucite syenites, which approach foyaite in composition. The pseudoleucite syenites consist of idiomorphic, more frequently ovoidal, pseudoleucites (30 to 40%) 2 to 6 cm in diameter set in a matrix of granular nepheline and alkali feldspar. The pseudo-

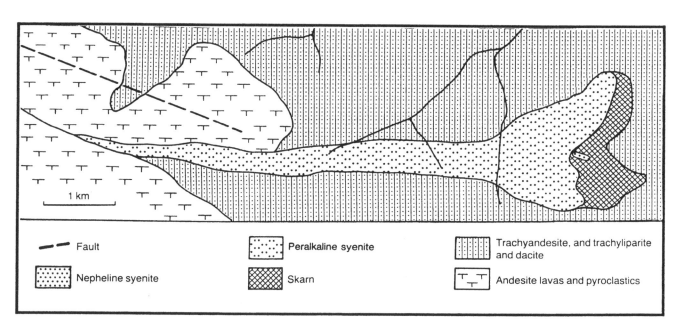

Fault

Nepheline syenite

Peralkaline syenite

Skarn

Trachyandesite, and trachyliparite and dacite

Andesite lavas and pyroclastics

Fig. 51. Bunduk (after Bagdasaryan, 1966, p. 264).

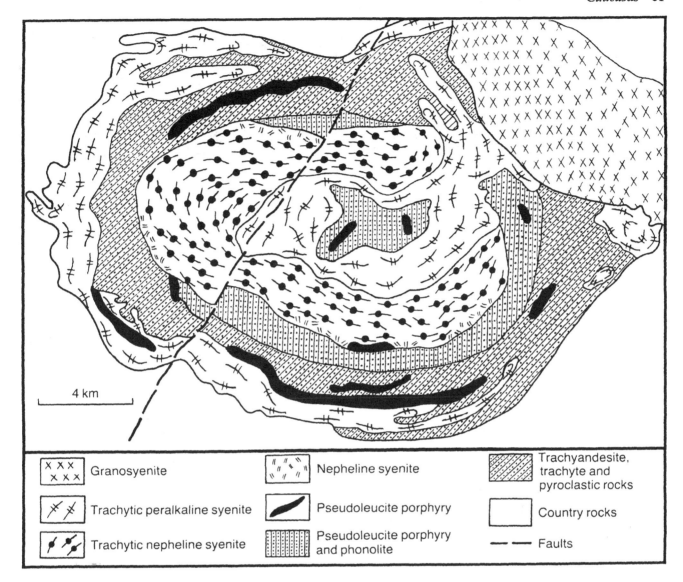

X X X / X X X	Granosyenite	// ' //	Nepheline syenite	Trachyandesite, trachyte and pyroclastic rocks
	Trachytic peralkaline syenite		Pseudoleucite porphyry	Country rocks
	Trachytic nepheline syenite		Pseudoleucite porphyry and phonolite	— — Faults

Fig. 52. Tezhsar (after Meliksetyan, 1971, Fig. 37).

leucite represents pseudomorphs having the typical intergrown texture of sodium-rich nepheline (25 to 35%) and sodium-depleted orthoclase ($Or_{90}Ab_{10}$). The central, conical intrusion is composed of peralkaline syenites, which grade in some places to peralkaline monzonites, nordmarkites and pulaskites. There are dykes of peralkaline and nepheline microsyenites, nepheline syenite porphyries, aplite, aplites dominantly composed of albite, tinguaite and bostonite, as well as feldspar and feldspathoid pegmatites.

Economic The nepheline syenites have been investigated for rare earth minerals, and there are rare earth minerals in fluorite–carbonate–biotite veins found in nepheline syenites (Bagdasaryan and Meliksetyan, 1966).

Age K–Ar determinations gave 37 to 39 Ma (Abovyan *et al.*, 1981; Bagdasaryan, 1966).

References Abovyan *et al.*, 1981; Bagdasaryan, 1956 and 1966; Dzhsrbashyan and Meliksetyan, 1974; Kochinyan, 1975; Kotlyar, 1939; Meliksetyan, 1971.

---

6 VEDI-AZATSKII

39°59'N; 45°00'E

This complex is developed within the Vedi-Azat volcanic

formation. It consists exclusively of volcanic and subvolcanic intrusions including sills, dykes, laccoliths and small stocks, amongst which, in a central position, are a number of volcanic necks that are circular in section and up to 1 km². The rocks of the suite comprise a series from alkaline olivine basalts through trachyte and trachyandesite to leucite syenite porphyry and leucite phonolite.

Economic Rare earth, copper, molybdenum and mercury mineralization is associated with the complex.

Age K–Ar determinations on rocks gave 12 Ma (Aslanyan, 1958).

References Aslanyan, 1958; Gabrielyan *et al.*, 1968; Ostroumova, 1967.

---

7 AIOTSDZOR

39°58'N; 45°01'E

This volcanic complex lies in the southwestern part of central Armenia within the Western Aiotsdzor district and extends over an area of about 150 km². The volcanic rocks have a thickness of 500 to 700 m and in vertical sections three series can be distinguished, (1) 220 to 300 m of andesitic basalts and tuff breccias, (2) 200 to

400 m of sanidine trachytes and tuffs and (3) about 100 m of dacitic lavas. In the lower part of the first series andesitic lavas and pyroclastics with fragments of trachyliparite and sanidine crystals are predominant; in the upper part tuff breccias of andesitic basalt with fragments of alkaline trachyte, sometimes containing pseudomorphs of leucite in the groundmass, are common. Also found within the first series are leucite and hauyne tephrites and alkali basalts. In the second series, which is composed predominantly of tuffaceous and other fragmental formations, the rocks are essentially trachytic and represented by porphyritic sanidine trachyliparites with a microlitic groundmass. In this series are concentrated subvolcanic bodies which are also of sanidine trachyliparite.

Age K–Ar determinations gave 21 to 24 Ma (Bagdasaryan and Meliksetyan, 1966).

References Bagdasaryan and Meliksetyan, 1966; Ostroumova, 1962.

---

8 ELPINSKII   39°27′N; 46°09′E

The Elpinskii complex is located within the Zavozhen-Akhavnadzorskaya tectonic depression in the south of Armenia. The earliest manifestations of igneous activity are represented by analcime basalts, the thickness of which is from 25 to 30 m. Higher in the section there occur analcime tephrites in the form of both lavas and explosive breccias which in their turn are succeeded by potassium-rich (leucitic) tephrites and dykes of hauyne-bearing basalt. Immediately above these rocks there occur trachyandesites and trachytes which are overlain by both effusive and subvolcanic trachytes and trachyrhyolites.

Age K–Ar determinations on whole rocks gave 12.5 to 14.5 Ma (Bazarova and Kazaryan, 1986.

References Bagdasaryan, 1966; Bazarova and Kazaryan, 1977 and 1986.

---

9 BARGUSHATSKII   38°57′N; 46°20′E
                                                           Fig. 53

The rocks of the Bargushatskii occurrence are located in the Bargushatskii mountain range and consist of a single 250 × 150 m mass as well as small lens-shaped and vein-like bodies, which occur in limestones that have been altered to skarns in a contact zone with leucocratic aplitic granites. The alkaline rocks including the smaller bodies, of which there are about 10, varying from 10–50 cm to 3–4 m across, comprise alkaline syenites, sviatonossite (andradite garnet-bearing syenite) and feldspathoidal rocks. The rock-forming minerals are microcline, albite, sodic plagioclase, quartz and aegirine. The accessory minerals include apatite, xenotime, magnetite, titanite, rutile, fluorite and anatase.

Age A K–Ar whole rock determination gave 35 Ma (Bagdasaryan, 1966).

References Bagdasaryan, 1966; Guyumdzhan, 1963; Perchuk, 1963; Tatevosyan, 1962.

---

10 SHVANIDZORSKII   39°00′N; 46°23′E

Shvanidzorskii is situated in the southeastern part of the Megrinskii granitoid pluton and is confined to a fracture zone striking northwest. The massif consists of three intrusive phases: (1) medium-grained melanocratic and mesocratic peralkaline syenites, (2) pegmatitic leuco-

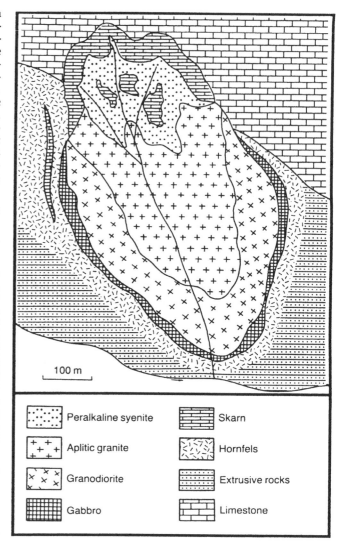

Peralkaline syenite

Aplitic granite

Granodiorite

Gabbro

Skarn

Hornfels

Extrusive rocks

Limestone

Fig. 53. Bargushatskii (after Tatevosyan, 1966, Fig. 67).

cratic syenites and (3) fine-grained nepheline syenites. There are also olivine-bearing gabbros of the essexite-theralite family. The most common rocks are the pegmatitic leucocratic syenites, which account for 85% of the intrusion. The principal rock-forming minerals are orthoclase–microperthite, oligoclase, nepheline, biotite and aegirine with accessory minerals including sodalite, monazite, pyrochlore, fluorite, cancrinite, calcite, ilmenite, apatite, zircon, titanite, baddeleyite and tourmaline. Fenitization of the country rocks (Devonian volcanics and diorites) is widely developed.

Age K–Ar and Rb–Sr determinations gave ages from 37 to 40 Ma (Bagdasaryan and Gukasyan, 1962; Gukasyan, 1963).

Economic Within the massif there are rare-earth pegmatites and carbonate-bearing veins with high contents of Th, REE, U and Pb (Bagdasaryan, 1954; Meliksetyan, 1963a, 1963b and 1971).

References Abovyan *et al.*, 1981; Adamyan, 1955; Bagdasaryan, 1947, 1954 and 1966; Gevorkyan, 1965; Gukasyan, 1963; Kochinyan, 1975; Kotlyar, 1939; Meliksetyan, 1963a, 1963b and 1971; Tatevosyan, 1966.

11 PKHRUTSKII   38°51′N; 48°10′E

This massif is an intrusion covering about 10 km². It is formed of a single intrusive phase, which is represented by nepheline–kalsilite monzonites, essexites and medium- and fine-grained syenites. The main rock-forming minerals are nepheline, kalsilite, orthoclase, andesine, augite, aegirine–augite and biotite. The rocks were strongly affected by metasomatic processes involving addition of sodalite and albite.

Age K–Ar and Rb–Sr determinations gave 37 to 40 Ma (Bagdasaryan, 1966).

References Adamyan, 1955; Bagdasaryan, 1947, 1954 and 1966; Gevorkyan, 1965; Kochinyan, 1975; Kotlyar, 1939; Meliksetyan, 1963b and 1971; Tatevosyan, 1966.

12 TALYSHSHKII   38°45′N; 48°22′E

The Talyshshkii folded zone is situated in the southeastern part of Azerbai′an and during the Eocene an extensive area of alkaline basaltic volcanic rocks were erupted. A series of leucite–sanidine tuffs and leucite tephrites and basanites, having a thickness of about 500 m, were ejected, during early, mid- and late Eocene times, in the southern part of the Talyshshkii zone; they extend into Iran. The leucite tuffs occur in the volcanic domes of Shandan-Kalasy and Green Hill. Mineralogically the tuffs consist of sanidine, plagioclase, leucite, augite, aegirine–augite, hornblende and volcanic glass. The leucite tephrites and basanites are found in the volcanoes of Geldar, Govery, Kiz-Kalashym Pashgra and Rasgov. They contain phenocrysts of plagioclase, leucite, olivine and augite set in a groundmass of glass, feldspar, augite, leucite and magnetite. The leucite tephrites also form dykes and sills including the Dimansky sill. The largest centre is the Kalakhanskaya subvolcanic intrusion, which forms a cone with a height of 1732 m; it is oval in plan. The rocks developed within Kalakhanskaya are trachyandesites, gabbro-teschenites, picrobasalts and leucite-bearing basalts. The gabbro-teschenites are composed of plagioclase (35–40%), augite (15–20%), biotite (3–5%), amphibole (1–6%), analcime (10–12%) and natrolite (2%) with accessory apatite, titanite and titanomagnetite.

Reference Azizbekov *et al.*, 1979.

Caucasus references

ABOVYAN, S.B., AGAMALYAN, V.A., ASLANYAN, A.T. et al. 1981. I.G. Magakyan (ed.) *Magmatic and metamorphic formations of the Armenian SSR.* Akademii Nauk Armyanskoi SSR, Erevan. 331 pp.

ADAMYAN, A.I. 1955. *Petrography of the alkaline rocks of the Megrinsky region Armenian SSR.* Akademii Nauk Armyanskoi SSR, Erevan. 102 pp.

ADAMYAN, A.I. and MKRTCHYAN, K.A. 1959. The Garnasar intrusion of alkaline syenites in the eastern part of the Khalabsky ridge. *Transactions of the Department of Geology and Mineral Deposits of Armenia, Erevan.* 2: 34–7.

ASLANYAN, A.T. 1958. *Regional geology of Armenia.* Aipetrat, Erevan. 430 pp.

AZIZBEKOV, S.A., BAGIROV, A.E., VELIEV, M.M., ISMAIL-ZADE, A.D., NIZHSERADZE, N.S., EMELYANOVA, E.N. and MAMEDOV, M.N. 1979. *Geology and volcanism of Talish.* Akademii Nauk Azerbaidzhanskoi SSR, Baku. 246 pp.

BAGDASARYAN, G.P. 1947. Nepheline syenites of the Pambaksky shield. *Doklady Akademii Nauk Armyanskoi SSR. Erevan,* 1: 19–35.

BAGDASARYAN, G.P. 1954. Alkaline pegmatites of central Armenia. *Doklady Akademii Nauk Armyanskoi SSR. Erevan,* 19(4): 117–22.

BAGDASARYAN, G.P. 1956. Petrography of alkaline effusive rocks. *Ivestiya Akademii Nauk Armyanskoi SSR, Seriya Geologiya. Erevan,* 2: 25–36.

BAGDASARYAN, G.P. 1966. Intrusive rocks of the Basumo-Pambaksky region. *Geology of the Armenian SSR. Petrography. Intrusive rocks.* 3: 256–308. Armenian Academy of Sciences, Erevan.

BAGDASARYAN, G.P. and GUKASYAN, P.X. 1962. The results of the absolute age determination of the separate magmatic complexes of the Armyanskaya SSR. Nauka, Moscow and Leningrad. 283–303.

BAGDASARYAN, G.P. and MELIKSETYAN, B.M. 1966. Genetic peculiarities of the Armenian alkaline rocks. *Ivestiya Akademii Nauk Armyanskoi SSR, Seriya Geologiya. Erevan,* 11: 82–100.

BAZAROVA, T.Yu. and KAZARYAN, G.A. 1977. Peculiarities of crystallization of the leucite phonolites of the Azat-Vedi river. *Doklady Akademii Nauk SSSR,* 6: 137–42.

BAZAROVA, T.Yu. and KAZARYAN, G.A. 1986. Neogene alkaline effusives of South Armenia and conditions for their formation. *Volcanology and Seismology,* 2: 34–45.

DZHSRBASHYAN, R.T. and MELIKSETYAN, B.M. 1974. Geochemical and petrogenetic peculiarities of the alkaline volcanic series of the Pambaksky ridge. *Magmatism and metallogenesis of the Armenian SR.* 54–67. Akademii Nauk Armyanskoi SSR, Erevan.

GABRIELYAN, A.A., BAGDASARYAN, G.P., KARAPETYAN, K.I., MELIKSETYAN, B.M., MELKONYAN, P.L. and MNATSAKANYAN, A.C. 1968. Main stages of geotectonic development and magmatic activity on the territory of the Armenian SSR. *Ivestiya Akademii Nauk Armyanskoi SSR, Seriya Nauk o Zemle, Erevan,* 21: 6–39.

GEVORKYAN, R.G. 1965. *Geochemical peculiarities and petrogenesis of the alkaline rocks of central Armenia.* Trudy, Moscow. 144 pp.

GUKASYAN, R.X. 1963. Determination of the absolute age by the Rb–Sr method on a sample from the Megrinsky pluton Armenian SSR. *Doklady Akademii Nauk Armyanskoi SSR, Seriya Geologiya. Erevan,* 36(3): 173–8.

GUYUMDZHAN, O.P. 1963. Paragenesis of alkaline metasomatites of the Bargushat ridge. *Izvestiya Akademii Nauk Armyanskoi SSR, Seriya Nauk o Zemla, Erevan.* 3: 31–9.

ISMAIL-ZADE, A.D., MALOV, Y.V. and BELEN′KIY, B.G. 1990. Systematics of Cainozoic volcanic formations by methods of multiple statistics. *Izvestiya Akademii Nauk Azerbaidzhanskoi SSR, Seriya Nauk o Zemle, Baku,* 2: 16–25.

KOCHINYAN, R.E. 1975. *Mineralogy and geochemistry of the alkaline rocks of the Pamback.* Akademii Nauk Armyanskoi SSR, Erevan. 175 pp.

*KOTLYAR, V.N. 1939. Discovery of the leucitic rocks in the Caucasus. *Sovetskaya Geologiya,* 4–5: 4–11.

LORDKIPANIDZE, M.B., ZAKARIADZE, G.S. and POPOLITOV, E.I. 1979. Volcanic evolution of the marginal and interarc basins. *Tectonophysics,* 57: 71–83.

MELIKSETYAN, B.M. 1963a. On the mineralogy, geochemistry and genesis of pegmatites of the Megrinsky pluton. *70-anniversary volume dedicated to K.N. Paffengolts,* 71–98. Akademii Nauk Armyanskoi SSR, Erevan.

MELIKSETYAN, B.M. 1963b. Mineralogical geochemical peculiarities of the alkaline rocks of the Magrinsky pluton. *Ivestiya Akademii Nauk Armyanskoi SSR, Erevan,* **2**: 57–80.

MELIKSETYAN, B.M. 1971. Mineralogy, geochemistry and petrological peculiarities of the Tezhsarskii complex. *Petrology of the intrusive complexes of the important regions of the Armenian SSR.* 117–308. Akademii Nauk Armyanskoi SSR, Erevan.

OSTROUMOVA, A.S. 1962. Triple volcanism of Western Daralagese (Armenia). Unpublished thesis. Leningrad. 197 pp.

OSTROUMOVA, A.S. 1967. A basalt-trachyte formation of the Small Caucasus. *Alkaline volcanological formations of folded regions.* 64–76. Nedra, Leningrad.

PERCHUK, L.L. 1963. Magmatic interchange of carbonate layers with the formation of nepheline syenites and other alkaline rocks: an example from the Dezhnev massif. *Physico-chemical problems of the formation of igneous rocks and ores* **2**: 160–81. Akademii Nauk SSSR, Moscow.

TATEVOSYAN, T.Sh. 1962. Conditions of formation of alkaline rocks of the Bargushatsk ridge (Armenian SSR), *Ivestiya Akademii Nauk Armyanskoi SSR, Seriya Geologo-geograficheskikh. Erevan,* **5**: 15–26.

TATEVOSYAN, T.Sh. 1966. Intrusive rocks of the Bargushatsk ridge. *Geology of the Armenian SSR. Petrography. Intrusive rocks,* **3**: 114–214. Akademii Nauk Armyanskoi SSR, Erevan.

* In English

THE URALS

Occurrences of alkaline rocks are known in both the central and southern Urals (Fig. 54), the massifs of the Ilmenogorskii-Vishnevogorskii complex being the largest and best studied. They stretch along a north–south line over a distance of about 150 km from the Vishnevogorskii complex in the north through the so-called intermediate zone, in the regions of Lakes Ishkul and Uveldy, to the Ilmenogorskii complex in the south (Fig. 56). The width of the outcrop of alkaline rocks varies from some 4–7 km down to 250–300 m. This group of intrusions is located in the narrowest part of the Urals, where the course of the folded structures changes sharply from northeasterly in the southern Urals, to north–south, and then to a northwesterly trend in the middle Urals. The occurrences of Ilmenogorskii and Vishnevogorskii are located in large anticlinal structures, while in the zone between these two occurrences the alkaline rock occurrences are much smaller.

Fig. 54. Distribution of alkaline igneous rocks and carbonatites in the Urals (after Levin, 1974, Fig. 1).

1 Kushva	4 Vishnevogorskii
2 Nyazepetrovsk	5 Ishkul
3 Berdyaush	6 Ilmenogorskii

1 KUSHVA
58°18'N; 59°47'E

Veins of nepheline (liebeneritic) syenites and nepheline syenite pegmatites up to 2 m thick are present in the intrusion, which is essentially of syenites. The nepheline syenites of the veins are variable in composition and structure with both trachytic and massively textured types. They contain up to 20–25% of nepheline, which may also be pseudomorphed by cancrinite, analcime, zeolites and spreustein, 65–70% of feldspar, including microperthite and albite-oligoclase, and 5–10% of aegirine–augite, salite, hastingsite and lepidomelane.

Accessory minerals, forming about 2.5% of the rocks, include melanite, titanite, orthite, magnetite and apatite.

Age K–Ar on syenites gave 369–389 Ma (Levin, 1974).

References Borodin et al., 1974; Levin, 1974; Levin et al., 1973.

2 NYAZEPETROVSK
56°01'N; 59°36'E

This complex consists of intrusive and extrusive peralkaline and mildly alkaline rocks. Nepheline and peralkaline syenites are encountered as small lens-like bodies 0.2–10 m long. The extrusive rocks are represented by trachybasalts and trachyandesites, including tuffs, trachytes and perhaps phonolites. The intrusive and extrusive rocks of Nyazepetrovsk are situated in a zone of approximately north–south-trending faults which cut sedimentary and volcanogenic rocks of Ordovician age. Related to the faults also are small areas and massifs of earlier dunite, peridotite, pyroxenite and gabbroic rocks. Cutting through pyroxenite and peridotite are nepheline and peralkaline syenites which are enriched in K–feldspar and nepheline and vary in composition to shonkinites and ijolites. Adjacent to veins of alkaline syenites cutting gabbroic rocks are reaction zones containing blue-green amphibole, K–feldspar, zeolites and calcite. Not infrequently the nepheline and peralkaline syenites have porphyritic textures with a trachytic groundmass. Feldspar (orthoclase and albite) comprises 65%, and spreustein and liebnerite, which have replaced nepheline, 27% of the nepheline syenites. Dark coloured minerals are represented by hastingsite, andradite-grossular, aegirine–salite and sometimes by biotite. Accessory minerals include titanite, apatite, fluorite, pyrite, chalcopyrite, magnetite, zircon and pyrochlore. Chemically, the predominance of potassium over sodium is typical. Shonkinites and ijolites have spotted textures and banded structures with K–feldspar developing large porphyroblasts in the former. Clinopyroxene, olivine, magnetite, amphibole and biotite are typical dark coloured constituents of these rocks. The peralkaline syenites are mineralogically similar but for the absence of nepheline. The petrochemistry of the rocks is described by Levin (1974) and Svyazhin and Levin (1971).

Age K–Ar on nepheline syenite gave 298–378 Ma (Svyazhin and Levin, 1971).

References Levin, 1974; Svyazhin and Levin, 1971; Zhilin and Seliverstov, 1971.

3 BERDYAUSH
55°10'N; 59°09'E
Fig. 55

This intrusion of 3.5 × 10 km lies within Precambrian limestones. Three intrusive phases comprise the complex: these are rapakivi granite, syenite and nepheline syenite. The intrusion has a concentric zonal structure with the nepheline syenites located in the centre (Fig. 55). However, mapping and prospecting work in the past few years, the map from which is not yet available, has indicated that the nepheline syenites do not form a continuous body, but comprise a series of dykes 0.4–0.5 km thick with lengths of up to 2.5 km. The nepheline syenites are fine-grained, bluish-grey rocks which are usually porphyritic and with a trachytic groundmass. All the main minerals are present as phenocrysts. The dark-coloured minerals are represented mainly by hastingsite and aegirine–salite with small quantities of magnetite and late biotite. On average the nepheline syenites contain 72% feldspar, 21% nepheline and 7% dark-coloured minerals. Not infrequently the nepheline syenites are altered and replaced by albite, zeolites and sericite. The marginal zones of nepheline syenite dykes often grade into alkaline syenite, but the alkaline syenites also form independent dykes. The alkaline syenites consist predominantly of alkali feldspar with small amounts of oligoclase or andesine also present. Mafic minerals are represented by amphibole, biotite and aegirine–salite and accessories, which are more diverse in the alkaline syenites than the nepheline syenites, are titanite, orthite and zircon. The genetic role of the rapakivi granites in the complex is a matter of discussion.

Age K–Ar on amphibole from nepheline syenite gave 1,370 Ma. K–Ar, Rb–Sr and U–Pb dating of the rapakivi granites gave 1300–1400 Ma (Ovchinnikov *et at.*, 1964; Levin, 1974).

References Levin, 1974; Ovchinnikov *et al.*, 1964; Zavaritsky, 1937.

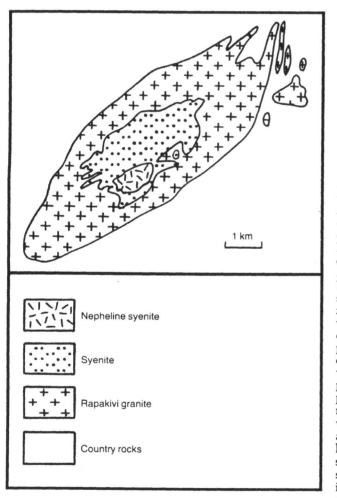

Fig. 55. Berdyaush (after Borodin et al., 1974, Fig. 13a).

4 VISHNEVOGORSKII 55°59′N; 60°34′E
Figs 56 and 57

The Vishnevogorskii intrusion forms the northern part of the Ilmenogorskii-Vishnevogorskii complex. This

massif is about 25 km long and up to 4 km wide and is located in the axial part of the Urals anticlinal structure. The northern part of the massif is wider and has the form of a steeply dipping dome which extends under metamorphic rocks of Riphean age. The northern part of the main intrusive body is accompanied by satellite intrusions of nepheline syenite which have thicknesses up to several hundreds of metres. The intrusion has contacts which are conformable with the enclosing rocks. In the southern part the complex is represented by veins and dykes of miaskitic nepheline syenite. Generally the southern part of Vishnevogorskii closely resembles the northern part of the Ilmenogorskii complex (No. 6 and Fig. 59). The Vishnevogorskii complex is composed mainly of miaskitic nepheline syenites with biotite, but there are nepheline syenites with amphibole and intermediate varieties with biotite and amphibole. Small bodies of biotite-bearing alkaline syenites are also found. The alkaline rocks of the complex have gneiss-like structures and textures with a clear preferred orientation of the feldspar. The nepheline syenites are frequently greatly albitized and partly replaced by calcite, especially in the central part of the massif, and also in the satellite intrusion at the northern end. In some places stockworks of calcite carbonatites occur. Numerous alkaline pegmatites, which are predominantly developed within fenites and close to the outer contacts of the intrusion, are a characteristic feature of the Vishnevogorskii intrusion. Miaskitic nepheline syenites occupy about half of the area of the intrusion. These vary from fine- to coarse-grained rocks with massive or banded structures which are reflected in the preferred orientation of biotite and other minerals. The main rock-forming minerals of the miaskites are nepheline, oligoclase, orthoclase, lepidomelane and magnetite. Secondary and accessory minerals include microcline, albite, biotite, pyrochlore, zircon, ilmenite, apatite, titanite, calcite, cancrinite, zeolites and sulphides. A highly variable mineralogy is typical of rocks near to the contact zone. Alkaline pegmatites are divided into two groups. Those of the first group are tabular in form, have thicknesses of up to 5–7 m and can be traced for distances of tens or hundreds of metres. They are generally located in the outer parts of the complex close to the contact and are composed of 95–98% microcline, albite, nepheline, biotite and natrolite. They also contain cancrinite, sodalite, analcime, calcite, magnetite, ilmenite, apatite, zircon and pyrochlore. The second group of pegmatites, which are situated in the central parts of the complex and cut across the gneissosity of the nepheline syenites, comprises microcline, nepheline, biotite, albite, natrolite, cancrinite and aegirine–augite which form 97–98% by volume. In the axial parts of the pegmatites nepheline predominates over microcline, while in the marginal parts microcline is predominant. Accessories include minerals of uranium, thorium and rare-earths, as well as vishnevite, for which this is the type locality, magnetite and calcite. Two stages of carbonatite emplacement have been identified. Early carbonatites, which constitute the main phase, are represented by calcite and silico-calcite types. They lie both within the complex and in the fenite aureole. Sometimes the bodies of carbonatite are concentrated to form complex zones 100–150 m across and stretching for many kilometres. In plan such zones consist of chains of flattened lens-like bodies which follow each other in an *en echelon* pattern. Albite–biotite–calcite and biotite–calcite varieties with apatite (up to 20%), ilmenite, titanite and zircon are the most abundant varieties. The veins which constitute the later carbonatite stage are narrow, up to 10 cm wide, and have variable compositions depending on the composition of the enclosing rock. For instance, in miaskites

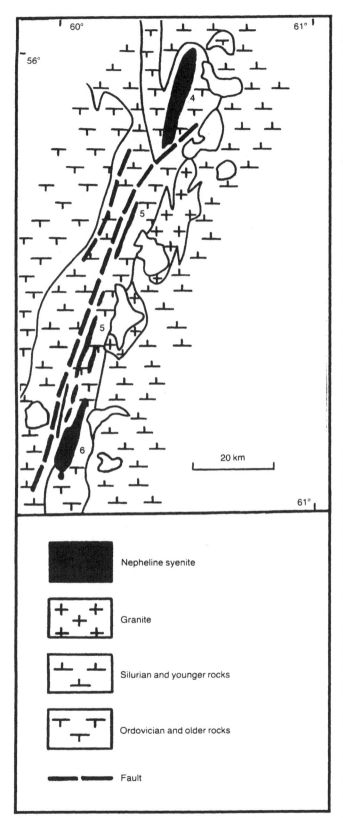

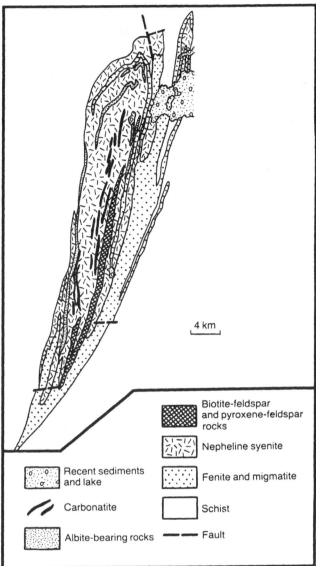

Fig. 57. Vishnevogorskii (after Ronenson, 1966, Fig. 4).

Fig. 56. The Ilmenogorskii-Vishnevogorskii occurrences including the Vishnevogorskii (No. 4), Ishkul (No. 5) and Ilmenogorskii (No. 6) occurrences (after Levin, 1974, Fig. 2 and Kramm et al., 1983, Fig. 1).

these veins contain calcite, apatite, ilmenite, orthite, betafite and thorite, whereas in fenites the veins consist of ankerite, calcite, aegirine, rutile, brookite and sphalerite. Ronenson (1966) suggested that the nepheline syenites are of anatectic origin, being the final stage of a metasomatic process whereby the enclosing rocks were transformed by alkaline fluids. Rb–Sr and U–Pb isochron studies (Kramm *et al.*, 1983; Chernyshev *et al.*, 1987) have, however, apparently established the mantle origin of the nepheline syenites of the Vishnevogorskii massif (Rb–Sr ratio 0.70304 ± 0.00060) and that there were two stages in the generation of the nepheline syenites. Calcite from carbonatites and alkaline rocks gave $^{18}O^0/_{00}$ values of +6.4 to +7.5, with some values up to +8.0. Silicates from miaskitic nepheline syenites gave $^{18}O^0/_{00}$ values of +5.4 to +5.5. Calcite from carbonatite gave $C^0/_{00}$ of −7.1 to −7.5 and sulphide −1.6 to −2.9 (Dontsova *et al.*, 1977; Kononova *et al.*, 1979; Grinenko *et al.*, 1970).

Age A Rb–Sr whole-rock isochron gave 478 ± 55 Ma but a mineral isochron 244 ± 8 Ma, which is taken as a recrystallization episode related to the regional Hercynian folding (Kramm *et al.*, 1983). K–Ar dating of

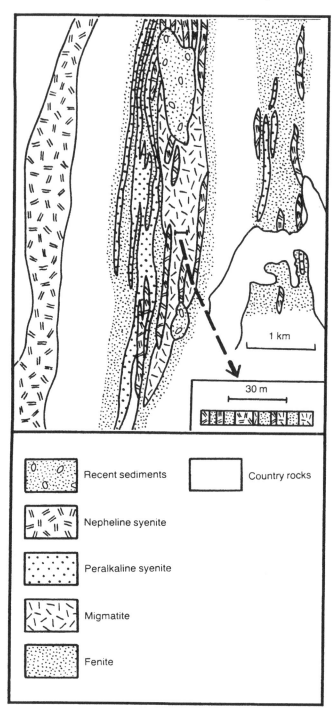

Fig. 58. Ishkul (after Kramm et al., 1983, Fig. 3). Inset diagram illustrates the complexity developed in one section by interdigitation of different rock types.

minerals from pegmatite gave the following results: nepheline 256 ± 10 and biotite 252 ± 10 Ma (Kononova and Shanin, 1971; Kononova and Shanin, 1973).

Economic Pegmatites have been mined to obtain raw materials for the production of ceramics (Eskova *et al.*, 1964).

References *Chernyshev *et al.*, 1987; Dontsova *et al.*, 1977; Eskova *et al.*, 1964; Grinenko *et al.*, 1970;

Kononova and Shanin, *1971 and 1973; Kononova *et al.*, 1979; *Kramm *et al.*, 1983; Levin, 1974; Ronenson, 1966.

5 ISHKUL
55°33′N; 60°21′E
Figs 56 and 58

The Ishkul occurrence is located near to the Ishkul and Uveldy lakes and is an elongate body extending north-south for 10 km and having a width of about 0.6 km. It lies within granites of the western sector of the Ilmenogorskii anticlinorium. The complex is situated in the so-called intermediate zone between the Vishnevogorskii and Ilmenogorskii occurrences (Fig. 56). The massif is composed predominantly of miaskitic nepheline syenites with lepidomelane; miaskitic hastingsite–nepheline syenites and peralkaline hastingsite syenites also occur. Both to the west and east of the complex are numerous sheets and veins of nepheline syenite and peralkaline syenite which vary in thickness from several centimetres to several metres and which lie within fenitized country rocks. The following sequences of rocks are typical: intrusive nepheline syenite–alkaline syenite–syenitic migmatite–fenite–unaltered gneiss. The biotite–nepheline syenite of this zone has Rb–Sr ratios of 0.70337 ± 0.00036 (Kramm *et al.*, 1983). Silicates from nepheline-feldspar migmatites gave $^{18}O^0/_{00}$ values of +8.2 (Kononova *et al.*, 1979).

Age A Rb–Sr isochron, on whole-rocks, gave 467 ± 105 Ma (Kramm *et al.*, 1983, Fig. 5) and U–Pb isochrons on zircon from carbonatites ages of 432 ± 12 and 261 ± 6 Ma (Chernyshev *et al.*,1987).

References *Chernyshev *et al.*, 1987; Kononova *et al.*, 1979; *Kramm *et al.*, 1983; Ronenson, 1966.

6 ILMENOGORSKII
55°04′N; 60°14′E
Figs 56 and 59

The Ilmenogorskii complex extends for 18 km in a north-northeasterly direction and has a maximum width 4.5 km. It is situated in the domed part of the Urals anticlinorium structure which expands southwards but is greatly compressed to the north. The contacts of the complex are complicated and accompanied by numerous injections of the alkaline rocks into the country rocks, which are fenitized. The massif is heterogeneous with the predominant miaskitic biotite-nepheline syenites occupying 38% of the area, amphibole miaskitic nepheline syenites 18% and biotite syenites 14%. In addition, about 20% of the area is occupied by biotite–nepheline syenites which interdigitate with syenites and biotite plagiosyenites, while 10% is taken up by amphibole–nepheline syenites which interdigitate with syenites and amphibole plagiosyenites. Dykes of miaskitic aplite are also found. Hundreds of pegmatite bodies have been discovered which include granite, miaskitic nepheline syenite and syenite pegmatites. The miaskitic nepheline syenites contain about 30% nepheline, orthoclase, oligoclase, albite and lepidomelane or biotite together with accessory magnetite, zircon, pyrochlore, titanite and ilmenite. The nepheline syenite pegmatites form concordant and, rarely, cross-cutting vein- like bodies up to 2 m thick which stretch for tens of metres, as well as small lenses. They occur mainly within the miaskitic nepheline syenites and rarely in the enclosing rocks. They are represented by two varieties: biotite–nepheline–microcline and muscovite–nepheline-microcline. Typically they also include cancrinite, sodalite, ilmenite, zircon, apatite, calcite, pyrochlore and zeolites. Amongst the nepheline syenites, restricted zones, up to 0.5 m across, of biotite–calcite rocks with apatite are sometimes found. Narrow calcite veins with apatite, titanite, magnetite, ilmenite, pyrochlore, orthite, fluorite and sulphides are more widespread. According to Ronenson (1966) the nepheline syenites are the result of anataxis and the final product of the metasomatic alteration of the enclosing rocks. Isotope studies of Kramm *et al.* (1983) and Chernyshev *et al.* (1987) indicated Rb–Sr ratios of 0.703290 ± 0.0023. Calcite from carbonatite gave $^{18}O^0/_{00}$ values of 6.6 to +7.0 and silicates from miaskitic nepheline syenite +5.6 (Kononova *et al.*, 1979). Melting experiments using miaskite samples from Ilmenogorskii in the presence of aqueous fluids are reported by Shchekina *et al.* (1985).

Economic Ilmenorutile and molybdenite ores are present and the complex is famous for its pegmatite veins with unique associations of minerals including aeschynite and betafite (Semenov *et al.*, 1974).

Age A whole-rock Rb–Sr isochron gave an age of 446 ± 13 Ma but for rock-forming minerals 244 ± 5 Ma and 245 ± 24 Ma (Kramm *et al.*, 1983). A U–Pb isochron on zircon from nepheline syenites gave 422 ± 10 and 261 ± 14 Ma (Chernyshev *et al.*, 1987).

References *Chernyshev *et al.*, 1987; Kononova *et al.*, 1979; *Kramm *et al.*, 1983; Levin, 1974; Ronenson, 1966; Semenov *et al.*, 1974; *Shchekina *et al.*, 1985.

Urals references

BORODIN, L.S., LAPIN, A.V. and PYATENKO, I.K. 1974. Alkaline provinces of Europe. *In* L.S. Borodin (ed.), *Principal provinces and formations of alkaline rocks.* 11–90. Nauka, Moscow.

*CHERNYSHEV, I.V., KONONOVA, V.A., KRAMM, U. and GRAUERT B. 1987. Isotope dating of Urals alkaline rocks in the light of U–Pb data for zircons. *Geochemistry International*, 24(10): 1–15.

DONTSOVA, E.I., KONONOVA, V.A. and KUZNETSOVA, L.D. 1977. Isotopic composition of oxygen of carbonatites and carbonatite-like rocks in connection with the sources of their material and the problem of ore content. *Geokhimiya*, 7: 963–75.

ESKOVA, E.M., ZABIN, A.G. and MUKHITDINOV, G.N. 1964. *Mineralogy and rare element geochemistry of the Vishnevy Gory.* Nauka, Moscow. 319 pp.

GRINENKO, L.N., KONONOVA, V.A. and GRINENKO, V.A. 1970. Isotopic composition of sulphur of sulphides from carbonatites. *Geokhimiya*, 1: 66–76.

*KONONOVA, V.A. and SHANIN, L.L. 1971. On possible application of nepheline for alkaline rock dating. *Bulletin Volcanologique*, 35: 1–14.

KONONOVA, V.A. and SHANIN, L.L. 1973. On the radiogenic argon in nepheline in connection with its application to the dating of alkaline complexes. *In* G.D. Afanas'ev (ed.) *Geological-radiological interpretation of discordance age data*. Nauka, Moscow. 40–52.

KONONOVA, V.A., DONTSOVA, E.I. and KUZNETSOVA, L.D. 1979. Isotopic composition of oxygen and strontium of the Ilmen-Vishnevy Gory alkaline complex and problems of miaskite genesis. *Geokhimiya*, 12: 1784–95.

*KRAMM, U., BLAXLAND, A.B., KONONOVA, V.A. and GRAUERT, B. 1983. Origin of the Ilmenogorsk-Vishnevogorsk nepheline syenites, Urals, USSR, and their time of emplacement during the history of the Ural fold belt: a Rb–Sr study. *Journal of Geology*, 91: 427–35.

LEVIN, V.Y. 1974. *The alkaline province of the*

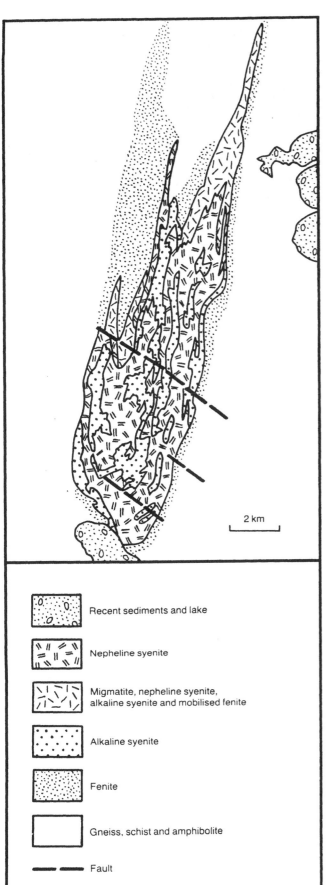

Fig. 59. Ilmenogorskii (after Levin, 1974, Fig. 3).

Recent sediments and lake

Nepheline syenite

Migmatite, nepheline syenite, alkaline syenite and mobilised fenite

Alkaline syenite

Fenite

Gneiss, schist and amphibolite

Fault

2 km

Ilmenogorsk-Vishnevy Gory (formation of the nepheline syenite of the Urals). Nauka, Moscow. 224 pp.

LEVIN, A.Y., PETROV, V.I. and LEVINA I.A. 1973. Experience of the facial and formation analyses of the nepheline syenite of the Urals. In Problems of the petrology of the Urals. *Proceedings of the Institute of Geology and Geochemistry, Sverdlovsk* 100: 134–54.

OVCHINNIKOV, L.N., DUNAEV, V.A. and KRASNOBAYEV, A.A. 1964. Materials on the absolute geochronology of the Urals. *Absolute age of the geological formations*. Nauka, Moscow. 157–71.

RONENSON, B.M. 1966. Origin of the miaskites and their relationship with the rare-metal ores. In A.I. Ginzburg (ed.) *Geology of the rare elements deposits* 28: Nedra. 174 pp.

SEMENOV, E.I., ESKOVA, E.M., KAPUSTIN, U.L. and HOM'AKOV, A.P. 1974. *Mineralogy of alkaline massifs and their deposits*. Nauka, Moscow. 246 pp.

*SHCHEKINA, T.I., GRAMINITSKIY, Ye.N. and YUDINTSEV, S.V. 1985. Experimental study of the melting of miascites from the Il'menskiye Gory massif. *Geochemistry International*, 22(3): 95–108.

SVYAZHIN, N.V. and LEVIN, V.Y. 1971. Alkaline magmatism of the Urals. *Magmatic formations, metamorphism and metalogenasis of the Urals* 4: 375–89. Academy of Sciences of the USSR, Urals Scientific Centre, Sverdlovsk.

ZAVARITSKY, A.N. 1937. *Petrography of the Berdyaush pluton*. ONTI. Leningrad and Moscow. 331 pp.

ZHILIN, I.V. and SELIVERSTOV, G.F. 1971. The trachybasalt–picrite association of the Nyazepetrovsk area (western slope of the Urals). *Doklady Akademii of the Academy of Sciences of the USSR*, 197: 686–8.

*In English

KAZAKHSTAN

Kazakhstan is a large country with an area of about 2.7 million km². More than 50 occurrences of alkaline rocks are known which are spread widely over this extensive territory. It is probable that a number of distinct provinces are represented within the country but for the purpose of the present work all occurrences have been combined as one national group. The geology of Kazakhstan is complex and involves blocks of the East European (Russian) and Scyptian-Turanian platforms and Caledonian, Hercynian and Alpine fold belts. The distribution of the occurrences of alkaline igneous rocks is shown on Fig. 60.

1 KUNDZINSKII   52°54′N; 64°27′E

This occurrence is elliptical in form and covers 2.5 km². It is composed of porphyritic solvsbergites consisting of orthoclase, aegirine, riebeckite and biotite.

Age K–Ar gave 252 Ma.

Reference Ksenofontov, 1966.

2 UBAGANSKII WEST   52°55′N; 64°54′E

The Ubaganskii intrusion has an elliptical form, elongated east–west, and is 2.5 km² in area. It is composed of solvsbergites containing phenocrysts of orthoclase, aegirine, riebeckite and biotite, each of which may comprise between 2% and 14% of the rock. Secondary albite and carbonate form 2.5–3.5% and accessories include magnetite and fluorite.

Age 252 Ma.

Reference Ksenofontov, 1966.

1 Kundzinskii	28 Tluembetskii
2 Ubaganskii West	29 Zhil′tauskii
3 Barchinskii	30 Polumesyats
4 Karly-kul′skii	31 Berkuty South
5 Krasnomaiskii	32 Akbiik
6 Pavlovskii	33 Kyzyl-Char
7 Kosistekskii	34 Kainarskii
8 Damdysaiskii	35 Kshi-Orda
9 Shinsaiskii	36 Korgantas
10 Ishimskii	37 Keregetas
11 Kentaldykskii	38 Manrak
12 Karasyrski	39 Bolektas
13 Borsuksai	40 Kandygatai
14 Zhusalinskii	41 Akkoitas
15 Karsakpai	42 Kurozekskii
16 Zhilanda and Kulan	43 Koksalinskii
17 Daubabinskoe	44 Arsalan
18 Kaindy	45 Donenzhal
19 Irisu	46 Batpak
20 Elinovskii	47 Kalgutinskii North
21 Askatinskii	48 Kalgutinskii South
22 Maiorskii	49 Iisorskii
23 Shanshal′skii	50 Biesimas
24 Berkutinskii	51 Bolshoi and Malyi Espe
25 Dyke field	52 Bakanasskaya
26 Birdzhankol′skii	53 Talgarskii
27 Karatal′ski	

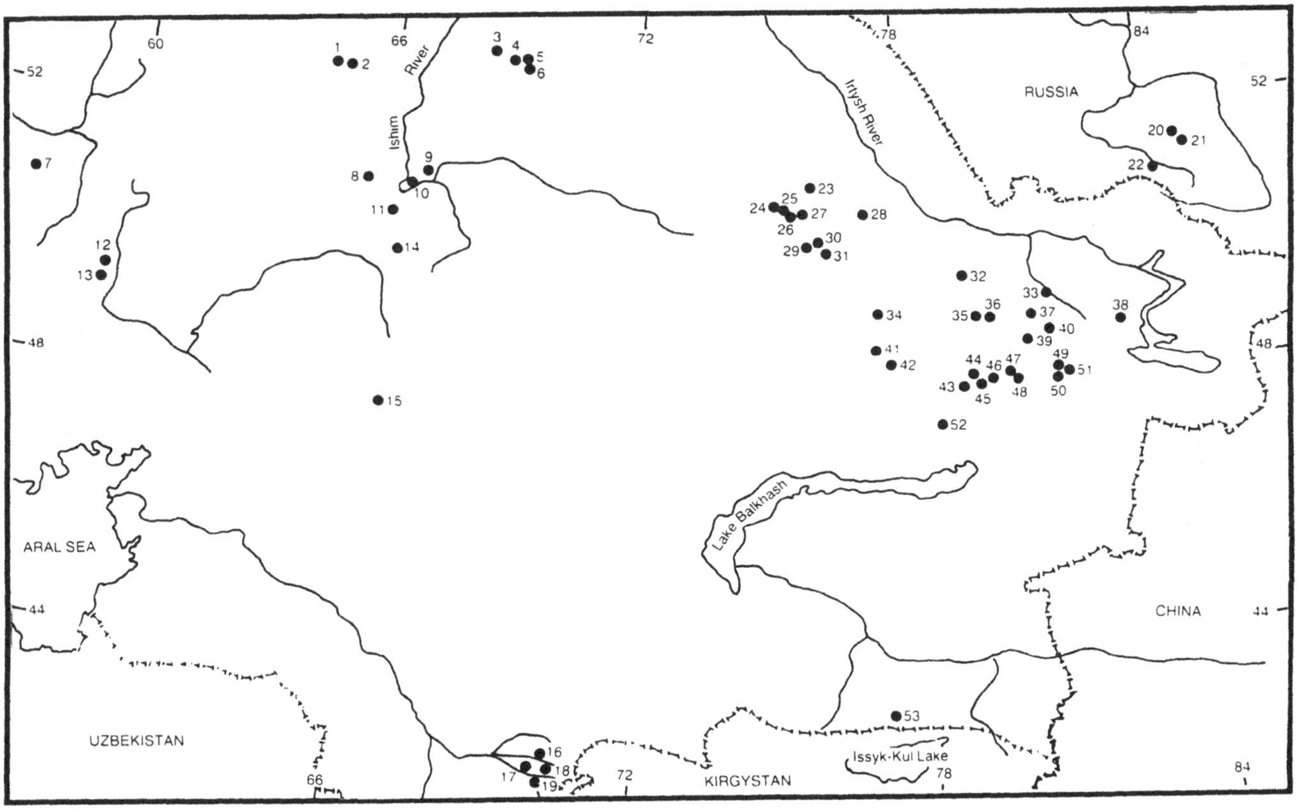

Fig. 60. Distribution of alkaline igneous rocks and carbonatites in Kazakhstan.

3 BARCHINSKII
53°15′N; 68°25′E
Fig. 61

This intrusion is situated to the southwest of the town of Kokchetav. It is 4.3 km long, but averages only 300–350 m in width, and is composed of pyroxenites, alkaline gabbroids, including olivine gabbro, malignite, sviato-nossite, pulaskite and shonkinite, nepheline syenites and syenite porphyries, as well as carbonatites. The carbonatites generally form veined and lense-like bodies situated among pyroxenites and metasomatically altered ultrabasic and nepheline-bearing rocks. Apart from calcite the carbonatites contain magnetite, phlogopite, forsterite, feldspar, pyroxene (diopside, augite–diopside and aegirine–diopside) and up to 10% apatite. The accessories are spinel, magnetite, perovskite, baddeleyite and brown zircon.

Age K–Ar gave 521–547 Ma (Nurlibayev, 1973).

Reference Nurlibayev, 1973.

4 KARLY-KUL'SKII
53°15′N; 68°35′E
Fig. 61

This is a faulted intrusion, lens-shaped in plan having a width of 100–800 m and an area of 2.5 km². It is composed of banded clinopyroxenites rich in apatite and biotite and sometimes containing melanite and K–feldspar. A subordinate role is played by melano-cratic and leucocratic mica shonkinites which sometimes contain pseudoleucite and altered olivine. The massif is cut by thin dykes of alkaline syenite and ledmorite, which are intensively granitized and altered to mica along the contacts. The intrusion has been intensively mylonitized and autometasomatically altered.

Age Based on geological data the intrusion is early Palaeozoic in age (Mikhailov and Orlova, 1971).

Reference Mikhailov and Orlova, 1971.

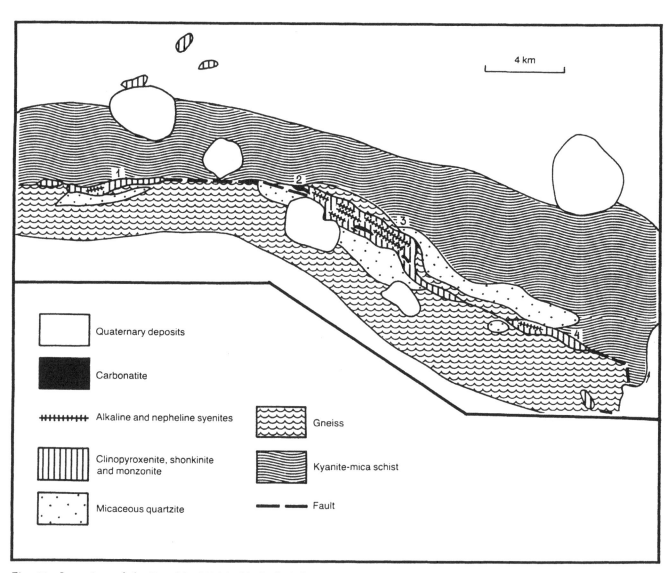

Fig. 61. *Location of the Barchinskii (1), Karly-kul'skii (2), Krasnomaiskii (3) and Pavlovskii (4) occurrences (after Nurlibayev, 1973, Fig. 4).*

5 KRASNOMAISKII
53°10′N; 68°50′E
Fig. 61

This massif is situated in the Krasnomaiskii deep fracture-zone which marks a tectonic contact between Archaean eclogitic formations and Proterozoic schists. The massif takes the form of a curved dyke and is 9 × 0.3–1.0 km with an area of 6.3 km². It is composed principally of biotite pyroxenites and nepheline syenites with subordinate malignite, sviatonossite, olivinite, biotite peridotite, zeolite, liebenerite and sodalite syenite and carbonatites. The intrusive sequence is: (1) olivinite and biotite perido-tite (2) pyroxenite (3) malignite and syenites (4) syenite porphyry (5) carbonatite. Within the ultrabasic rocks there are veins of syenite pegmatite consisting of ortho-clase, albite, apatite, mica and carbonate. In the alkaline and nepheline syenites are found veins of syenite pegma-tite and syenite porphyry as well as schlieren, veins and patches of carbonatite and apatite. Hydrothermal veins of zeolite, carbonate and fluorite occur in almost all the rocks of the intrusion but they are most typical of the pyroxenites, alkaline and nepheline syenites. Biotitization and phlogopitization (vermiculitization) of peridotite and pyroxenites and microclinization and liebeneritization of alkaline and nepheline syenites are broadly developed within the complex.

Age K–Ar gave 500–563 Ma (Nurlibayev, 1973).

Reference Nurlibayev, 1973.

6 PAVLOVSKII
53°07′N; 69°00′E
(Shagalinskii, South Krasnomaiskii)
Fig. 61

Pavlovskii has the form of a dyke and is about 4 km long. The principal rock type, occupying about two-thirds of the intrusion, is lepidomelane–apatite pyroxe-nite with 5–15% apatite; there are streaks and patches enriched in melanite. The pyroxenites are cut by veins and dykes, from 2–50 cm to 1–5 m wide, of sviatonos-site, malignite and alkaline and nepheline syenites. In the nepheline syenites the nepheline is sometimes replaced by zeolite, cancrinite and liebenerite. Zones of weather-ing up to 20–30 m deep, which has produced 9–20% hydro-mica and vermiculite, are developed over the pyroxenites.

Economic Apatite and vermiculite concentrations occur.

Reference Nurlibayev, 1973.

7 KOSISTEKSKII
50°44′N; 58°00′E
Fig. 62

The Kosistekskii complex is situated to the northeast of the town of Aktyubinsk. It is highly irregular in shape with a separate part of the complex lying several kilo-metres to the southeast; it has an area of about 60 km². The marginal contacts have steep dips of 70–90° and it is probably a thick, sheet-like intrusion the upper parts of which have not been eroded extensively. Alkaline rocks occupy about two-thirds of the area of the com-plex, the principal part of which is composed of quartz, biotite, biotite–amphibole and pyroxene granosyenites and alkaline granites. Nepheline syenites, which are divided into fine-, medium- and coarse-grained varieties, contain sodalite, which has replaced nepheline, while the feldspars are extensively albitized.

Reference Nurlibayev, 1973.

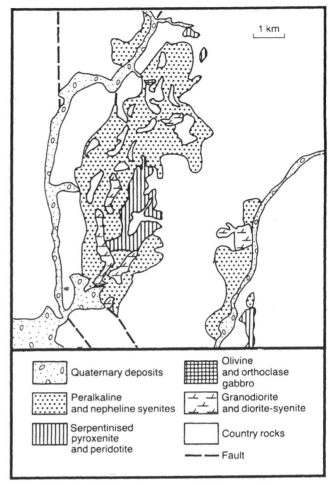

Quaternary deposits		Olivine and orthoclase gabbro
Peralkaline and nepheline syenites		Granodiorite and diorite-syenite
Serpentinised pyroxenite and peridotite		Country rocks
		Fault

Fig. 62. Kosistekskii (after Nurlibayev, 1973, Fig. 31).

8 DAMDYSAISKII
51°17′N; 65°33′E

This occurrence has an area of about 30 km² and prin-cipally comprises shonkinites with subordinate perido-tites. The shonkinites are composed of augite, bronzite, barkevikite, biotite and orthoclase. The accessory minerals are magnetite, ilmenite, titanite and apatite. Although this intrusion is not strictly alkaline in char-acter, traditionally it has been included with the alkaline series.

Age Geological data indicate an Ordovician age.

Reference Nurlibayev, 1973.

9 SHINSAISKII
51°25′N; 67°00′E

The 7.5 × 2.5 km Shinsaiskii complex cuts through schists, quartzites, limestones and argillites. Circular and zonal structures are typical of the massif. The peri-pheral areas are composed of pyroxenites, in which the pyroxene is aegirine–augite, while the nucleus is composed of peralkaline syenites, kentallenite–essexite, orthoclase gabbro and liebenerite syenites.

Reference Nurlibayev, 1973.

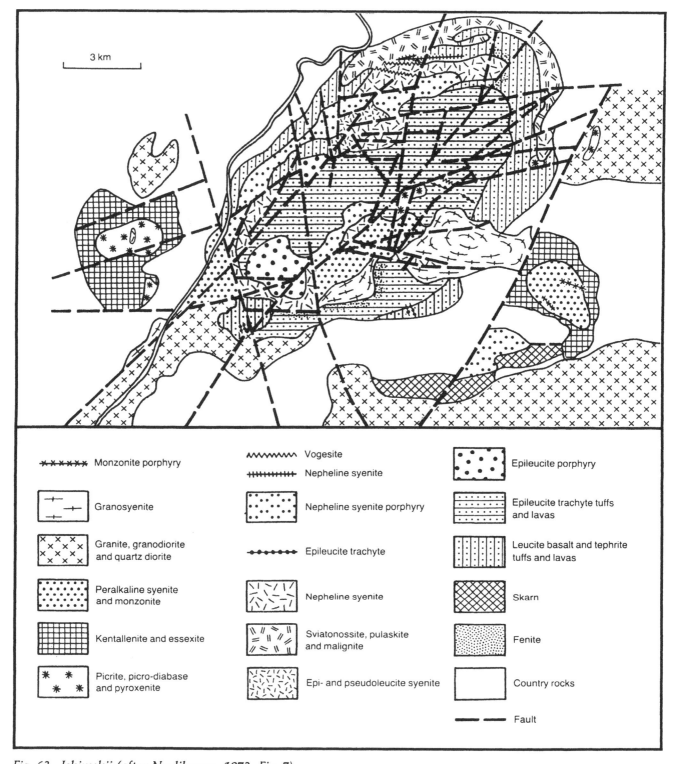

Fig. 63. Ishimskii (after Nurlibayev, 1973, Fig. 7).

10 ISHIMSKII
(Kubasadyrskii)

51°17'N; 66°33'E
Fig. 63

This massif is located on the periphery of Cuba-Sadir mountain in Central Kazakhstan. It is a volcanic complex which is composed of extrusive and intrusive rocks. Among the extrusive and subvolcanic facies are leucite-bearing tephrites, augite porphyrites, leucite (pseudoleucite) porphyries, two phases of pseudoleucite trachyte porphyry and dykes of pseudoleucite tinguaite. Within the leucite suite phenocrysts are generally represented by pseudoleucite, which may reach three centimetres in diameter; rarely K–feldspar is present as phenocrysts. The fine-grained groundmass consists of microlites of feldspar, nepheline and altered glass. The earliest rocks of the intrusive facies consists of coarse- and medium-grained nepheline syenites within which is a weakly developed system of veins that is confined to the marginal areas of the complex. They consist of K–feldspar, nepheline, hastingsite, aegirine–augite, titanium, garnet and biotite. Sometimes a little andesine is present. Secondary minerals include liebenerite, cancrinite and analcime with accessory fluorite, apatite, titanite, zircon and orthite; corundum has been identified in one specimen. The next intrusive phase is represented by a series of kentallenites, congressites, essexites, essexite porphyries, monzonites, biotite and amphibole alkaline syenites and granosyenites. Also broadly developed within the complex are liebenerite syenites and liebeneritized alkaline effusive rocks, as well as fenites at the outer contacts. A noteworthy feature of the main rocks of the complex is the presence of pyroxene together with abundant plagioclase, features indicative of the alkaline gabbroid association. A discussion of the chemistry of the complex, as a representative of miaskitic ultrapotassic rocks, will be found in Mineyeva (1972).

Age K–Ar gave 390–430 Ma (Nurlibayev and Panchenko, 1968).

References *Mineyeva, 1972; Nurlibayev and Panchenko, 1968; Nurlibayev *et al.*, 1965; Zavaritsky, 1936.

---

11 KENTALDYKSKII   50°50′N; 66°15′E

This intrusion covers 9 × 11 km and near to it there are three dyke-shaped bodies disposed in an arcuate form. The complex includes peridotites, alkaline pyroxenites, nepheline syenites, shonkinites and essexites.

Reference Nurlibayev, 1973.

---

12 KARASYRSKII   49°28′N; 59°50′E
Fig. 64

The Karasyrskii intrusion is situated on the eastern side of the Mugod'arsky anticlinorium amongst Precambrian gneisses. It is a stock-shaped body lying in an area of amphibole, amphibole–biotite and biotite syenites which are accompanied by dykes of nepheline and liebenerite syenites, syenite porphyry and pegmatites and quartz syenites. Sodium metasomatism has occurred and results typically in the complete alteration of K-feldspar to white albitites. The principal ferromagnesian rockforming mineral is ferrohastingsite (from 4 to 15%) but aegirine is present in some rocks.

Reference Bekbotayev, 1968.

---

13 BORSUKSAI   49°45′N; 59°20′E
Figs 65 and 66

Borsuksai is situated amongst Precambrian gneisses and metamorphosed shales in the eastern part of the Mugodzharsky mega-anticlinorium, where it coincides with the intersection of north–south- and east–west-trending faults. Two rather different geological accounts and maps of Borsuksai are Nurlibayev (1973) (Fig. 65) and one by Semenov *et al.* (1974) (Fig. 66). According to Nurlibayev (1973) the intrusion extends over about 50 km² and is emplaced in Precambrian biotite and granite gneisses and mid-Palaeozoic diorites, gabbros and gabbro-amphibolites. He distinguishes five intrusive stages the first of which is represented by dykes and sheets of hastingsite–biotite syenite and liebenerite syenite, which were followed by nepheline-, aegirine-, biotite-, sodalite- and amphibole-bearing syenites. The third phase comprises muscovite and nepheline-microcline pegmatites and syenitic aplites, which were followed by injection of veins of syenite porphyry. During the fifth stage alaskitic granite, granite aplite and granite pegmatite were intruded. The country rocks immediately adjacent to the complex are fenitized and

zones rich in sodalite and quartz-feldspar metasomites were developed while quartz and calcite veins containing fluorite and sulphides were injected.

Semenov *et al.* (1974) present a simpler picture of an intrusion with a core of biotite syenite and a periphery of hastingsite syenites, which rocks are cut by small intrusions and veins of nepheline syenite. Dykes of nepheline syenite from 0.1 to 30 m thick are described as extending for up to 10 km from the complex and are accompanied by dykes of peralkaline syenite, granosyenite, syenite porphyry, syenite aplite, bostonite and sviatonossite (andradite-bearing syenites). Pegmatitic areas, up to several metres long, occur in the central parts of some nepheline syenite dykes. The enclosing shales and gneisses near the contact of the central complex are fenitized and xenoliths of the country rocks are present in the central parts of the massif. The nepheline syenites are fine-grained, often trachytic rocks which contain about 20–40% of nepheline, 5–20% of coloured minerals (biotite, hastingsite and aegirine), microcline and albite; biotite–hastingsite-bearing varieties are the most widespread. Secondary and accessory minerals found in these rocks include fluorite, cancrinite, sodalite, calcite, natrolite, pyrochlore, zircon, orthite, monazite, bastnaesite, britholite, ancylite, thorite and sulphides. Peralkaline syenites contain microcline, oligoclase, biotite, aegirine, riebeckite and hastingsite. Accessories include zircon, rutile, columbite, titanite, apatite, garnet and Fe–Ti oxides with secondary albite, muscovite, natrolite and calcite. Miaskitic nepheline syenite pegmatites form sets of concordant veins 2–50 m thick and hundreds of metres long in the country rocks. Generally they are unzoned but infrequently they have two zones, the central one of which displays large block structures. Albitites are widely distributed and form zones hundreds of metres long and up to 2 m thick in the marginal part of vein bodies, and less frequently in rocks of the massif itself. The albitites are enriched in rare-metal minerals including columbite, zircon, rutile and thorite. Interesting discoveries of concentrically zoned aggregates of rare-earth minerals have been made and these include monazite–britholite–orthite, britholite (including the variety lessingite)–orthite and bastnaesite–britholite-orthite.

Age K–Ar dating indicates an age of 300 Ma (Levin *et al.*, 1973).

References Borodin *et al.*, 1974; Levin, 1974; Levin *et al.*, 1973; Nurlibayev, 1973; Semenov *et al.*, 1974.

---

14 ZHUSALINSKII   50°14′N; 66°26′E

The Zhusalinskii intrusion is situated in North Ala-Tayu. It comprises two small massifs of alkaline gabbroids. The northern intrusion is some 450 m in length, the southern, 120 m. The southern intrusion consists of shonkinite porphyries, with rather less olivine shonkinites. The northern massif is composed mostly of essexite and shonkinite, but there is some mica peridotite.

Age K–Ar on biotite gave 360 Ma (Nurlibayev and Panchenko, 1968).

Reference Nurlibayev and Panchenko, 1968.

---

15 KARSAKPAI   47°53′N; 66°25′E
Fig. 67

Karsakpai is situated in the central area of the Mai-Tubinskii anticlinorium and lies amongst Middle Proterozoic crystalline schists, amphibolites and granite gneisses. The complex is about 6 × 4 km and has a

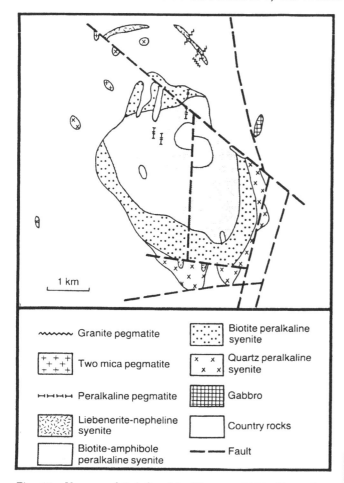

Fig. 64. *Karasyrskii (after Nurlibayev, 1973, Fig. 37).*

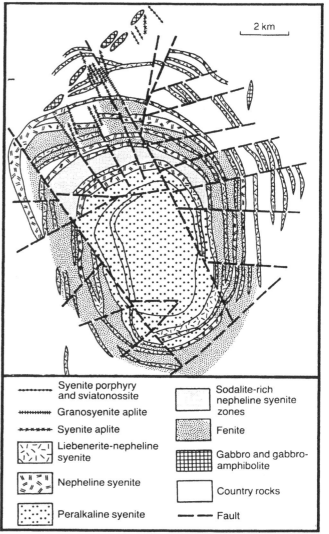

Fig. 65. *Borsuksai (after Nurlibayev, 1973, Fig. 35).*

concentric ring structure. It is composed principally of alkaline melanocratic syenites, with rather less abundant leucocratic ones; liebenerite syenites and nepheline syenites are subordinate. The complex was generated by the intrusion of the following phases: (1) monzonite and syenite–diorite; (2) albite–hastingsite syenite; (3) feldspar-bearing urtite; (4) melanocratic alkaline syenite (lepidomelane–hastingsite, augite–hastingsite and lepidomelane); (5) nepheline syenite; (6) dykes including trachytic syenites, syenite porphyries and syenite pegmatites. Albitized and liebeneritized syenites occur throughout the complex.

Economic Rare earth mineralization is typically associated with albitization of syenites and is represented by zircon, baddeleyite, apatite, titanite, orthite and thorite (Nurlibayev and Panchenko, 1968).

Age Geological evidence indicates a lower Devonian age (Nurlibayev and Panchenko, 1968).

Reference Nurlibayev and Panchenko, 1968.

---

16 ZHILANDA AND KULAN   42°37′N; 70°23′E

Zhilanda is a small stock of about 1 km² consisting of subalkaline gabbroids (monzonites) and peralkaline syenites. Khulan is situated near Zhilanda and is very similar in form and size. It also is composed of subalkaline gabbroids (monzonites) and peralkaline syenites.

Age K–Ar gave 260–300 Ma (Nurlibayev, 1973).

Reference Nurlibayev, 1973.

17 DAUBABINSKOE   42°28′N; 70°07′E
Fig. 68

Located in southern Kazakhstan, Daubabinskoe, together with Kaindy (No. 18) and Irisu (No. 19), lies on an east–west-trending deep fracture zone in the Precambrian basement which has been detected geophysically. There are two volcanic sequences within the complex, that have been divided into three fields by faulting. The lowest part of the succession (500 m) is composed of alkaline leucite and analcime trachybasalts and tuffs, together with lavas, breccias and tuffs of leucite tephrite as well as some minor horizons of picritic lavas. The central part of the succession (450 m) consists of leucite-bearing rocks including leucite tephrite and leucitite. These are agglomerates, breccias, tuffs and more rarely lavas. The uppermost section (400 m) is composed of agglomerates and tuffs of trachyandesite and latite that are underlain by agglomerates of leucite–biotite phonolite and trachyte. Both among the extrusive rocks and in the vicinity there are necks and dykes of trachybasalt, leucite tephrite and melanocratic trachyte, as well as stocks of shonkinite porphyry and biotite syenite porphyry. A post-magmatic series of apatite–magnetite

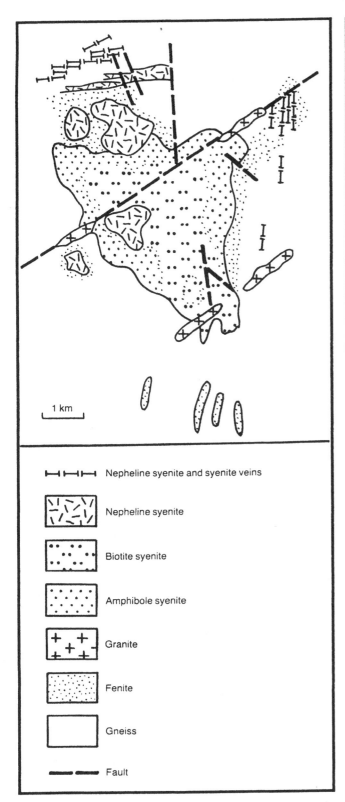

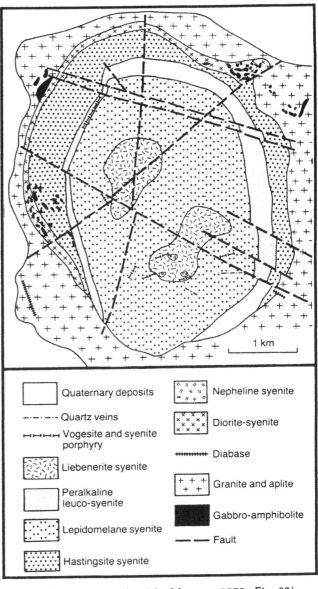

Fig. 67. *Karsakpai (after Nurlibayev, 1973, Fig. 32).*

legend for Fig. 67:
- Quaternary deposits
- Quartz veins
- Vogesite and syenite porphyry
- Liebenerite syenite
- Peralkaline leuco-syenite
- Lepidomelane syenite
- Hastingsite syenite
- Nepheline syenite
- Diorite-syenite
- Diabase
- Granite and aplite
- Gabbro-amphibolite
- Fault

legend for Fig. 66:
- Nepheline syenite and syenite veins
- Nepheline syenite
- Biotite syenite
- Amphibole syenite
- Granite
- Fenite
- Gneiss
- Fault

Fig. 66. *Borsuksai (after Semenov et al., 1974, Fig. 31).*

rocks (ores) are developed as small dykes. There are zones, some linear, of propilitization with hematite and copper sulphide mineralization distributed in the central and western parts of the field. In the southwestern area the alkaline basaltoids have been affected by an intensive post-magmatic zeolitization.

Age The extrusive rocks give 322 Ma by K–Ar (Eremeev, 1984).

References Eremeev, 1984; Nurlibayev, 1973; Orlova, 1959a.

18 KAINDY

42°21′N; 70°35′E
Fig. 69

The Kaindy complex is situated in the Talassky Alatau and has a rather irregular configuration. It appears to be a pipe-like stock with a zonal structure, the central part of which is composed of alkaline pyroxenites which contain pseudoleucite. There are younger intrusions of monzonite, melanocratic syenite and pseudoleucite syenite, while there are also further pipe-like intrusions of pyroxenites which are characterized by their small size (50–150 m diameter) and steep, inward-dipping margins (75–90°). The igneous activity was brought to a close with the emplacement of pegmatites and stocks of nepheline syenite.

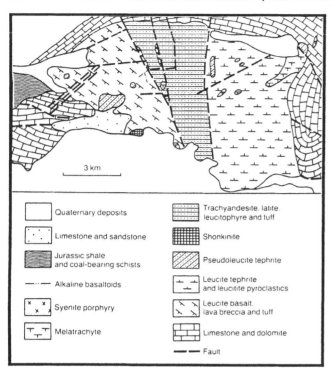

Fig. 68. Daubabinskoe volcanic field (after Nurlibayev, 1973, Fig. 26).

Age Geological evidence points to a mid-Carboniferous to early Permian age (Nurlibayev, 1976).

References Nikolayev, 1935; Nurlibayev 1973 and 1976; Orlova, 1959.

---

19 IRISU   42°20'N; 70°27'E
Fig. 70

Irisu is an intrusion of assymetrical form which is concentrically zoned and emplaced into limestones. The area is more than 25 km² (9 × 3 km), but only the more easterly 8 km² outcrops at the surface. It has a complicated structure being composed of subvolcanic stocks and ring intrusions that change from the periphery to the centre. Three stages have been distinguished, the rocks of which comprise, from youngest to oldest: (1) leucite basalts and leucitite necks; (2) a subvolcanic stage including nepheline porphyry, shonkinite porphyry and phonolite; (3) a plutonic stage consisting of koswite (olivine clinopyroxenite), leucite and orthoclase pyroxenite (first phase), monzonite, shonkinite, melanite alkaline syenites (second phase), pyroxene–biotite syenite, nepheline syenite, fergusite (third phase) and syenite porphyry (fourth phase). Detailed investigations of the structure of the Irisu massif confirm that all the alkaline rocks can be related to up to three intrusive centres, that appear to be centred about the stock of leucite basalts. The most widely distributed rocks in the complex are pseudoleucite syenite, fergusite, pseudoleucite porphyry and pseudoleucite pyroxenite which are composed of feldspar–zeolite, feldspar–analcime and feldspar–liebenerite aggregates of psuedo-and epileucite, K–feldspar, pyroxene (diopside, diopside–augite and augite), biotite (lepidomelane), amphibole, nepheline and plagioclase. The accessories are magnetite, apatite, titanite, melanite and fluorite. K–feldspathization is well developed, especially along fissure zones, in which primary rocks may be feldspathized to the composition

of syenites, while the development of biotite is also typical in Irisu rocks.

Economic Deposits of chalcopyrite–magnetite ores and vermiculite and alkaline kaolinite deposits are genetically linked with the alkaline rocks (Nurlibayev, 1973, 1976).

References Nurlibayev, 1973 and 1976; Orlova, 1959.

---

20 ELINOVSKII   51°31'N; 84°35'E
Fig. 71

Elinovskii is a long, narrow granitic body of some 400 × 50 m lying within terrigenous lower Devonian sandstones and limestones and lower Silurian limestones. In the southeastern part of the contact they are intensively altered to skarns. The main intrusive phase is a medium-grained riebeckite granite that gradually merges into medium- to coarse-grained peralkaline alaskites. In the country rocks there are rare dykes of granosyenite and alaskite as well as aplitic granites.

Reference Ermolov *et al.*, 1988.

---

21 ASKATINSKII   51°22'N; 84°41'E
Fig. 72

This occurrence is situated between mid-Palaeozoic sedimentary rocks and granitoids of upper Palaeozoic age. Peralkaline granites, which comprise most of the intrusion, form a stock the form of which is complicated by numerous faults. The principal rock types are riebeckite and biotite–riebeckite granites, which sometimes grade into alaskites. The marginal zone is composed of porphyritic granites and granite porphyries with micropegmatitic granites and aplites forming dykes and veins.

Reference Ermolov *et al.*, 1988.

---

22 MAIORSKII   51°02'N; 84°02'E
Fig. 73

The Maiorskii peralkaline granite massif extends along the southwestern margin of the huge Permian Talitskii granite pluton and covers about 5 km². Biotite granites of the southeastern margin of the Talitskii granite are cut by dykes of peralkaline fine-grained granite and granite porphyry. A second small body of peralkaline granite is situated on the northwestern margin of the Talitskii massif and both this and the Maiorskii intrusion are composed of fine-grained, sometimes miarolitic, riebeckite granites.

Age Post-Permian.

Reference Ermolov *et al.*, 1988.

---

23 SHANSHAL'SKII   51°11'N; 76°00'E
Fig. 74

This intrusion covers nearly 30 km² and is composed of porphyritic and medium-grained peralkaline granites which are cut by vein-like bodies and dykes of peralkaline granite porphyry. The country rocks have been altered with the development of quartz and muscovite. The granites consist of microcline, albite, quartz, arfvedsonite, riebeckite and biotite. They are rich in accessory minerals including fluorite, zircon, including a metamict variety, orthite, monazite, ferrothorite, topaz, ilmenite, titanite, apatite, pyrite, baryte, garnet, rutile, anatase, galena, wulfenite and scheelite.

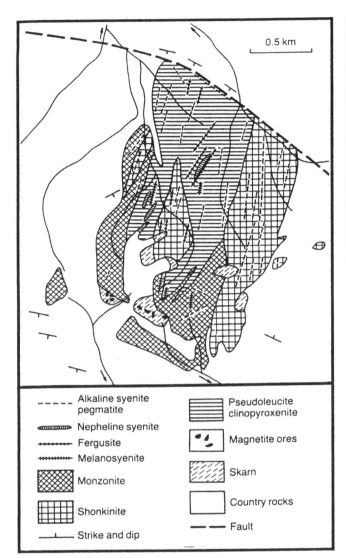

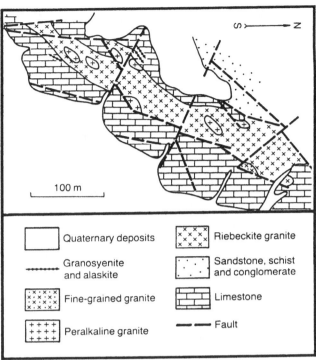

Fig. 71. *Elinovskii (after Ermolov et al., 1988, Fig. 9).*

Fig. 69. *Kaindy (after Nikolayev, 1935, Fig. 1).*

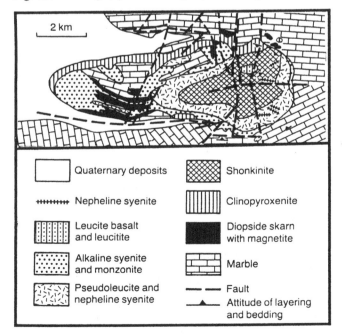

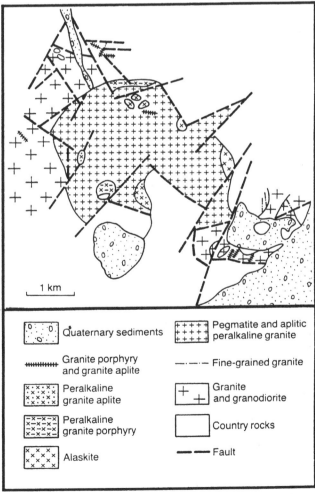

Fig. 70. *Irisu (after Eremeev, 1984, Fig. 8).*

Fig. 72. *Askatinskii (after Ermolov et al., 1988, Fig. 10).*

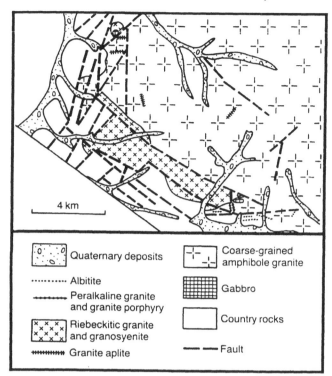

Fig. 73. *Maiorskii (after Ermolov et al., 1988, Fig. 8).*

Age Late Carboniferous to early Permian (Monich *et al.*, 1965b).

References Monich *et al.*, 1965a, 1965b.

---

24 BERKUTINSKII
(Berkuty)

50°54′N; 75°02′E
Fig. 75

This stock has an area of about 6 km² and is elongated in an east–west direction; it is composed of peralkaline syenites, granosyenites and granites.

References Monich *et al.*, 1965a, 1965b.

---

25 DYKE FIELD

50°50′N; 75°15′E
Fig. 75

Fifteen kilometres to the southeast of the Berkutinskii intrusion there is a field of dykes and dyke-shaped bodies which are situated to the northeast of the town of Zsheltan. It stretches for 10–12 km with a width of 3–4 km; individual dykes are from 0.1 to 0.3 km thick. The dykes are of fine-grained arfvedsonite–riebeckite syenite and granosyenite and also contain aegirine and accessories including fluorite. There are some peralkaline syenites which are cut by veins of sodalite–nepheline syenite in which the nepheline comprises 20–45% and sodalite plus cancrinite 5–25%.

Age Later Palaeozoic.

References Monich *et al.*, 1965a, 1965b; Sevryugin and Semenova, 1960.

---

26 BIRDZHANKOL'SKII

50°50′N; 75°19′E
Fig. 75

This small, 1 km² intrusion is a stock-like body composed of peralkaline syenites and granosyenites with subordinate dykes of nepheline syenite. The syenites

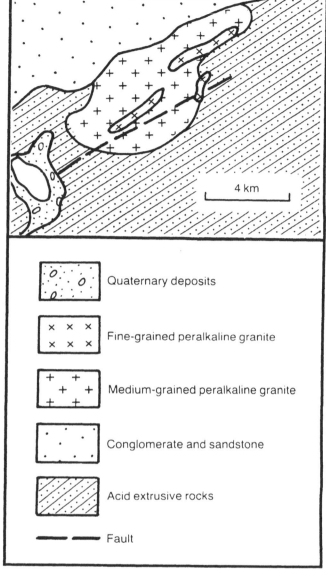

Fig. 74. *Shanshal'skii (from Geological Map of the USSR, 1960. Sheet M-43-X, 1:200,000).*

and granosyenites are composed of microcline, albite-oligoclase, arfvedsonite, hastingsite, aegirine, lepidomelane and melanite; accessories are zircon, fluorite, apatite, titanite and magnetite. In the nepheline syenites there are 0–45% of nepheline, 6–28% cancrinite, aegirine, hastingsite, lepidomelane and melanite with accessory zircon, apatite, titanite, fluorite, garnet, ilmenite and magnetite.

References Monich *et al.*, 1965a, 1965b.

---

27 KARATAL'SKII

50°52′N; 75°23′E
Fig. 76

This somewhat elongated stock of 2 × 0.7–1.0 km lies within sandstones. The intrusion is concentric, as indicated by the mineral fabric, and comprises three intrusive phases, namely (1) peralkaline syenites, which form the periphery, (2) nepheline syenites, which are confined to the centre, and (3) syenite porphyries and aplites and small pegmatitic bodies of sodalite, feldspar and hastingsite.

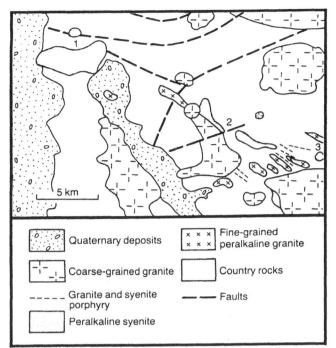

Fig. 75. *Location of the Berkutinskii intrusion (1), dyke field (2) and Birdzhankol'skii intrusion (3) (from Geological Map of the USSR, 1960. Sheet M-43-X, 1:200,000).*

Age K–Ar determinations gave 250–280 Ma (Nurlibayev, 1973).

Reference Nurlibayev, 1973.

---

28 TLEUMBETSKII
50°46′N; 77°02′E
Fig. 77

The Tleumbetskii intrusion is situated in the region of Chingiz-Tarbogotay. It is a large complex with an area of 175 km². The earliest rocks are porphyritic medium-grained riebeckite–arfvedsonite–biotite granites and syenites which are cut by small stocks and dykes of peralkaline granosyenite and granite porphyry. Alteration of these rocks is expressed by the development of albite and chlorite and zones of greisen. The rocks are exceptionally rich in accessory minerals including fluorite, zircon, orthite, monazite, ferrothorite, topaz, ilmenite, titanite, apatite, pyrite, baryte, columbite and garnet with more rarely scheelite, rutile, anatase, galena and wulfenite.

Economic There is sulphide, rare-earth phosphate and rare-earth greisen-type mineralization (Nurlibayev, 1973).

Age Geological evidence indicates an early Permian age (Monich *et al.*, 1965).

References Monich *et al.*, 1965a, 1965b; Nurlibayev, 1973; Ziryanov, 1964.

---

29 ZHILTAUSKII
50°15′N; 75°44′E
Fig. 78

This stock-like intrusion of 12 km² is composed predominantly of medium-grained porphyritic peralkaline granites which cut porphyritic peralkaline syenites and quartz syenite porphyries. The associated dykes are of peralkaline granite porphyry. The rock-forming minerals

are anorthoclase, albite, quartz, hastingsite, arfvedsonite and aegirine. Autometamorphism is manifested by the development of albite and fluorite, sometimes by kaolin and by a greisenization. Accessory minerals are zircon, ilmenite, apatite, fluorite, garnet, monazite, anatase, titanite and fluorite.

Age Geological evidence indicates a late Palaeozoic age.

Reference Ziryanov, 1964.

---

30 POLUMESYATS
(Kokon)
50°27′N; 76°05′E
Fig. 79

Having a crescentic outline and an area of about 1 km² the Polumesyats intrusion is composed of coarse- and medium-grained arfvedsonite–riebeckite granites with a subordinate role played by fine-grained riebeckite granite and peralkaline granosyenites. Dykes of granosyenite porphyry and peralkaline granosyenite porphyry occur. The rocks are rich in accessories including zircon, orthite, monazite, thorite and fluorite.

Age Geological data indicate a late Palaeozoic age.

References Geological map of the USSR 1:200,000, 1960, Sheet M–43–XVII; Nurlibayev, 1973.

---

31 BERKUTY SOUTH
50°19′N; 76°13′E

Berkuty South is situated in the Chingiz-Tarbagotay and covers approximately 11 km². It is a symmetrical body composed of coarse- and medium-grained porphyritic arfvedsonite and riebeckite granites. The accessory minerals are zircon, malacon (a variety of zircon), orthite, monazite, thorite and fluorite. The country rocks are hornblende and muscovite schists and quartzites.

Age Early Permian from geological data (Sevryugin and Semenova, 1960).

References Sevryugin and Semenova, 1960; Ziryanov, 1969.

---

32 AKBIIK
49°45′N; 79°18′E

The Akbiik intrusion cuts through Carboniferous volcanic rocks and small intrusive bodies of early upper Palaeozoic age. The massif is 18 × 4 km and composed of intrusions of biotite and riebeckite–biotite granites. Aegirine–riebeckite and riebeckite granites are located in the southwestern part of the massif but are of subordinate development.

Reference Ziryanov, 1969.

---

33 KYZYL-CHAR
49°24′N; 81°18′E

This is a dyke-like intrusion of riebeckite granite some 2 km in length lying along the northern margin of the Kandygataiskii granite massif. The Kyzyl-Char granite consists of 55–60% alkali feldspar, 25–30% quartz, 5–7% riebeckite and 2–3% biotite. There are also several dykes of riebeckite granite porphyry lying northeast of this locality.

Age Geological evidence suggests a late Palaeozoic age.

Reference Nokolskii and Sokratov, 1965.

---

34 KAINARSKII
49°17′N; 77°20′E

Kainarskii extends for 16 km north–south, has an average width of 7 km and an area of about 100 km². Both

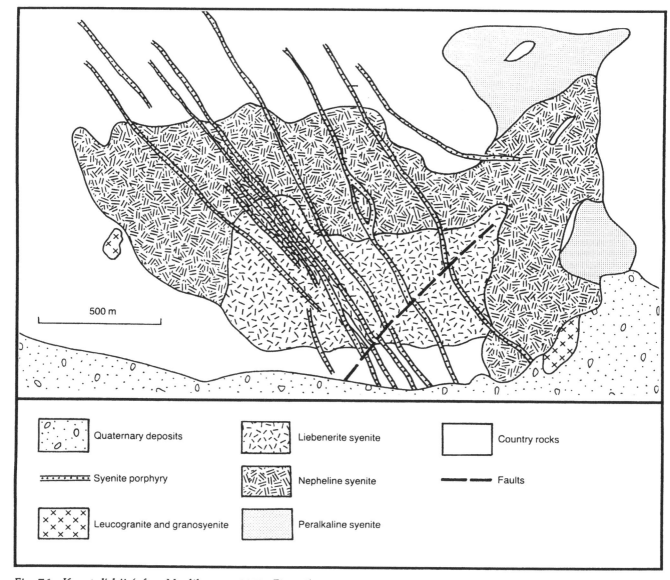

Fig. 76. Karatal'skii (after Nurlibayev, 1973, Fig. 33).

biotite and peralkaline riebeckite and aegirine–riebeckite granites are present with the former being in greater abundance and the latter forming linear, narrow strips parallel to the length of the intrusion and corresponding to fracture zones and zones of intense faulting. A hypabyssal phase is represented by dykes of granite, aplite, quartz syenite porphyry and microdiorite.

Reference Ziryanov, 1969.

35 KSHI-ORDA   49°06′N; 79°29′E
(Abachevskii)   Fig. 80

The Kshi-Orda complex cuts through syenites, diorites and monzonites of the Bakshory complex, which are probably Silurian in age. It occupies about 10 km² and forms a partial ring structure which extends along a northwesterly line. The contacts dip at 60–70°. Within the massif there are systems of steep fractures which control the igneous structures. Zonality is strongly marked in the complex with the outermost zone composed of pink liebenerite syenites. The next zone inwards comprises hastingsite–nepheline syenites, which have a circular distribution and gradually grade into nepheline

syenites via a zone of nepheline syenites in which muscovite is developed. The central part of the complex is composed of albitized nepheline syenites. Dykes that are developed along circular and radial fissure systems are from 15–20 cm in thickness and extend for 100–200 m. The dykes are nepheline syenites with aplitic textures and occasional aegirine granites.

Reference Ziryanov, 1969

36 KORGANTAS   49°03′N; 79°50′E
                                     Fig. 81

This 10 × 4 km intrusion consists essentially of peralkaline granites. An extensive system of dykes is developed the first phase of which consists of aplites, granites, bostonites and spherulitic porphyries of both normal and peralkaline compositions. The dykes of a second phase comprise quartz and felsitic porphyries, diorite porphyries, microdiorites and diabases. Skarns are developed at the contact with a gabbro-diorite intrusion.

Reference Ziryanov, 1969.

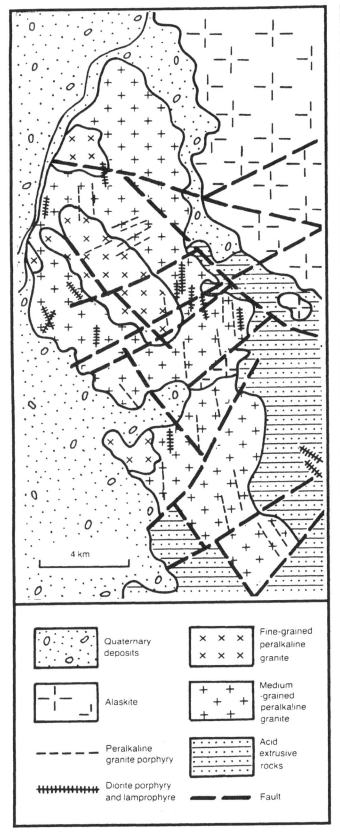

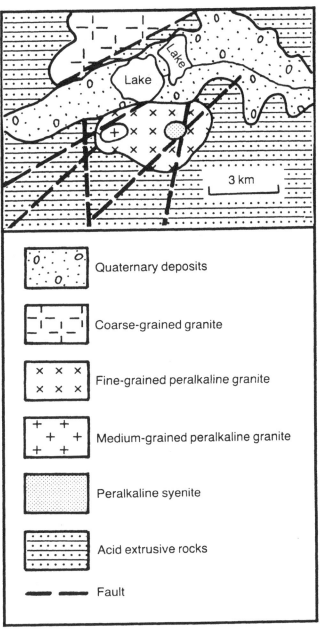

Fig. 78. Zhil'tauskii (from Geological Map of the USSR, 1960. Sheet M-43-XVI, 1:200,000).

37 KEREGETAS   49°03′N; 80°42′E
                                            Fig. 82

Situated in the Keregetas Mountains, the intrusion lies within extrusive deposits of Carboniferous age and Upper Palaeozoic granitoids. It is somewhat elongate and dyke-like in form, the length being 9 km but the width varying from only 100 to 850 m. It consists of biotite, riebeckite, aegirine–riebeckite and astrophyllite–riebeckite granites. In the central part of the intrusion in astrophyllite–riebeckite granites there is a zircon mineralization (Ziryanov, 1969).

Reference Ziryanov, 1969.

Fig. 77. Tleumbetskii (after Ermolov et al., 1988, Fig. 22).

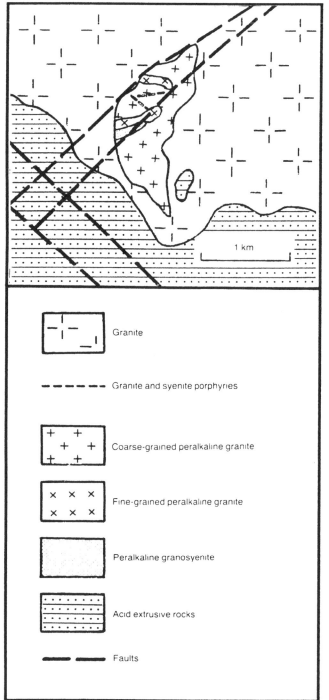

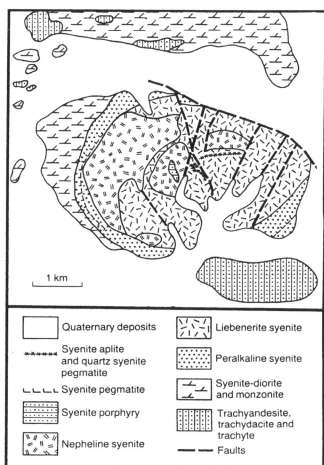

Fig. 80. Kshi-Orda (after Nurlibayev, 1973, Fig. 34).

Fig. 79. Polumesyats (from Geological Map of the USSR, 1960. Sheet M-43-XVII, 1:200,000).

38 MANRAK
48°50′N; 82°52′E
Fig. 83

This intrusion consists of several varieties of peralkaline granite emplaced in Carboniferous sedimentary rocks. The western granite is 1.5 × 3 km, but partly concealed by Quaternary deposits, and composed of aegirine–riebeckite microgranite porphyry which is partly layered, as indicated by concentrations of quartz and K–feldspar and melanocratic phases. At the contact with upper Palaeozoic deposits the rock passes into a cryptocrystalline granite porphyry. The base of the intrusion descends

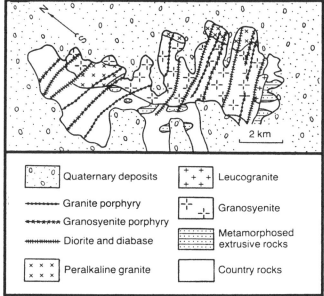

Fig. 81. Korgantas (after Ermolov et al., 1988, Fig. 3).

to the west beneath and conformably with the bedding of the country rocks. The thickness of the western part of the sill is estimated at a few hundred metres. The eastern granite is in contact with Carboniferous basalts.

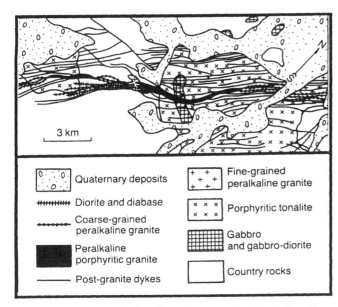

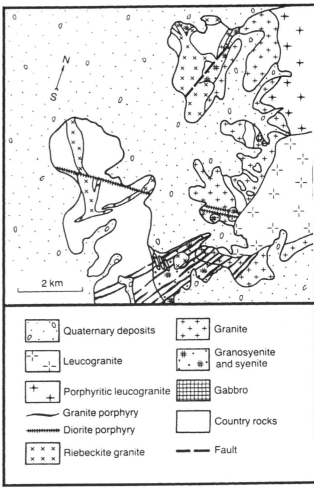

Fig. 82. Keregetas (after Ermolov et al., 1988, Fig. 4).

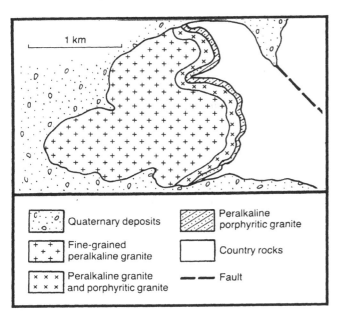

Fig. 83. Manrak (after Ermolov et al., 1988, Fig. 7).

It extends over 0.3 × 2 km and is composed of bluish-grey massive riebeckite–aegirine granite porphyry.

Age Geological evidence indicates post Upper Palaeozoic.

Reference Ermolov *et al.*, 1988.

39 BOLEKTAS 48°39'N; 80°34'E
Fig. 84

This occurrence is represented by a dyke-like body and a rectangular intrusions of about 4.6 × 0.5 km. It cuts through terrigenous sediments with subordinate horizons of lavas and pyroclastics of basic compositon. To the south of the main intrusion there is a small, intrusive body that is associated with a northeasterly-trending fracture zone, which is probably a satellite of the massif. The massif is composed principally of riebeckite granites, but biotite granites are observed in the northern

Fig. 84. Bolektas (after Ermolov et al., 1988, Fig. 5).

part of the main intrusion. There are some distinct areas of biotite–riebeckite granite that are characterized by the secondary nature of the biotite. There are dykes of granite, aplite, quartz porphyry and granodiorite porphyry.

References Ermolov *et al.*, 1988; Ziryanov, 1969.

40 KANDYGATAI 48°48'N; 81°02'E

Extending over 22 km², Kandygatai is composed of riebeckite granites, which sometimes contain astrophyllite. Accessories include zircon, columbite and apatite. There are dykes of peralkaline microgranite.

Age Geological evidence indicates an early Permian age.

Reference Ermolov *et al.*, 1988.

41 AKKOITAS 48°41'N; 77°13'E

A granite intrusion of 9 × 3.5–4 km, Akkoitas possibly is sill-like in form. Biotite granites are the most abundant rocks with less abundant areas of riebeckite and aegirine–riebeckite granites. Aegirine–riebeckite granites are mostly restricted to the upper part of the intrusion, where they form dome-shaped outcrops containing xenoliths of country rock. There is a very strongly developed system of dykes.

Reference Ziryanov, 1969.

42 KUROZEKSKII   48°30′N; 77°35′E

Intruded into Silurian sandstones, this is a typical faulted intrusion extending for 8.5 km in a northwesterly direction but having a maximum width of only 750 m. There are peralkaline varieties of granite which have developed in the central part of the massif.

Reference Ziryanov, 1969.

43 KOKSALINSKII   48°03′N; 79°09′E

Koksalinskii is emplaced in Ordovician and Devonian extrusive formations. It has an elongate, dyke-like form and occupies an area of about 15 km². Biotite granites are developed in the northern and southern parts of the massif and in the central area are riebeckite and aegirine-riebeckite varieties that gradually grade into the biotite granite. Along the eastern margin there is a narrow strip of riebeckite and hastingsite granites. There are early dykes of granite and aplite and later ones of syenite and diorite porphyry.

Reference Ziryanov, 1969.

44 ARSALAN   48°11′N; 79°18′E

This granite intrusion is pear-shaped and covers 25 km². It cuts through volcanogenic Devonian deposits and granodiorites. The inner structure of the intrusion is very complex because of the presence of zones of aegirine-riebeckite granite that have developed metasomatically from primary biotite granite. Early aplite and microgranite dykes are followed by dykes of diorite porphyry and microdiorite.

Reference Ziryanov, 1969.

45 DONENZHAL   48°06′N; 79°29′E

This 15 km² body is intruded into Cambrian and Ordovician sediments and tuffs. It comprises biotite, riebeckite–biotite, aegirine–biotite and muscovite-riebeckite syenites that grade gradually into each other. Pegmatites are also typical for the massif. Dykes are not well developed in the intrusion but there are some early aplitic syenite and bostonite dykes, while a second stage is represented by dykes of microdiorite and porphyry.

Reference Ziryanov, 1969.

46 BATPAK   48°07′N; 79°46′E

The 30 km² Batpak intrusion is situated in the Chingis-Tarbagatai region and is composed of peralkaline quartz syenites. These consist of microcline perthite, albite-oligoclase, quartz, riebeckite, aegirine and biotite with an abundance of accessory minerals including magnetite, ilmenite, titanite, zirkelite, orthite, apatite, zircon, rutile, monazite, thorite and fluorite. Dykes are of aplite and granite porphyry.

Reference Putalova, 1968.

47 KALGUTINSKII NORTH   48°10′N; 80°10′E

This intrusion, of about 30 km², is composed of riebeckite–aegirine syenites and granosyenites and is cut by small bodies of porphyritic leucogranite. The country rocks are late Palaeozoic granitoids which are cut by a northeasterly-trending system of basic and syenite dykes.

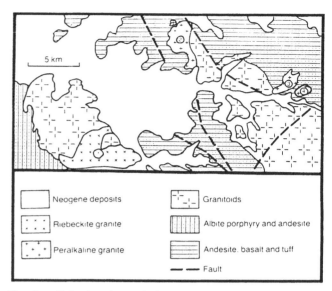

Fig. 85. *Relationship of the Biesimas (1), Iisorskii (2), Bolshoi Espe (3) and Malyi Espe (4) intrusions (from Geological Map of the USSR, 1960. Sheet M-43-XXXIV, 1:200,000).*

Age Geological data indicate a late Palaeozoic age.

Reference Ermolov *et al.*, 1988.

48 KALGUTINSKII SOUTH   48°07′N; 80°12′E

The Kalgutinskii South intrusion covers about 6 km². It comprises syenites and quartz syenites composed of microcline, albite–oligoclase, aegirine, riebeckite and accessory apatite, titanite, zircon and magnetite.

Age Geological evidence indicates a late Palaeozoic age.

References Borukaev, 1962; Ermolov *et al.*, 1988.

49 IISORSKII   48°11′N; 81°19′E
                                                       Fig. 85

The 7.5 km² Iisorskii intrusion lies on the northwestern margin of a much larger granitoid massif of the same age. The relationship of the Iisorskii peralkaline granite, which contains riebeckite and annite, to the more extensive granitoids is not clear, but many investigators consider the peralkaline granites to be an alkaline facies generated by alkali metasomatism and zirconium-niobium mineralization.

Economic Zr–Nb mineralization is known in this occurrence.

Age Late Palaeozoic – on geological evidence (Ziryanov, 1964).

References Ermolov *et al.*, 1988; Ziryanov, 1964.

50 BIESIMAS   48°05′N; 81°14′E
                                                       Fig. 85

The 17 km² Biesimas peralkaline intrusion is part of a larger, late Palaeozoic granite massif. It is composed of medium-grained porphyritic and fine-grained riebeckite granites which contain small amounts of aegirine and arfvedsonite and accessory zircon, bastnaesite, thorite, magnetite and fluorite. There are pegmatites and dykes of peralkaline granite porphyry and aplite.

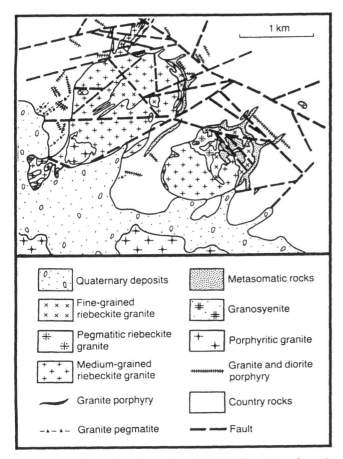

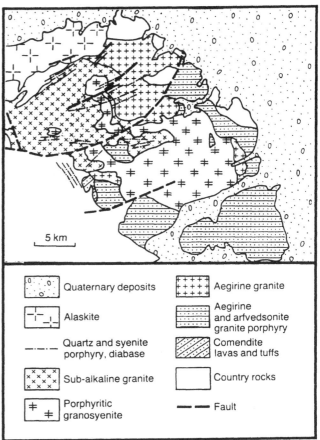

Fig. 86. Bolshoi (northwest) and Malyi Espe (southeast) (after Ermolov et al., 1988, Fig. 6).

Fig. 87. Bakanasskaya (after Ermolov et al., 1988, Fig. 11).

Age 257 Ma by K–Ar.

Reference Ziryanov, 1964.

51 BOLSHOI ESPE and MALYI ESPE

48°09′N; 81°27′E
Figs 85 and 86

These two adjacent bodies, of about 4 km² and 2 km², are both stock-like in form and both are composed of albite–riebeckite granites of a range of grain sizes.

Age Whole rock K–Ar determinations indicated 247 Ma (Ermolov *et al.*, 1988).

Reference Ermolov *et al.*, 1988.

52 BAKANASSKAYA

47°31′N; 78°33′E
Fig. 87

This approximately circular, 30 km diameter extrusive-intrusive complex has been emplaced into extrusive volcanic and sedimentary formations of late Palaeozoic age. The complex is formed of the rocks of three granitoid suites which, from oldest to youngest, are (1) comendites, comenditic tuffs and peralkaline granites, (2) granosyenites and leucogranites and (3) alaskitic granites. The rocks of the first, peralkaline suite comprise mainly an external ring while within the complex they are present as large fragments within the granitoids of the second suite. The oldest rocks of the peralkaline suite are comendites and comenditic tuffs, which comprise phenocrysts of alkali feldspar and micropheno-

crysts of aegirine, plus or minus quartz, in a glassy base. These were followed by emplacement of aegirine and arfvedsonite–aegirine granite porphyries, then coarse-grained miarolitic aegirine granites and finally by fine- and medium-grained, massive and miarolitic aegirine granites. The granitoids of the second suite build more than 50% of the Bakanasskaya complex. They differ from the granites of the earliest suite by being generally less alkaline, by the taxitic textures of the mafic minerals, by the abundance of inclusions of earlier rocks and by the abundance of titanite, apatite and fluorite. The rocks comprising the second suite, from oldest to youngest, are biotite–amphibole porphyritic granosyenites and grano-syenite porphyries, coarse-grained biotite–amphibole porphyritic granites, fine-grained porphyritic arfved-sonite and biotite granites and dykes of diabase, syenite and granite porphyry. Development of the complex was completed with the intrusion of the alaskites of the third suite which lie along the northwestern margin of the complex.

Age Geological evidence indicates a Permian age.

Reference Ermolov *et al.*, 1988.

53 TALGARSKII

43°05′N; 77°11′E

This large intrusion, covering more than 100 km², is composed of riebeckite and hastingsite granites and alkaline leucogranites.

Age K–Ar on whole rocks gave 349–365 Ma (Monich *et al.*, 1965b).

References Monich *et al.*, 1965a and 1965b; Nurlibayev, 1973.

Kazakhstan references

BEKBOTAYEV, A.T. 1968. Alkaline rocks of the Karasyrskii massif. *Kazakhstan Academy of Sciences, Geology Series*, **2**: 44–8.

BORUKAEV, R.A. (ed.) 1962. Geological map of the USSR. 57–8. *State Scientific Technical Publishing House, Moscow.*

EREMEEV, N.V. 1984. *Volcanic-plutonic complexes of potassic alkaline rocks*. Nauka, Moscow. 134 pp.

ERMOLOV, P.V., VLADIMIROV, A.G. and TIKHO-MIROVA, N.I. 1988. *The petrology of silica over-saturated agpaitic alkaline rocks*. Nauka, Novosibirsk. 86 pp.

KSENOFONTOV, O.K. 1966. On alkaline intrusions of the Mugodyar and Turgaisky flexures. *Scientific notes of the Hertzen Institute, Leningrad, Nedra*, **290**: 140–61.

MIKHAILOV, N.P. and ORLOVA, M.P. 1971. Alkaline ultrabasic formation (Krasnomaysky intrusive complex). *In* N.P. Mikhailov (ed.) *Petrography of Central Kazakhstan* 2: 298–313. Nedra, Moscow.

*MINEYEVA, I.G. 1972. Principal geochemical characteristics of potassic alkalic rocks. *Geochemistry International*, **9**: 173–9.

MONICH, V.K., ABDRAKHMANOV, K.A., NURLI-BAYEV, A.N., ZHYR'ANOV, V.N., STAROV, V.I., NARSEEV, A.A., BUGAEZ, A.N. and SEMENOV, Yu, A. 1965a. Alkaline intrusions of Kazakhstan and their age correlation. *Abstracts of the First Kazakhstan Petrographic Conference. Kazakhstan Academy of Science, Alma-Ata.* 19–22.

MONICH, V.K., IVANOV, A.I., KOMLEV, L.V., LAPICHEV, G.F. and SEMENOV, T.P. 1965b. Age of magmatic and metamorphic rocks of the East Kazakhstan geochronological scale. *Magmatism and metamorphism of East Kazakhstan.* 27–8. Akademii Nauk Kazakhstan SSR, Alma-Ata.

NIKOLAYEV, V.A. 1935. *Alkaline rocks of the Kaindy River in Talassky Alatau.* CNIGRI, Leningrad and Moscow. 11: 121 pp.

NOKOLSKII, A.P. and SOKRATOV, G.I. 1965. *Geological map of the USSR. M–44–XXIII. 1:200,000. Explanatory Notes.* 45–6. Gosgeoltekhizdat, Moscow.

NURLIBAYEV, A.N. 1973. *Alkaline rocks of Kazakhstan and their ore deposits.* Nauka, Alma-Ata. 296 pp.

NURLIBAYEV, A.N. 1976. *Magmatism of north Kazakhstan.* Akademii Nauk Kazakhstan SSR, Alma-Ata. 249 pp.

NURLIBAYEV, A.N. and PANCHENKO, A.G. 1968. *Alkaline and subalkaline rocks of Northern Kazakhstan and Kirgizia.* Izd-vo Nauka Kazakskoi SSR, Alma-Ata. 100–7.

NURLIBAYEV, A.N., PANCHENKO, A.G. and MONICH, V.K. 1965. New data on the geology of the Kubasadyrsky massif of alkaline rocks. *Izvestia Akademii Nauk Kazakhstan SSR, Alma-Ata. Seriya Geologicheskaya*, **1**: 57–62.

ORLOVA, M.P. 1959a. Alkaline basaltoids of the River Daubaba (Tholass Alatau). *Information Collection, VSEGEI, Leningrad*, **16**: 87–95.

ORLOVA, M.P. 1959b. Alkaline gabbroid intrusions of the north-west part of the Tholoss Alatau Range. *Scientific Note, Leningrad University, Geology Series*, **291**: 91–121.

PUTALOVA, R.B. 1968. Accessory minerals of alkaline rocks of the Prechingiz area. *In* J.E. Smorchkov (ed.) *Accessory minerals of the eruptive rocks.* 218–23. Nauka, Moscow.

SEVRYUGIN, N.A. and BESPALOV, V.F. 1960. *Geological map of the USSR. Sheet M–43–X. 1:200,000. Explanatory notes.* Gosgeoltehizdat, Moscow.

SEVRYUGIN, N.A. and SEMENOVA, T.P. 1960. The determination of the age of the eruptive rocks of the Degelen-Chingiz region by the Ar method. *Trudy Kazakhstan IMS*, **3**: 91–9.

ZAVARITSKY, A.N. 1936. Alkaline mountain rocks of Ishim. *Works of the Petrographical Institute of the Academy of Sciences USSR, Moscow*, 47–105.

ZIRYANOV, V.N. 1964. Nepheline syenites of the middle Prichingiz. *Izvestia Akademii Nauk Kazakhstan SSR, Alma-Ata. Seriya Geologicheskaya*, **5**: 57–67.

ZIRYANOV, V.N. 1969. *Petrology of the metasomatically altered granitoids and alkaline rocks of the Chingizsky zone.* Nauka, Moscow. 160 pp.

* In English

CENTRAL ASIA
(UZBEKISTAN, KIRGYSTAN, TADZIKISTAN)

The term Central Asia covers a vast area extending south from Kazakhstan to the border with Afghanistan and China. Westwards it extends into Uzbekistan and to the east takes in much of the Tyan-Shan range of mountains. A group of alkaline intrusions also lies close to the Chinese border in the Pamirs (Fig. 89).

1 Kolbashinskii	24 Isfairamskii
2 Tokailuashu	25 Dgamandgar
3 Sandyk	26 Skalystyi
4 Kyzyl-Ompul	27 Matchinskii
5 Chonashu	28 Aikel
6 Shamatorskii	29 Dara-Pioz
7 Kokdzharskii	30 Utren
8 Kaichinskii	31 Matchinskii South
9 Ailagir	32 Tutekskii
10 Sarisai	33 Kulpskii
11 Surtekin	34 Khodzhaachkan
12 Uzunbulak	35 Dzhilisu
13 Tozbulak	36 Nedostupnyi
14 Aksai	37 Yarkhich
15 Akhbasai	38 Turpi
16 Kaznokskii	39 Pamir pipes – Northern
17 Rokshifo-Sabakhskii	40 Pamir pipes – Southern
18 Karaganskii	41 Pamir dykes – Northern Belt
19 Saliepskii	
20 Sokhskii	42 Pamir dykes – Southern Belt
21 Surmetashskii	
22 Chekindy	43 Dunkeldykskii
23 Kichikalaiskii West	

1 KOLBASHINSKII
(Karabalty-Kol'bashi)

42° 20′N; 73°44′E

Kolbashinskii is a volcanic field situated in the Kirgiz shield of the northern Tyan-Shan west of Lake Issyk-Kul. Leucite-bearing basaltoids form an east–west-trending zone 40–45 km long and 5 km wide. They comprise the so-called 'Kolbashinskaya suite' that is divided into three series which, from the base upwards, consist of (1) leucite-bearing basaltoids, including leucite tephrites, (2) alkaline trachytes and (3) tuff-breccias and tuffs of peralkaline trachyte.

Age K–Ar indicated 270–309 Ma (Bagdasarov et al., 1974).

References Bagdasarov et al., 1974; Goretskaya et al., 1972.

2 TQKAILUASHU
(Bulak-Ashu)

42°10′N; 73°46′E
Fig. 88

The 17 km² sub-volcanic intrusion of Tokailuashu is emplaced in late Ordovician sediments and is located south of the Kolbashinskii volcanic field (No. 1). The rocks consist of augite, aegirine–augite and aegirine peralkaline syenites, melasyenite, syenite porphyry, monzonites and shonkinites. Dyke rocks include peralkaline syenite pegmatite and lamprophyres. The country rocks are granitoids and volcanics.

References Bagdasarov et al., 1974; Goretskaya et al., 1972.

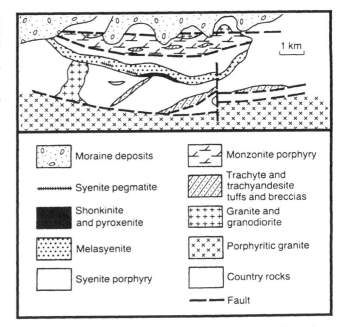

Fig. 88. Tokailuashu (after Dolzhenko, 1968, Fig. 1).

Legend:
- Moraine deposits
- Syenite pegmatite
- Shonkinite and pyroxenite
- Melasyenite
- Syenite porphyry
- Monzonite porphyry
- Trachyte and trachyandesite tuffs and breccias
- Granite and granodiorite
- Porphyritic granite
- Country rocks
- Fault

3 SANDYK
(Chechekty)

42°09′N; 74°59′E
Fig. 90

The Sandyk massif is situated in the D'umbgolsky Mountain Range 90 km to the west of the Kyzyl-Ompul intrusion (No. 4) in a northwesterly-trending fault zone. It extends over about 80 km² but much of the northern part is obscured by recent deposits. The country rocks are Caledonian granitoids of the Susamyrskii batholith. The complex is composed of both subalkaline and alkaline rocks, the most extensive being rocks of the subalkaline series, which occupy 75% of the area, which are monzonites and mesocratic and leucocratic syenites. These are coarse-grained, porphyritic and sometimes trachytic textured rocks containing alkali feldspar, basic plagioclase, pyroxene and biotite. Through the series from monzonite to leucocratic syenite the plagioclase decreases from 40% to 5%, diopside–augite from 50% to 5% and biotite from 15% to 2–3% while the content of orthoclase increases from 30% to 90%. Some syenites contain 5–10% quartz and 2–10% hornblende; accessory minerals are titanite, zircon, magnetite and apatite. The rocks of the alkaline series, which are located mostly in the northern part of the complex in a centre known as Chechekty (Fig. 90), cut across the subalkaline rocks. They are coarse-grained, leucocratic nepheline syenites, composed mostly of orthoclase (60–80%) and nepheline (10–30%) with hornblende, augite and biotite; accessories include titanite, zircon, magnetite and thorianite. There is some pseudoleucite, which displays dactylotypic textures, in the nepheline syenites. The presence of up to 10% sodic plagioclase and the prevalence of ferrohastingsite over the other dark-coloured minerals is typical for the alkaline syenites.

References Borodin, 1974; Zlobin, 1960.

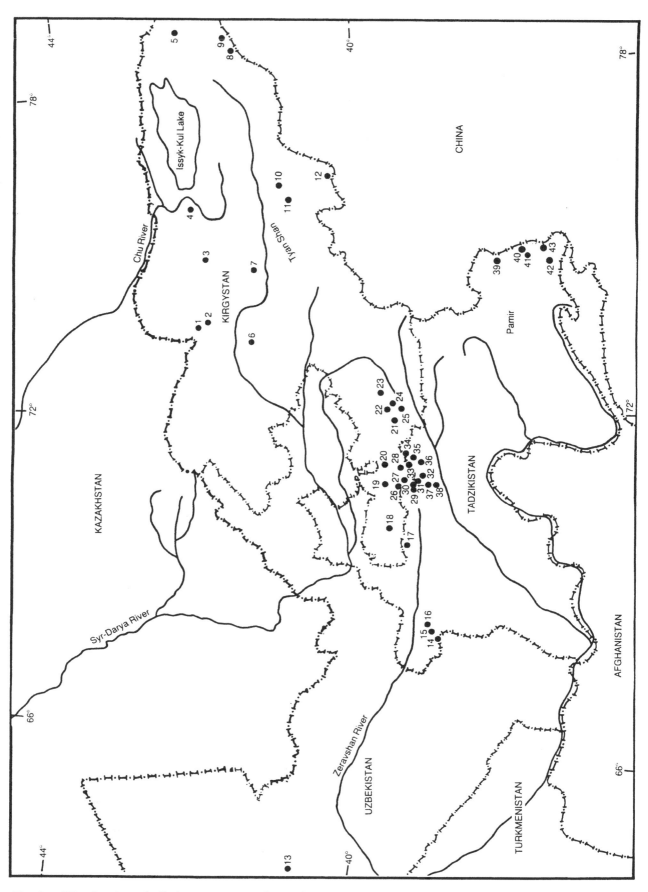

Fig. 89. Distribution of alkaline igneous rocks in the area of Central Asia.

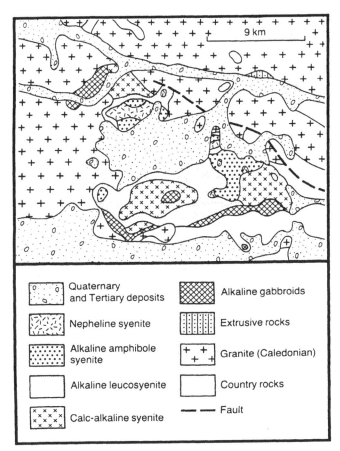

Fig. 90. Sandyk (after Zlobin, 1960, Fig. 1).

Quaternary and Tertiary deposits	Alkaline gabbroids
Nepheline syenite	Extrusive rocks
Alkaline amphibole syenite	Granite (Caledonian)
Alkaline leucosyenite	Country rocks
Calc-alkaline syenite	— — — Fault

The second major intrusive phase consists of grano-syenites, which contain biotite and amphibole, while the third phase comprises granites with biotite only.

Economic Quartz-baryte veins with sulphide mineralization are associated with the massif, as are deposits of rare-metal and rare-earth minerals (Dodonova *et al.*, 1984).

Age Geological evidence indicates a Permian age (Gavrilin, 1964).

References Dodonova *et al.*, 1984; Gavrilin, 1964; Izraileva and Turovsky, 1958.

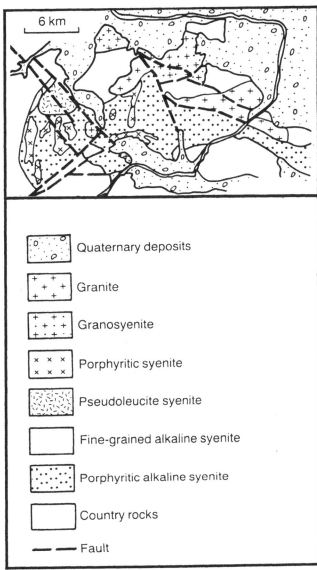

Quaternary deposits	
Granite	
Granosyenite	
Porphyritic syenite	
Pseudoleucite syenite	
Fine-grained alkaline syenite	
Porphyritic alkaline syenite	
Country rocks	
— — — Fault	

Fig. 91. Kyzyl-Ompul (after Gavrilin, 1964, Fig. 3).

4 KYZYL-OMPUL
(Ortotokoiskii)

42°21'N; 75°55'E
Fig. 91

This occurrence is situated in the central part of the northern Tyan-Shan to the west of Lake Issyk-Kul. It is a large, elongate, complex body, triangular in plan with an area of about 350 km². It is emplaced into Palaeozoic schists and sandstones and Permian extrusive rocks and is composed of the rocks of three intrusive phases, the first of which is alkaline. Among the rocks of the first phase four sub-phases can be distinguished. The first of these consists of porphyritic quartz-bearing and per-alkaline syenites (pulaskites). The main rock-forming minerals of these rocks are orthoclase, andesine, diop-side, biotite, alkali amphibole and sometimes quartz; accessories are apatite, titanite, titanomagnetite and zircon. During the second sub-phase small bodies of leucocratic syenite were intruded, the principal minerals of which are orthoclase, andesine, diopside, amphibole and more rarely quartz. During the third sub-phase small volumes of nepheline and pseudoleucite syenites were emplaced. The nepheline syenites are composed of orthoclase, diopside, amphibole, biotite and nepheline, which is restricted to the mesostasis. The pseudoleucite syenites contain phenocrysts of nepheline, pseudoleucite and barkevikite in a groundmass of orthoclase, nephe-line, barkevikite, biotite, diopside and a little olivine. The accessories are titanite, apatite, zircon, titanomag-netite, fluorite and pyrite. The fourth subphase is repre-sented by evenly textured biotite quartz-bearing syenites which form small, vein-like bodies. The syenite intru-sions are crossed by numerous dykes and veins of lamprophyre, monzonite, syenite, granite and aplite.

5 CHONASHU
(Irtashskii)

42°22'N; 79°04'E

Situated within the deep fracture zone which divides the north and central Tyan-Shan, the Chonashu intrusion extends over 10 km in a northeasterly direction but has a width of only 1–3 km. The country rocks to the south are Precambrian granitoids and to the north phyllites, schists and andesitic tuffs. The complex includes gabbros, diorites and pyroxenites with alkaline and nepheline syenites forming a series of vein-shaped bodies. The nepheline syenites are unevenly grained, gneiss-like rocks, with a pegmatitic facies, composed of albite, microcline, nepheline, associated with which are sodalite and cancrinite, and biotite. Sometimes the nepheline syenites pass into nepheline-bearing or alkaline syenites 80–90% of which may comprise large crystals of albite, biotite and actinolite. Similar accessories occur in all the peralkaline rocks of the complex, namely titanite, garnet, zircon, fluorite, catapleite, corundum and apatite.

Age K–Ar on biotite gave 324 Ma.

Reference Borodin, 1974.

---

6 SHAMATORSKII
41°35′N; 73°26′E
Fig. 92

Situated in the central Tyan-Shan, Shamatorskii has an area of about 80 km² and is a steeply dipping, lens-shaped body with a length 25 km and width up to 7 km. The intrusion, the inner part of which is layered, is composed of biotite clinopyroxenite, biotite melagabbro and subordinate essexite, monzonite and nepheline syenite. The layering within the complex conforms with the contacts and dips at steep angles. The intrusion is crossed by dykes of syenite aplite, nepheline syenite pegmatite, camptonite and spessartite. At the contacts with the country rocks are narrow zones of fenite and hornblende rocks.

Age K–Ar determinations on biotite from essexite gave 252 Ma.

Reference Kushev, 1960.

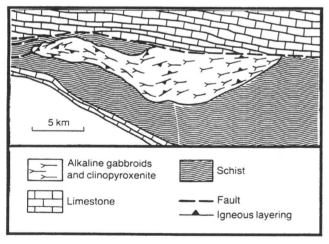

Fig. 92. Shamatorskii (after Kushev, 1960, Fig. 1).

---

7 KOKDZHARSKII
41°32′N; 74°46′E

This is a dyke field of limburgites, camptonites, bekinkinites, monchiquites and analcime basaltoids. The dykes are 2–50 m in width and 200 m or more in length.

Age Permian.

Reference Zubtsov, 1967.

8 KAICHINSKII
41°39′N; 78°41′E

Kaichinskii is situated on the northern slope of the Kok-Shaalsk Mountain ridge (Tyan-Shan) and a huge fracture zone. It cuts through upper Silurian schists, sandstones and limestones which form a large anticlinorium. Pyroxene and cordierite–biotite are developed in sandstones and schists adjacent to the steep contacts and limestones are altered to wollastonite skarns. The 20 km² intrusion has an oval form with the longer axis trending northeastwards parallel to the strike of the folding of the country rocks. It comprises four intrusive phases: (1) melteigite and ijolite, (2) coarse-grained aegirine–augite syenites, which are the major phase, (3) trachytic and giant-grained peralkaline syenites and (4) tourmaline-bearing granites. There are dykes of alkaline and nepheline syenites, pegmatites and carbonatites. The melteigite and ijolite occur in the western part of the intrusion as xenoliths up to 60 × 20 m in aegirine–augite syenites. They are also found in the marginal parts of the intrusion as small veins and sills and among the contact altered sandstones and schists. The aegirine–augite syenites occupy 80% of the intrusion and are traversed by the trachytic and giant-grained alkaline syenites. Alkaline and nepheline syenite pegmatites are concentrated in the outer parts of the intrusion and form dykes up to 10 m thick and 2 km long. Carbonatites in the western section of the complex form veins and dykes with a width up to 0.8 km.

Age Whole rock K–Ar determinations gave 314–330 Ma (Purkin, 1968).

References Kayumov and Karabaev, 1981; Purkin, 1968.

---

9 AILAGIR
41°43′N; 78°58′E

Ailagir lies within upper Silurian limestones and schists and upper Carboniferous biotite–amphibole granites and granosyenites. It is a northeasterly-trending elongate intrusion with steep to vertical contacts composed of coarse-grained porphyritic aegirine–augite syenites.

Reference Kayumov and Karabaev, 1981.

---

10 SARISAI
41°06′N; 76°17′E

This is a dyke-shaped body of small size, intruded into Carboniferous limestones and schists along a northeasterly-trending fault. It comprises aegirine–augite–nepheline syenites and mariupolites.

Reference Purkin, 1968.

---

11 SURTEKIN
40°59′N; 76°02′E
Fig. 93

The Surtekin intrusion is situated on the southern slope of the Atbashi mountain ridge of the Tyan-Shan. It has an area of 20 km² and lies within schists and limestones which are metamorphosed up to 400 m from the contact. The contacts dip steeply at 40–80° towards the centre. The intrusion was formed in four phases involving (1) monzonite, shonkinite and essexite, (2) peralkaline syenite, (3) nepheline syenite and (4) quartz syenite. Monzonite and essexite are concentrated in the southeastern contact zone. Nepheline syenite occupies 75% of the area of the intrusion with biotite–amphibole-bearing varieties predominant; many are trachytic in texture. Volumetrically of less importance are urtite and aegirine–nepheline syenite. Alkaline syenites are intimately connected with the nepheline syenites, the intrusive contacts between them being generally gradational.

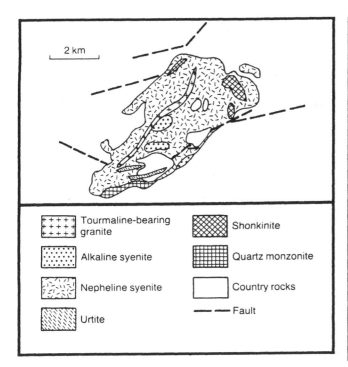

Fig. 93. Surtekin (after Kayumov and Karabaev, 1981, Fig. 6).

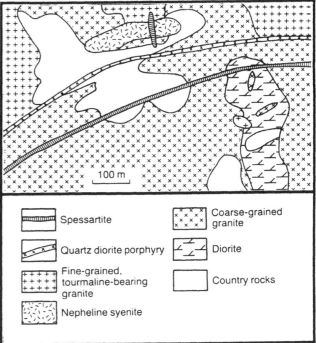

Fig. 94. Tozbulak (northeast) (after Kayumov and Karabaev, 1981, Fig. 10a).

The marginal parts of the nepheline syenites are strongly albitized such that sometimes they are altered into pure albitites. Quartz syenites (shonkinites), in the form of dyke-shaped and stock-like bodies, cut the nepheline syenites. Dykes of nepheline syenite pegmatite and tinguaite are concentrated in the marginal areas of the intrusion where they are sometimes albitized.

Age Whole rock K–Ar determinations gave 203–272 Ma (Purkin, 1968).

References Kayumov and Karabaev, 1981; Purkin, 1968.

---

12 UZUNBULAK 40°26′N; 76°25′E

Uzunbulak is a dyke-like intrusion of only 0.8 km² situated on the southern slope of the Atbashi mountain ridge (Tyan-Shan) where it cuts through Silurian and Devonian schists and limestones. It consists of medium- and coarse-grained cancrinite syenites with small veins of alkaline pegmatite.

Reference Kayumov and Karabaev, 1981.

---

13 TOZBULAK 40°48′N; 63°36′E
  Figs 94 and 95

This intrusion is located in Uzbekistan, far to the west of the rest of the province (Fig. 89). It has an area of 52 km²; but the two maps (Figs 94 and 95) show only small segments in the northeastern and southeastern parts of the occurrence. It is located within sedimentary rocks, amongst which limestones are predominant, and subordinate extrusive igneous rocks. The country rocks are folded into a large anticline in the axis of which the intrusion was emplaced. Four intrusive phases are present, namely: (1) diorite and quartz diorite, (2) coarse-grained biotite granite and granodiorite, (3) peralkaline and nepheline syenites and (4) fine-grained tourmalinized granite. Dyke rocks associated with the syenites are nepheline syenite pegmatite and peralkaline aplite.

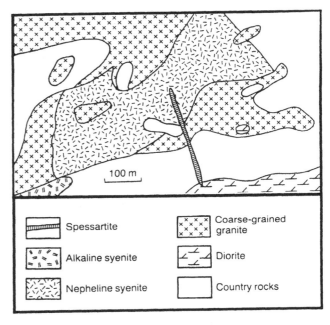

Fig. 95 Tozbulak (southeast) (after Kayumov and Karabaev, 1981, Fig. 10c).

Reference Kayumov and Karabaev, 1981.

---

14 AKSAI 39°01′N; 68°01′E

The 2 km² Aksai intrusion lies within Carboniferous sediments and consists of nepheline, nepheline–analcime and cancrinite syenites and syenite porphyries. The mafic minerals are aegirine, arfvedsonite and biotite.

The intrusion is accompanied by teschenite, camptonite, trachyte and essexite dykes.

References Ivanova *et al.*, 1940; Shinkarev, 1966.

---

15 AKHBASAI   39°08′N; 68°12′E

This occurrence is situated on the southern slope of the Zeravshansky mountain ridge and takes the form of a stratified body extending in a northeasterly direction for 7 km but having a width of only 500–600 m. The rocks of the intrusion comprise two phases, the first of which consists of syenites and quartz syenites that form a strip in the marginal contact zone. They have been altered under chlorite facies conditions and albitized. The second phase, which is predominant, is represented by foyaites and phonolites. Fine-grained foyaites are not widely developed and generally form xenolith-like bodies. In the central part of the massif the foyaites have a pegmatitic grain size, while pegmatite veins from several centimetres to 1–2 metres across are also broadly developed. Dyke rocks, including tinguaite and liebenerite and nepheline microsyenite, occur within the limits of the intrusion. Subvolcanic bodies of liebenerite phonolite are also present.

Age K–Ar on foyaite gave 209–275 Ma (Abdusalomov and Dusmatov, 1978).

References Abdusalomov and Dusmatov, 1978; Kuddusov *et al.*, 1980.

---

16 KAZNOKSKII   39°10′N; 68°15′E

This is a lens-shaped, 2–2.5 km² body intruded into Devonian and Carboniferous rocks. It is composed of nepheline and cancrinite syenites.

Age Geological evidence indicates an Upper Palaeozoic age.

Reference Ivanova *et al.*, 1940.

---

17 ROKSHIFO-SABAKHSKII   39°33′N; 69°40′E

Situated on the northern slope of the Turkestan mountain ridge, east of Samarkand, Rokshifo-Sabakhskii is a conformable, 20 km² intrusion, which has steep contacts. The following succession of rocks occurs: (1) gabbro-diorites and diorites, which are met as xenoliths in the northeastern part of the intrusion, (2) quartz and peralkaline syenites are the dominant rock types and are situated in the central area of the intrusion, (3) nepheline syenites, which are located in the marginal parts of the intrusion and (4) sodalite and cancrinite syenites. There are concentrations of fluorite, apatite, zircon, garnet, titanite, pyrite, arsenopyrite and molybdenite (Dusmatov and Salikhov, 1964).

Age The intrusion has been dated at 190–237 Ma by K–Ar. Albitization was dated at 204 Ma and the development of sodalite and cancrinite in syenites at 190–204 Ma (Melnichenko and Dusmatov, 1974).

References Baratov, 1966; Baratov *et al.*, 1970b; Dusmatov and Salikhov, 1964; Melnichenko and Dusmatov, 1974.

---

18 KARAGANSKII   39°45′N; 70°00′E

Karaganskii is a 6 km² intrusion composed of nepheline and alkaline syenites which are commonly altered with extensive replacement by sodalite, cancrinite and liebe-

nerite. Nepheline comprises about 30% of the rocks and accessories include zircon, garnet, apatite, monazite, anatase, tantalo-niobates and magnetite. There are dykes and veins of syenite aplite, albitite, peralkaline syenite pegmatite and lamprophyre.

Economic There is a sulphide mineralization of copper, lead, zinc, arsenic and molybdenum; concentrations of Nb, Ta and rare earth minerals are associated with the alkaline syenites.

Age Late Palaeozoic – on geological evidence.

Reference Shinkarev, 1966.

---

19 SALIEPSKII   39°45′N; 70°48′E

This stock of about 4 km² is composed of subalkaline gabbroids (monzonite and gabbro-monzonite) and peralkaline syenites. There are dykes of lamprophyre.

Age Geological evidence indicates a Permian age.

Reference Shinkarev, 1966.

---

20 SOKHSKII   39°47′N; 71°07′E
(Zardalek)   Fig. 96

Lying on the right bank of the central section of the Sokh River the Sokhskii intrusion is over 100 km² in area. It forms an ethmolith (lopolith) with steep (70–80°) contacts which is located in the nucleus of an anticline of Silurian limestones which is overturned to the south. The form of the intrusion is well constrained by the geometry of lineations within it. The complex consists of (1) nepheline- and garnet-bearing trachytic textured syenites, (2) garnet-bearing trachytic syenites free of nepheline and (3) monzonites. The first two groups have many features in common and comprise the larger part of the massif whereas the monzonites either form bodies adjacent to the margins or form xenolith-like masses within the complex. There are gradations between all the

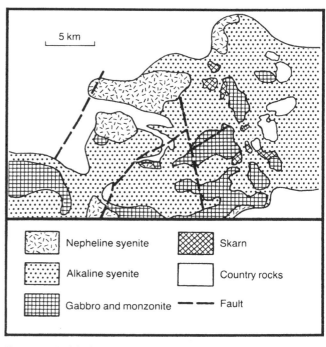

Fig. 96. Sokhskii (after Kayumov and Karabaev, 1981, Fig. 4).

rock types. The rock-forming minerals of the alkaline rocks are nepheline, microcline, plagioclase (andesine–labradorite), hornblende, biotite and garnet (grossular–andradite) with accessory zircon, apatite, fluorite, titanite and magnetite. The monzonites consist of diopside, hornblende, plagioclase and microcline with a mixture of analcime and muscovite. There are syenite and garnet-bearing nepheline syenite dykes. At the contacts the country rocks are phlogopitized and altered to marbles and skarns.

Reference Lyashkevich, 1963.

---

21 SURMETASHSKII   39°45′N; 71°54′E

Surmetashskii is a ring intrusion which is elliptical in plan. The country rocks are limestones, dolomites and schists. There are two principal intrusive phases: alkaline quartz-bearing syenites and tourmaline-bearing granites. Nepheline syenites are concentrated in the south of the intrusion with the central areas occupied by alkaline olivine-bearing melano- and leucocratic syenites. These rocks gradually pass into quartz-bearing syenites, in which veins and irregular patches of pegmatite and dykes of hybrid syenite are developed as well as skarn xenoliths.

References Kayumov and Karabaev, 1981; Shinkarev, 1966.

---

22 CHEKINDY   39°45′N; 72°11′E

Situated on the left bank of the Isfairamsaya River 1 km below the confluence with the Vostochnii Kichikalay River, Chekindy is compositionally and structurally similar to the Sokhskii (No. 20) intrusion. The first phase comprises monzonites and melanocratic syenites with garnet and diopside–augite. The second, and last, phase consists of nepheline syenites with sodalite and hauyne pegmatites up to 2 metres thick and up to several kilometres in length.

Reference Kayumov and Karabaev, 1981.

---

23 KICHIKALAISKII WEST   39°51′N; 72°25′E

Conglomerates and schists are the country rocks of the West Kichikalaiskii intrusion, the first igneous phase of which is represented by aegirine–augite syenites. Close to the margin of the intrusion they grade into garnet-bearing, melanocratic, fine-grained syenites and shonkinites. Feldspathoidal syenites comprise the last intrusive phase and these have been affected by processes of albitization and liebeneritization. The intrusion is encircled by a 200 m wide zone of hornblendites.

References Kayumov and Karabaev, 1981; Shinkarev, 1966.

---

24 ISFAIRAMSKII   39°44′N; 72°12′E
                                        Fig. 97

This intrusion is situated in the central part of the Alayski Mountains and occupies an area of about 20 km². It lies on the northern side of an anticlinorium composed of marbles of Carboniferous age. In the north it breaks through lower Carboniferous schists, conglomerates and sandstones. At the contact with the limestones in places garnet and pyroxene skarns have been produced which contain scheelite. There are numerous satellite intrusions composed of granodiorite, granite and syenite. Four principal groups of rocks make up the

intrusion: granosyenites, quartz diorites and monzonites, trachytic subalkaline syenites, alkaline granites and nepheline syenites. The granosyenites are developed in the southern part of the complex and the subalkaline syenites in the central part where they cut granosyenites and are represented by pyroxene–amphibole and biotite-amphibole varieties. The alkaline granites, which are leucocratic rocks, cut the syenites. Nepheline syenites are confined to the southwestern part of the complex where they form dykes up to 15 m in length; their relationship with the subalkaline syenites is not clear. Granite and syenite aplite dykes occur.

References Kayumov and Karabaev, 1981; Shinkarev, 1966.

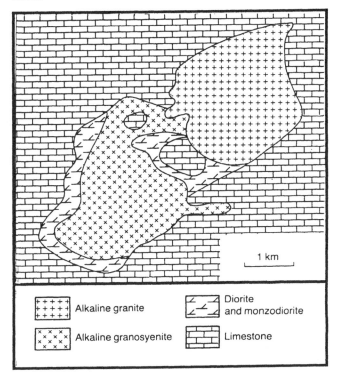

Fig. 97. *Isfairamskii (after Kayumov and Karabaev, 1981, Fig. 5).*

---

25 DGAMANDGAR   39°32′N; 72°10′E
                                        Fig. 98

Only 0.2 km², this intrusion cuts Permian granites and sedimentary rocks of Upper Carboniferous age. There is a broad skarn aureole. It has a layered structure and consists of nepheline syenites, which are often markedly porphyritic, peralkaline syenites and ijolites. The syenites are cut by syenite porphyry veins up to a metre across.

Reference Kayumov and Karabaev, 1981.

---

26 SKALYSTYI   39°37′N; 70°41′E

The complex is situated at the margin of the Alaisky shield in the core of an acute synclinal structure. It covers about 54 km² and is symmetrical in form but has a complicated structure; it is probably a ring complex. In the central part are nepheline syenites which change towards the periphery to peralkaline and subordinate quartz syenites. In the contact zone there is extensive cataclasis and development of quartz rocks.

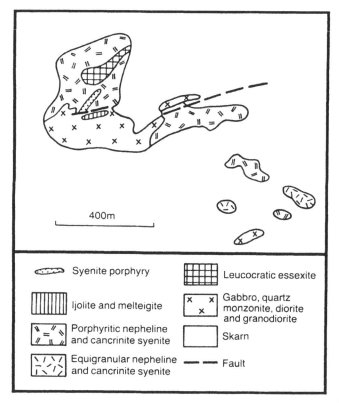

Fig. 98. Dgamandgar (after Kayumov and Karabaev, 1981, Fig. 9).

Age Permian, on geological evidence.

Reference Shinkarev, 1966.

27 MATCHINSKII   39°32'N; 70°47'E
Fig. 99

The 30 km² Matchinskii intrusion is in contact with Silurian psammites and comprises three phases. These are (1) aegirine–augite and biotite–nepheline syenites, (2) peralkaline pyroxenite and quartz syenites and (3) leucocratic granites. The nepheline syenites, which occupy the western part of the intrusion, are represented by huge xenoliths within the alkaline syenites. The aegirine–augite syenites are concentrated in the central part of the intrusion and grade into quartz syenites towards the margins, the contacts with biotite syenites indicating that there was remelting of the latter. The alkaline syenites vary from mesocratic to leucocratic varieties. There are areas of quartz–tourmaline rock within the quartz syenites, in which the tourmaline may form distinct layers, while there has been extensive albitization and carbonatization of the alkaline syenites. The leucocratic granites form veins in the syenites. The intrusion is cut by dykes of biotite and nepheline syenite and granite. The country rocks are intensively altered with the production of alkali amphibole, pyroxene and tourmaline. Rocks referred to as carbonatites have been produced at the eastern end of the intrusion as a result of carbonatization.

Economic Rare-earth minerals occur in carbonatite veins cutting schists close to the margins of the intrusion.

Reference Kayumov and Karabaev, 1981.

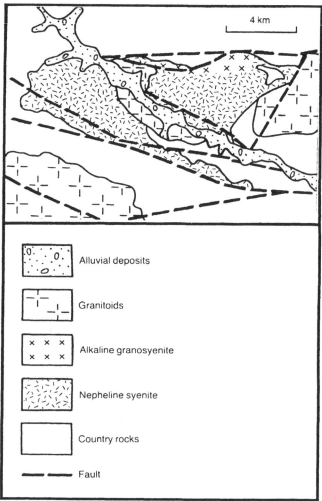

Fig. 99. Matchinskii (after Gavrilin, 1963, Fig. 1).

28 AIKEL   39°37'N; 69°36'E

Aikel is a symmetrical intrusion of about 14 km² lying near Lake Aikel. It is composed of subalkaline gabbroids (monzonites and syenodiorites) and syenites. It is traversed by approximately east–west-trending dykes of syenite aplite, syenite pegmatite, nepheline syenite (lujavrite) and leucite absarokite.

Age Permian.

Reference Shinkarev, 1966.

29 DARA-PIOZ   39°25'N; 70°44'E
(Dara-i-Pioz)

This intrusion comprises two fragments: the Upper massif and the Middle massif, both of which lie in limestones and terrigenous sediments of Carboniferous age. The Upper massif is a circular structure of 16 km², the outer part of which consists of tourmaline-bearing granosyenites and peralkaline granites and the central part of quartz-aegirine syenites. Granite and syenite pegmatites are broadly developed. The Middle massif is an oval body of 9.5 km², from the periphery to the centre of which a narrow zone of normal granite grades into peralkaline syenite, amongst which aegirine granite (sviatonossite) occurs. The centre of the massif consists of small bodies of garnet–nepheline syenite. The intrusion is crossed by lamprophyre dykes and pegmatites containing such rare minerals as sogdianite, tienshanite, searlesite, neptunite and hyalotekite (Grew *et al.*, 1994), for some of which this is the type locality.

Economic Alkali rare earth-rich minerals occur (Dusmatov, 1970).

References Dusmatov, 1970; *Grew *et al.*, 1994.

---

30 UTREN   39°28′N; 70°42′E

The 40 km² Utren intrusion is located in the core of a syncline (Perchuk, 1964) in lower Silurian schists and sandstones. The intrusion is divided into northern and southern sectors by xenoliths of schist up to 300 m across. The margins of the intrusion consist of quartz syenite, which also veins the country rocks, and this merges into trachytic nepheline syenites in the central parts. Biotite-, amphibole- and pyroxene-bearing varieties of nepheline syenite can be distinguished, as can sodalite and cancrinite types. The intrusion is cut by biotite granite dykes which often contain tourmaline.

References Kayumov and Karabaev, 1981; Perchuk, 1964.

---

31 MATCHINSKII SOUTH   39°27′N; 70°43′E

Lying 8 km to the south of Matchinskii (No. 27) this is a 20 km² intrusion composed of nepheline and peralkaline syenites which is structurally similar to Matchinskii. In the north and northeast it is cut by peralkaline granites.

Age Permian.

Reference Shinkarev, 1966.

---

32 TUTEKSKII   39°23′N; 70°50′E

The Tutekskii intrusion is situated high on the Alai mountain ridge. It is a stock-like body of 23 km² cutting Silurian mica and alkali amphibole schists. The central part of the intrusion is composed of medium- and coarse-grained biotite–nepheline syenites and pyroxenites. Zones of fenitization are developed at the margins of the intrusion by interaction with sedimentary rocks, and particularly with volcanics. The alkaline rocks are crossed by dykes of nepheline syenite pegmatite.

Economic Veins of fluorite are developed in limestones at the margins.

Age K–Ar gave 22–58 Ma (Baratov *et al.*, 1969).

References Baratov *et al.*, 1969; Kayumov and Karabaev, 1981; Melnichenko and Dustmatov, 1974.

---

33 KULPSKII   39°36′N; 71°07′E

The Kulpskii intrusion is located in the core of an anticline of Silurian and Carboniferous limestones, schists and phyllites. The intrusion is formed of three phases, namely nepheline syenites, syenites and quartz syenites, and tourmaline-bearing granites. The rocks of the first two phases contain numerous xenoliths of schist. The nepheline syenites occur in the northwestern and northeastern parts of the intrusion, as well as in east–west-trending zones in the central and southern parts of the massif. They are mostly biotite-bearing varieties with less abundant amphibole and amphibole-pyroxene types; there are also sodalite-and canrinite-bearing varieties. The syenites and quartz syenites form

an elongate body in the central part of the intrusion within which pyroxene-, biotite- and tourmaline-bearing variants have been distinguished. Fine-grained, leucocratic granites with patches of tourmaline form dykes in the central and eastern parts of the intrusion. Many minerals, including albite, sodalite, cancrinite, tourmaline and quartz, are considered to be the result of secondary processes.

Reference Kayumov and Karabaev, 1981.

---

34 KHODZHAACHKAN   39°37′N; 71°14′E
Fig. 100

The 50 km² intrusion lies in the core of a very large east–west-trending anticline, has vertical contacts and sends numerous apophyses into the surrounding schists. Syenite gneisses, so called because of their gneissose structure, monzonites and alkaline syenites as well as melanite–nepheline syenite, shonkinite, cancrinite syenite and amphibole–nepheline syenite grade into each other and comprise the first intrusive phase. Aegirine–augite–nepheline syenites occur in the eastern part of the intrusion and melanite syenites and shonkinites form schlieren some metres in length. Quartz syenites, of which there are biotite-, biotite–amphibole- and pyroxene-bearing varieties, occur as dyke-shaped bodies in the southwestern part of the intrusion and cut across the nepheline syenites and country rock schists. Fine-grained granites form sills and dyke-shaped bodies up to 3 m thick in the central part of the complex. Among the granites there are biotite- and tourmaline-bearing varieties and they are cut by dykes of syenite aplite and albitite, and pegmatites and veinlets of calcite and wollastonite. In the marginal contact zones the schists are intensively altered, including albitized, and in places changed into syenite gneisses; limestones are changed to wollastonite skarn.

References Kayumov and Karabaev, 1981; Perchuk, 1964.

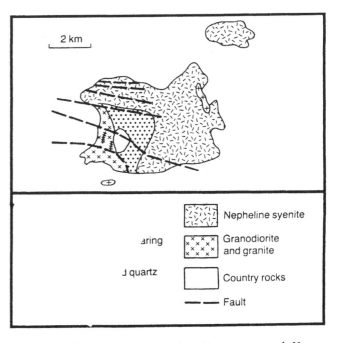

Fig. 100. Khodzhaachkan (after Kayumov and Karabaev, 1981, Fig. 8).

35 DZHILISU
39°35'N; 71°12'E
Fig. 101

Located in the nucleus of a syncline of psammites and granites, the Dzhilisu intrusion is composed principally of biotite- and aegirine–augite–nepheline syenites with gradational contacts. In the upper part close to the roof are heterogeneously grained syenites with dykes of nepheline and alkaline syenite aplite. Further into the intrusion there are evenly grained nepheline syenites and pegmatite bodies. In the upper parts of the intrusion particularly, there has been extensive albitization that involved the syenites and marginal schists.

References Kayumov and Karabaev, 1981; Shinkarev, 1966.

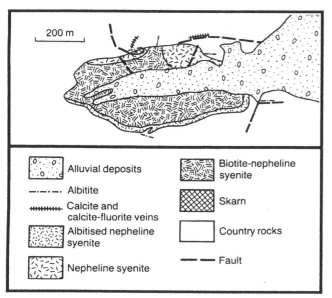

Fig. 101. *Dzhilisu (after Kayumov and Karabaev, 1981, Fig. 7).*

36 NEDOSTUPNYI
39°29'N; 71°11'E

This is a small (8 km²) intrusion lying in a fracture zone located in the nucleus of an east–west-trending anticline composed of Silurian schists. There are two principal phases, the first of which is diorite, the second nepheline syenite. The latter has assimilated country rock schists. Sharp contacts between the diorite and syenite have not been observed. There is a gradual transition between nepheline syenites and alkaline syenites, the latter having been produced by assimilation of schist by the former, and between the alkaline syenites and a marginal facies, which gives the intrusion a distinct zonality. The diorites are developed in a narrow zone in the north-western part of the intrusion and contain biotite and hornblende and large amounts of apatite and titanite. There are many veins of alkaline syenite.

Reference Lyashkevich, 1959.

37 YARKHICH
39°16'N; 70°42'E

The 50 km², east–west-trending, steeply inclined Yarkhich intrusion is composed of nepheline and peralkaline syenites, which grade, via quartz syenites, to hastingsite granites in the north. The nepheline syenites are traversed by numerous stocks and dykes of porphyritic granite and syenite aplite. Apophyses of the alkaline rocks and zones of hornfelses extend into the country rocks. The rock-forming minerals of the nepheline and alkaline syenites are microcline–perthite, oligoclase-albite, nepheline, sodalite, lepidomelane, aegirine and amphibole; accessories are orthite, apatite, zircon, titanite and magnetite.

Age K–Ar gave 230 Ma (Shinkarev, 1966).

References Goretskaya *et al.*, 1957; Shinkarev, 1966.

38 TURPI
39°09'N; 70°43'E
Fig. 102

This occurrence is 16 km² in area and situated in the core of an anticline composed of schists, gneisses and marbles with the immediate country rocks massive biotite and amphibole granites. Of the three phases comprising the intrusion the first consists of alkaline syenites, which form the central part of the massif. These are hastingsite and biotite–hastingsite syenites, the earliest of which are melanocratic rocks that are cut by mesocratic and leucocratic varieties. Pegmatites and quartz–albite and calcite veins are associated with the syenites which are widely albitized and zeolitized. The second phase, which intrudes the syenites, consists of foyaites with aegirine, sodalite and cancrinite. There is a pegmatitic facies and they are veined by albitites. The nepheline is partly altered to liebenerite and fluorite is widely developed. The third igneous phase is represented by dyke-shaped bodies of lepidomelane foyaite and biotite–hastingsite syenite, that are cut by dykes of medium-grained biotite–quartz syenite and alkaline granite.

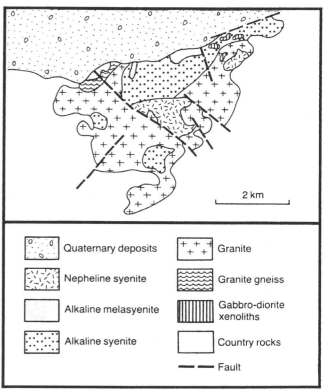

Fig. 102. *Turpi (after Lyashkevich, 1961, Fig. 1).*

Age K–Ar determinations gave dates of 218–268 Ma for the first phase of intrusion, 233–250 Ma for the second phase and 190–195 Ma for the third phase (Baratov *et al.*, 1978).

References Baratov *et al.*, 1978; Lyashkevich, 1959 and 1961.

---

39 PAMIR PIPES – NORTHERN   37°48′N; 74°55′E
Fig. 103

This northern group of pipes is related to a regional fault that separates geological zones of the central and southwestern Pamirs. The explosion pipes and plugs, with some dykes, which were discovered by Dmitriev (1976), form a 15 km long chain. The pipes are up to 10 m in diameter, are commonly multiple and contain tuffs and breccias as well as massive rocks. The 'Leucite pipe' illustrated in Fig. 103 has a border zone of porphyritic pyroxene–sanidine syenite and fergusite with the central part occupied by fergusite explosion tuffs and pseudoleucite phonolite.

Age K–Ar on K–feldspar gave 14 Ma.

Reference Dmitriev, 1976.

---

40 PAMIR PIPES – SOUTHERN   37°48′N; 74°57′E

This southeastern group of pipes is situated in the region of the Dun-Keldik and Agad'an-D'ilga Rivers. The pipes are generally 100–150 × 30–70 m and comprise fergusite porphyry and alkaline syenite porphyries; they often contain country rock xenoliths.

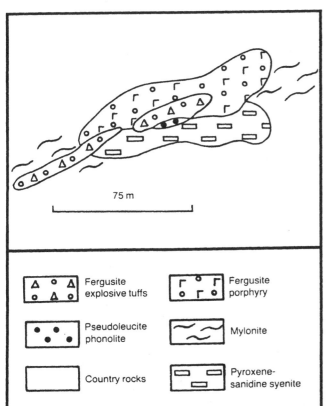

Age K–Ar on leucite and sanidine gave 14–23 Ma.

Reference Dmitriev, 1976.

---

41 PAMIR DYKES – NORTHERN BELT
38°05′N; 74°39′E
Fig. 104

More than 20 dykes extend over 100–120 km and have an arcuate distribution which is controlled by faulting. The dykes are from 1 to 8 m thick and comprise grorudite and a range of alkaline syenites and porphyries and minette and other lamprophyres. The dykes can be multiple, and a common type, illustrated by Fig. 104, contains fergusite, pseudoleucite tinguaite and syenite.

Age Neogene.

Reference Dmitriev, 1976.

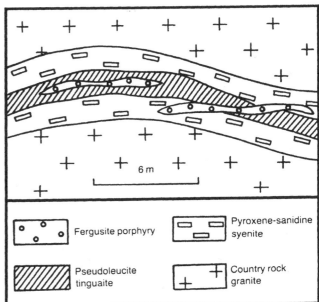

Fig. 104. Dyke at the head of the Agad'in-D'ilga River; one of northern belt of dykes in Pamirs (after Dmitriev, 1976, Fig. 4).

---

42 PAMIR DYKES – SOUTHERN BELT
37°44′N; 74°51′E

The southern belt of dykes in the Pamirs is located near Dunkeldik Lake to the north of the Dunkeldykskii intrusion. The dykes extend over some 25 km with individual dykes 300–500 m in length and generally 3–5 m thick. They comprise pseudoleucite tinguaite, sanidine-bearing syenites, bostonite and rarely fergusite.

Age Neogene.

Reference Dmitriev, 1976.

---

43 DUNKELDYKSKII   37°47′N; 74°57′E

Located on the watershed of the Sarikolsky Ridge the Dunkeldykskii massif is elliptical in outline and 2.5 × 1 km. It has vertical contacts and over 300 m can be seen in vertical exposure. The intrusion is zonal with

Fig. 103. 'Leucite pipe' – one of the group of Northern Pipes in the Pamirs (after Dmitriev, 1976, Fig. 5).

in the centre pseudoleucite rocks including syenite, borolanite, fergusite and fergusite porphyry. The outer part comprises alkaline syenites and porphyries. All the rocks of the intrusion are cut by dykes, including multiple types, and bodies of granosyenite porphyry, the greatest concentrations being in the southwestern and central parts. There are abundant xenoliths of country rocks throughout.

Age K–Ar on K–feldspar gave 14–16 Ma.

Reference Dmitriev, 1976.

Central Asia references

ABDUSALOMOV, F.N. and DUSMATOV, V.D. 1978. On the age of the Akhbasaisky massif foyaites. *Doklady Akademii Nauk Tadzhikskoi SSSR*, 21(8): 40–2.

BAGDASAROV, E.A., ORLOVA, M.P. and KOZYREV, V.I. 1974. Rare and trace elements of alkaline gabbroids and basalts of the North Tyan-Shan. *Zapiski Vsesoyuznogo Mineralogicheskogo Obshchestva.*, 6: 682–94.

BARATOV, R.B. 1966. *Intrusive complexes of the southern slope of the Gissarsky ridge and its ore formation.* Donish, Dushanbe. 90 pp.

BARATOV, R.B., DUSMATOV, V.D. and MELNICHENKO, A.K. 1969. First data on the K–Ar age of the nepheline syenites of the Tutek-Devonasuzhsky massif (Alaisky ridge) *Doklady of the Academy of Sciences, Tadzhikistan SSR*, 12(12): 41–3.

BARATOV, R.B., KUTENETS, V.A. and MADII, L.A. 1970a. On the formation of the intrusive complexes of the eastern Karagenin (central Tadzhikistan). *Doklady Akademii Nauk Tadzhikskoi SSSR*, 191(6): 54–5.

BARATOV, R.B., MELNICHENKO, A.K. and DUSMATOV, V.D. 1970b. On the alkaline rocks of the Rokshifo-Sabakhsky massif (Turkestansky ridge). *Doklady Akademii Nauk Tadzhikskoi SSSR*, 12(4): 21–6.

BARATOV, R.B., AKRAMOV, A.N., MELNICHENKO, A.K. and DUSMATOV, V.D. 1978. New data on the absolute age of the Turpi massif of alkaline rocks. *Izvestiya Akademii Nauk Tadzhikskoi SSSR. Otdelenie Geologo-khimicheskikh i Tekhnicheskikh nauk*, 2: 63–71.

BEKBOTAEV, A.T. 1968. *Alkaline rocks of Kirgizia and Kazakhstan.* Ilim, Frunze. 72 pp.

BORODIN, L.S. 1974. *The principle provinces and formations of alkaline rocks.* Izdatel'stvo, Nauka, Moscow. 376 pp.

DMITRIEV, E.A. 1976. *Cainozoic potassic alkaline rocks of East Pamir.* Donish, Dushanbe. 170 pp.

DODONOVA, T.A., POMAZKOV, K.D. and POMAZKOV, Ya.K. 1984. *Endogenic magmatic formations of Kirgizia.* 1: 152–9. Ilim, Frunze.

DOLZHENKO, V.N. 1968. Geologic-petrographic characteristics of the syenite massif of Bulak-Ashu. *Alkaline rocks of Kirgizia and Kazakhstan.* 87–99. Ilim, Frunze.

DUSMATOV, V.D. 1970. Mineralogy of the Dara-Pioz massif (Tadzikistan). *In Questions of the geology of Tadzhikistan*, 131–6. Geological Institute of Tadzhikistan, Dushanbe.

DUSMATOV, V.D. and SALIKHOV, D.N. 1964. Nepheline syenites of the River Tagoby-Sabakh (southern slope of the Turkestan ridge). *Trudy Instituta Geologii. Akademiya Nauk Tadzhikskoi SSR. Dushanbe*, 8:, 118–31.

GAVRILIN, R.D. 1963. The Matchinskii massif of syenite-granite. *Doklady Akademii Nauk SSSR*, 148(2): 403–5.

GAVRILIN, R.D. 1964. Geological composition of the complex syenite-granite massif of Kyzyl-Ompul (North Tyan-Shan). *Izvestiya Akademii Nauk SSSR. Geological Series*, 3: 69–83.

GORETSKAYA, E.I., DODONOVA, T.A. and LESKOV, S.A. 1972. Volcanogenic formations of orogenic stages. *Geology of the USSR*, 25(2): 99–134. Nedra, Moscow.

GORETSKAYA, E.I., UGLOV, S.M. and LESKOV, S.A. 1957. Materials on geology and metallogeny of South Gizzar. *Doklady Akademii Nauk Tadzhikskoi SSR*, 9: 68–72.

*GREW, E.S., YATES, M.G., BELAKOVSKIY, D.I., ROUSE, R.C., SU, S-C. and MARQUEZ, N. 1994. Hyalotekite from reedmergnerite-bearing peralkaline pegmatite, Dara-i-Pioz, Tajikistan and from Mn skarn, Langban, Varmland, Sweden: a new look at an old mineral. *Mineralogical Magazine*, 58: 285–97.

IL'INSKY, G.A. 1970. *Mineralogy of the Turkestan-Alay alkaline intrusives.* Leningrad University, Leningrad. 166 pp.

IVANOVA, T.N., GAVRILOVA, V.N. and UNKSOV, V.A. 1940. Intrusions of the north-western part of the Zeravshano-Pissarsky mountain system. *In* V.A. Unksov (ed.) *Geology and ore deposits of Tadzhikistan*, 61–75. Donish, Dushanbe.

IZRAILEVA, P.M. and TUROVSKY, S.D. 1958. The Ortotokoisky massif of giant-grained and giant-porphyritic syenites. *Guidebook around North Kirgizia. Trudy of the 2nd Petrographical Conference.* Academy of Sciences, Kirgiz SSR. Frunze. 11–25.

KAYUMOV, A.K. and KARABAEV, K.K. 1981. *Alkaline magmatism and ore-formation of the south Tyan-Shan.* Uzbek Academy of Science Publishers, Tashkent. 135 pp.

KUDDUSOV, Kh.K, ABDUSALOMOV, F.M. and DUSMATOV, V.D. 1980. Magnetometry of the alkaline rocks of the Akhbaisky massif (central Tadzhikistan). *Izvestiya Akademii Nauk Tadzhikskoi SSR, Otdelenie Geologo-khimicheskikh i Technicheskikh nauk*, 78(4): 49–70.

KUSHEV, V.G. 1960. Some data on alkaline rocks of the western part of the Shamatorsky intrusion. *Vestnik Leningradskogo Gosudarstvennogo Universiteta, Seriya Geologiya, Geografiya*, 1(6): 31–42.

LYASHKEVICH, Z.M. 1959. Petrography of the alkaline massifs of the western part of the Alaisky ridge. PhD Thesis, University of Lvov. 301 pp.

LYASHKEVICH, Z.M. 1961. Peculiarities of the Turpi alkaline massif. *Tadzhik Academy of Sciences, Geology-Chemistry Series*, 2: 49–70.

LYASHKEVICH, Z.M. 1963. New data on the formation of the Sokhsky alkaline massif (Alaisky ridge). *Geology and Prospecting*, 8: 69–78.

MELNICHENKO, A.K. and DUSMATOV, V.D. 1974. Formation time of the alkaline rocks of the Gissaro-Alay according to the data of geochronometrical observations (south Tyan-Shan). *New data on absolute geochronology* (XVII session), 330–41. Nauka, Moscow.

PERCHUK, L.L. 1964. *Physico-chemical petrology of granitoid and alkaline intrusives of the central Turkestan-Alay.* Nauka, Moscow. 159 pp.

PURKIN, M.M. 1968. Alkaline rocks of the Kok-Shaala ridge system. *In* A.K. Kayumov (ed.) *Alkaline rocks of Kirgiziya and Kazakhstan.* 47–87. Ilim, Frunze.

SHINKAREV, N.F. 1966. *Upper Palaeozoic magmatism of Turkestan-Alay.* Leningrad University Publishers, Leningrad. 152 pp.

ZLOBIN, B.I. 1960. A petrographical account and the petrochemistry of the alkaline rocks of the Sandyk intrusion. *Izvestiya Akademii Nauk SSSR, Seriya Geologicheskaya*, **2**: 91–104.

ZUBTSOV, E.I. (ed.) 1967. *Geological map of the USSR. 1:200,000* (Compiled by A.A. Luik), Nedra, Moscow.

* In English

TAIMYR PENINSULA

The Taimyr Peninsula (Fig. 105) is located on Arctic coast of Russia and lies immediately north of the Maimecha-Kotui Province (Figs 1 and 110).

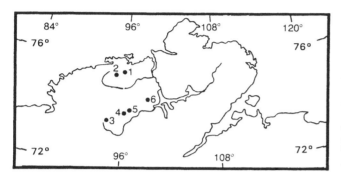

Fig. 105. Distribution of alkaline igneous rocks in the Taimyr Peninsula.

1 Kuropatochei	4 Angidritovyi
2 Vysokoi	5 Lunnyi
3 Kyidinskii	6 Fad'yu-Kuda

1 KUROPATOCHEI   75°31′N; 95°54′E

This intrusion is located in the vicinity of the rivers Kolomeitseva and Mamont to the north of the Birranga ridge. It lies within a granitoid batholith of Lower Devonian age in the contact zone of leucocratic granites. It is oval in plan, covers about 8 km² and is composed of coarse-grained syenite, syenite porphyry, trachyte and alkaline syenites with nepheline, although the quantity of nepheline (up to 5%) in the last is low.

Age The K–Ar method gave 200 Ma.

Reference Daminova, 1963.

2 VYSOKOI   75°26′N; 94°47′E

Vysokoi lies west of Kuropatochei and has been emplaced in similar leucocratic granites within the same Devonian granite batholith. The intrusion extends over about 4 km² and has a zonal structure grading from syenites in the centre through syenite porphyries to trachytes. Quartz syenites are present in the western contact zone. The syenites consist of orthoclase, andesine, augitic pyroxene, hornblende, biotite, titanite, titanomagnetite and apatite, sometimes with grains of marialite and analcime; there is rare zircon and fluorite. A dyke series comprises quartz porphyries and mica lamprophyres.

Age K–Ar indicated an age of 196 Ma.

Reference Daminova, 1963.

3 KYIDINSKII   73°45′N; 94°10′E
Fig. 106

This intrusion is emplaced in sedimentary rocks of Permian age. There were two periods of intrusion, the first involving syenites, which were affected by a folding episode, and the second essexites which cut the syenites and folded structures. The main minerals of the syenites are K–feldspar (50–70%), plagioclase (10–30%), quartz (0–15%), alkali pyroxene and amphibole (2–10%) with accessory nepheline, magnetite, apatite, titanite, zircon, orthite, fluorite, eudialyte and riebeckite. The riebeckite occurs in leucocratic facies of the quartz syenites. In the syenites in which eudialyte occurs it is an interstitial phase. The main minerals of the essexites are plagioclase (50–60%), pyroxene (20–30%), K–feldspar (10–25%) and olivine (0–7%); among the accessories are nepheline, apatite, magnetite, pyrite, titanite and zircon.

As well as Kyidinskii, there are two other intrusions along the upper reaches of the River Dikara-Bigay (right tributary of the Upper Taimyra River) which have the same composition and structure.

Reference Ravich and Chaika, 1959.

4 ANGIDRITOVYI   74°10′N; 96°50′E

This occurrence is situated in the southern part of the Birranga Ridge in the upper part of a small river, the Angidritovyi. It has a ring form and occupies only some 250 m². It is composed of nepheline syenites which may be equigranular or porphyritic. A typical composition of the nepheline syenites is 50% nepheline, 30% albite, 9% K–feldspar, 6% analcime and zeolites, 3% aegirine and aegirine–augite and 1% schorlomite.

Age 199–205 Ma by the K–Ar method.

References Daminova, 1963; Egorov and Surina, 1980.

5 LUNNYI   74°31′N; 99°21′E
Fig. 107

Lunnyi is composed of fluorite-bearing carbonatites at the borders of which are developed leucocratic alkaline metasomatic rocks. The carbonatite forms small bodies, the majority only tens of metres in diameter, but the largest of which, located near the Lunnyi spring, is 250 × 120 m. The carbonatites are medium-grained, massive rocks, containing 60–80% carbonates which are mainly calcite, with rare ankerite, dolomite and siderite. Other minerals present include baryte, fluorite, hematite, flourite and fluorcarbonates of the rare earths. Albite, phlogopite, apatite and magnetite are found in some varieties.

Reference Gulin, 1970.

6 FAD'YU-KUDA   74°33′N; 99°50′E
(Shchelochnoi)   Fig. 108

The intrusion is located near the River Fad'yu-Kuda and is a curved, lens-shaped body which extends along a zone of faulting for 2 km and has a width of from 400 to 650 m. The area of the intrusion is 1 km² and it is stock-like in form. It is composed of nepheline syenites of varying grain size but coarse- and medium-grained varieties are predominant. The principal mineralogy is nepheline, albite, microcline and aegirine with accessory melanite, titanite, apatite, zircon, rutile, sphalerite, magnetite and fluorite. Autometasomatic processes were extensive and caused the development of albite and cancrinite in the nepheline syenites.

Economic Pyrite and baryte mineralization is developed at the contacts. Pyrite ores, consisting of 60–70% pyrite,

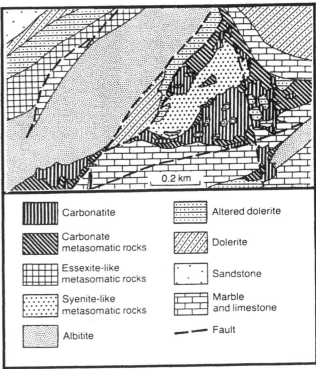

Fig. 107. Lunnyi (after Gulin, 1970, Fig. 2).

generally form a zone having a thickness of about 2 m, but wider zones also occur.

Reference Ravich and Chaika, 1959.

Taimyr Peninsula references

DAMINOVA, A.M. 1963. Alkaline rocks of central Taimyr. *Trudy, Patricia Lumumba University of Peoples Friendship, Moscow*, 3(1): 3–48.

EGOROV, L.S. and SURINA, N.P. 1980. Nepheline syenites of the Shelochnogo spring intrusion syenites in the Central Taimyr. *Alkaline magmatism and apatite-bearing rocks of the north of Siberia.* Scientific-Investigatory Institute of Arctic Geology, Leningrad. 155–74.

GULIN, S.A. 1970. On the formation of alkaline and carbonate metasomatites of the central Taimyr. *Carbonatites and alkaline rocks of the North of Siberia.* Scientific-Investigatory Institute of Arctic Geology, Leningrad. 170–84.

RAVICH, M.G. and CHAIKA, L.A. 1959. Small intrusions of the Birranga shield. *Trudy Nauchno-Issledovatel'skogo Instituta Geologii Arktiki, Leningrad*, 88: 1–149.

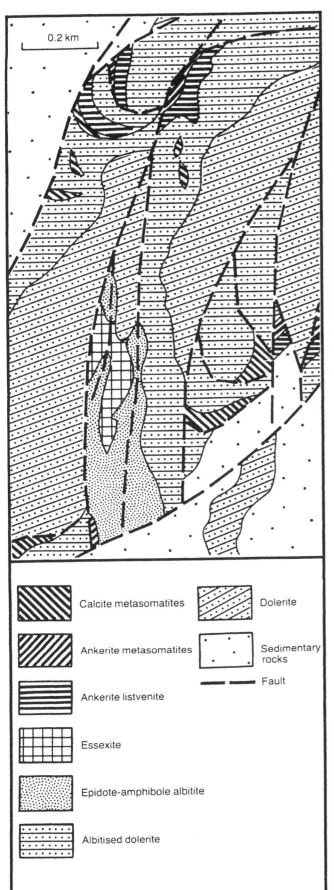

Fig. 106. Kyidinskii (after Gulin, 1970, Fig. 1).

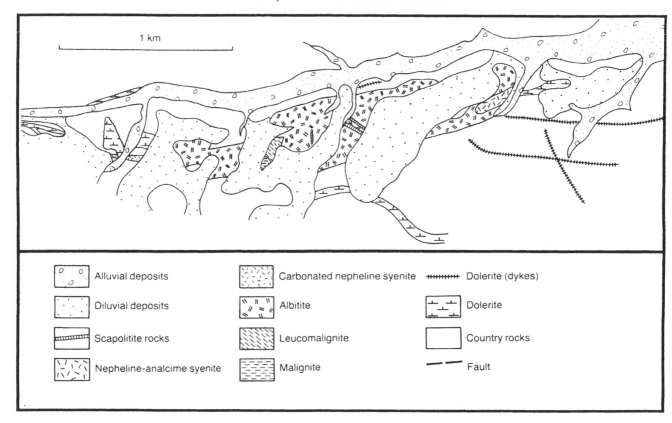

Fig. 108. Fad'yu-Kuda (after Ravich and Chaika, 1959, Fig. 21).

MAIMECHA-KOTUI

The area of development of the alkaline-ultrabasic rocks of the Maimecha-Kotui province comprises the western slope of the Anabar uplift (shield), from the margin of the middle Siberian platform in the north to the upper reaches of the Kotui River in the south. The alkaline rocks are concentrated close to the boundary of the platform with the Taimyr Depression in a zone which is intensely faulted. The province is one of the most extensive alkaline provinces in the world, extending over an area of 220 × 350 km (Figs 109 and 110). North of Maimecha-Kotui are the occurrences of the Taimyr Peninsula and to the northeast those of the Anabar Province (Fig. 110).

The first finds of alkaline rocks date back to 1937–38 and in 1940 G.G. Moor predicted the existance of a province of alkaline-ultrabasic rocks in the north of central Siberia. In 1943–44 Yu.M. Sheinman (1947) discovered the massifs of Guli, Bor-Uryakh, Dalbykha and Changit and towards the end of the 1940s and the beginning of 1950s Moor (1957) found some large complexes on the banks of the Kotui River. This unique province has now been investigated by numerous geologists, much of whose work is referred to in the following descriptions, but the fullest contribution has been that of Egorov, amongst whose numerous publications is a comprehensive review volume (Egorov, 1991). There is a brief review, in English, in Butakova (1974).

This province is particularly noteworthy not only for the abundance of carbonatites but for the association of many of them with ultramafic rocks. The rock types represented include dunite, pyroxenite, ijolite, melteigite, jacupirangite, phoscorite, a range of melilite-bearing rocks and numerous and varied carbonatites. It contains by far the greatest number and diversity of complexes of this type which, apart from those of the Kola Peninsula, are rare in the rest of the world. The Guli intrusion, which is probably the largest alkaline complex in the world, is a particularly fine example of this genre. The general geology of the province is shown in Fig. 109.

1 Extrusives of Maimecha-Kotui	19 West satellite of Changit *Dalbykha group of intrusions*
2 Guli	20 Kyndyn
3 Debkogo	21 Urukit
4 Nemakit and satellite plugs	22 Dalbykha-North
5 Odikhincha	23 Dalbykha
6 Churbuka	24 Dalbykha-South
7 Sona	25 Bykhyt-North
8 Sona-West	26 Bykhyt-West
9 Kugda	27 Bykhyt-East
10 SW satellite of Kugda	28 Chara
11 Ary-Mas	29 Saga Chara
12 Krestyakh	30 Bor-Uryakh
13 Sedete	31 Kara-Meni
14 Atyrdyakh	32 D'ogd'oo
15 Romanikha	33 Magan
16 Eastern satellite of Romanikha	34 Yraas
17 Changit	35 Essei
18 South satellite of Changit	

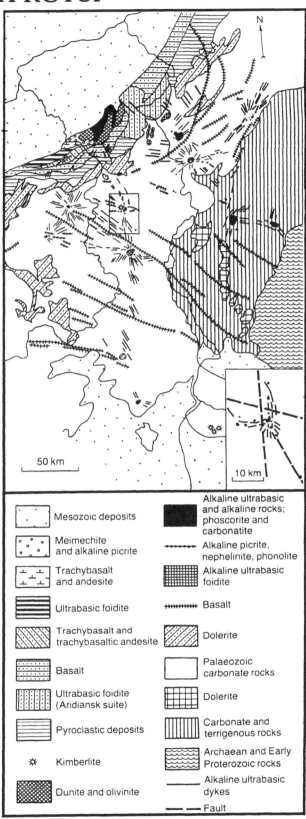

Fig. 109. General geology of the Maimecha-Kotui province (after Egorov, 1991, Fig. 2). The inset diagram shows the Dalbykha group of intrusions, which are located in the centre of the province.

106

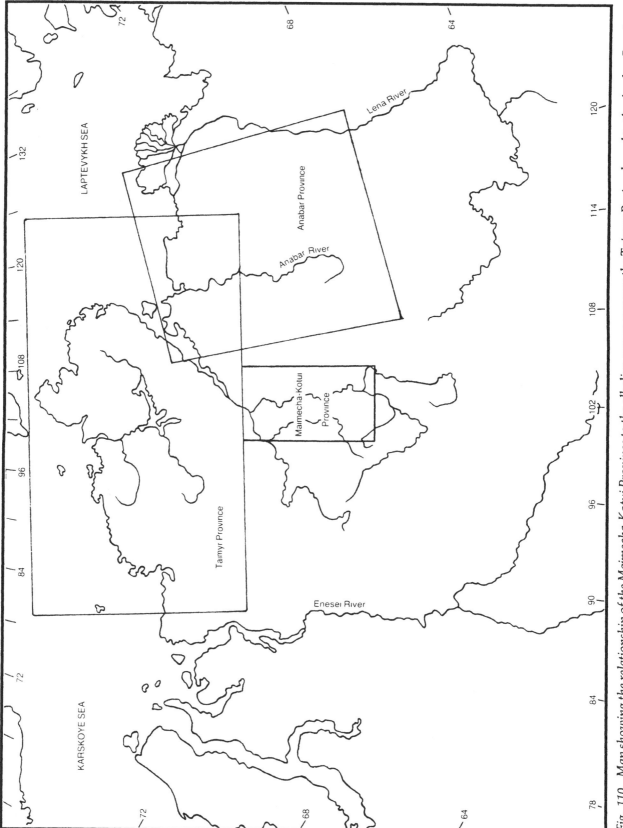

Fig. 110. *Map showing the relationship of the Maimecha-Kotui Province to the alkaline occurrences on the Taimyr Peninsula and to the Anabar Province. The areas illustrated in Figs 105 (Taimyr), 109 (Maimecha-Kotui) and 136 (Anabar) are indicated.*

1 EXTRUSIVES OF MAIMECHA-KOTUI

Western area 70°49′N; 95°55′E
Eastern area 71°14′N; 102°20′E
Figs 109 and 111

Extrusive volcanic rocks are developed extensively in the northern part of the Maimecha-Kotui province along the margin of the mid-Siberian platform with the Taimyr Depression (Fig. 109) in the vicinity of the Guli complex (No. 2). The overall sequence is a thick one and a number of distinct suites can be recognized (Fig. 111, inset diagram). The Aridiansk and Pravoboyarsky suites lie on middle Palaeozoic rocks, which are predominantly limestones. The Pravoboyarsky rocks, from 250 to 400 m thick, are basic pyroclastics and the 250–500 m thick Aridiansk suite comprises ultramafic alkaline rocks. At the base of the Aridiansk succession there is 25–50 m of alkaline ultrabasic tuffs above which are flows of melanephelinite, melilitite, olivine melanephelinite, limburgite and alkaline picrite with occasional layers of tuffs and gravels. Above these two suites are rocks of the Kogotsky suite, the lower sequence (300–400 m) of which consists of basalts with the upper 350–500 m thick sequence formed mainly of trachybasalts and trachyandesitic basalts, but also including flows of dacite, trachyte, andesite, olivine melanephelinite and nephelinite. The uppermost Delcan suite is divided into two sequences. The 545 m thick lower sequence consists of melanephelinite and melilitite, most of which contain olivine, mela-analcimite, limburgite, augitite, alkaline picrite, basanite and tephrite. The upper sequence is 300–500 m in thickness and sharply divided from the lower by a horizon of trachytes. It consists principally of rocks of the trachybasalt–trachyte series but there are also flows of ultrabasic foidites, subalkaline picrite, andesite, trachyrhyolite and trachytic ignimbrites. At the mouth of the Delcan River the Delcan suite is overlain by high Mg ultrabasic lavas and tuffs (meimichites) which extend for more than 20 km along the right bank of the Meimecha River. Analyses of the melilite-bearing rocks, including REE and other trace elements, are presented and discussed by Gladkikh (1991).

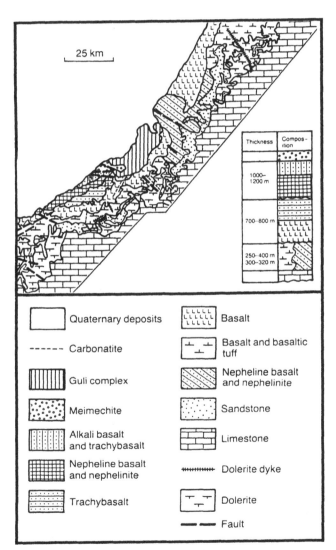

Fig. 111. *Distribution of extrusive rocks in the Maimecha-Kotui province. The inset stratigraphic column indicates the relationship of the principal suites of extrusive rocks which are, from the base upwards, (a) Aridiansk and Pravoboyarsky, (b) Kogotsky and (c) Delcan suites, the last being overlain by meimechitic lavas and tuffs.*

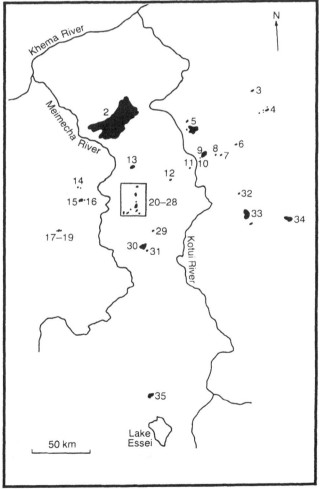

Fig. 112. *The distribution of occurrences of alkaline igneous rocks and carbonatites in the Maimecha-Kotui province. The rectangle encloses the Dalbykha group of intrusions, which are shown in greater detail in Fig. 125.*

Age K–Ar determinations on biotite gave 188–297 Ma (Egorov, 1991) and Rb–Sr and Sm–Nd isochrons on meimechite 239 ± 61 Ma (Kogarko et al., 1988).

References Egorov, 1991; *Gladkikh, 1991; Kogarko et al., 1988.

---

2 GULI
(Gulinskii)

70°57′N; 101°26′E

Figs 113 and 114

This complex occupies a large area between the Maimecha and Kotui Rivers at the boundary of the Siberian platform with the Hatanga trough. It has an oval form of 35 × 45 km and, including the two-thirds obscured by Quaternary deposits, has an area of 1500–1600 km² (Egorov, 1991). Geophysical evidence indicates near vertical contacts and hence probably a stock-like form (Egorov,1989). The complex was discovered by Sheinman (1947) in 1943 and at different periods was studied by Butakova (1956), Egorov (1991), Prochorova, Evzikova and Michailova (Prochorova et al., 1966), Vasiliev and Zolotuchin (1975), Zhabin (1965) and Kostyuk (1974). The country rocks of the Guli complex are volcanics (Figs 109 and 111), which have been described above (No. 1), and which include an extensive area of meimechites. The Guli massif, like many of the other alkaline-ultrabasic intrusives of the province, is a complex multi-stage pluton, as indicated in Table 5.

Table 5 Principal stages and rock types in the Guli complex

Intrusive stage	Sub-stage	Rock types
Sixth	Fourth	Dolomite carbonatite
	Third	Fine-grained calcite carbonatite
	Second	Coarse-grained calcite carbonatite
	First	Phoscorite and ore forsteritite
Fifth	Second	Micro-shonkinite and solvsbergite
	First	Peralkaline syenite
Fourth		Ijolite and ijolite pegmatite
Third	Third	Jacupirangite and melteigite
	Second	Melanephelinite, olivine melanephelinite, nepheline picrite, biotite–pyroxene picrite
	First	Melteigite, malignite, shonkinite
Second		Melilite rocks
First	Second	Ore pyroxenite, porphyritic olivine pyroxenite and peridotite
	First	Dunite

The predominant rocks of the complex are dunites, which occupy about 60% of the total area, and a range of types of melanocratic alkaline rocks, which extend over about 30%. All the other rock types, including melilitolite, ijolite, alkaline syenite and carbonatite, occupy less than 10% of the area of the complex. The earliest rocks of the intrusion (first stage) are dunites which form a curved area having a width of 9–10 km, which can be traced for 40 km. They are essentially olivine rocks with small amounts, less than 5% by volume, of clinopyroxene, titanomagnetite and chromite; accessory minerals include perovskite, phlogopite, calcite, clinohumite and spinel. During the second sub-stage the dunite intrusives were cut by numerous bodies of ore pyroxenite that are composed mainly of pyroxene and titanomagnetite, which form about 10% of the volume of the dunites; apatite and titanite are accessory. In the southwestern part of the dunite body the ore pyroxenite grades into texturally similar porphyritic trachytic peridotites and olivine pyroxenites, which have a lower content of titanomagnetite. Nepheline, biotite–phlogopite, perovskite, apatite and titanite are present as minor constituents. After the formation of the ultrabasic rocks a second stage of evolution of the pluton is marked by the intrusion of melilitic rocks. The melilite magma gave rise to a large (0.6 × 5 km) partial ring body in the southern structural centre of the massif and three smaller stocks 8 km to the southeast of the main ring. The composition of the rocks – melilitolite and kugdite – is characterized by the predominance of melilite over olivine, which in melilitolite is up to 90% of the volume. The accessory minerals are pyroxene, nepheline and titanomagnetite.

During the third intrusive phase a series of alkaline mafic and ultramafic rocks, which are compositionally similar, were emplaced in the following succession: sub-stage 1 melteigite–malignite–shonkinite; sub-stage 2 melanephelinite–alkaline picrite, and sub-stage 3 jacupirangite–melteigite. The most widespread rocks of this phase are the jacupirangites and melteigites which are characterized by the presence of pyroxene, in jacupirangite up to 90%, and nepheline, which in melteigite is from 10% up to 35–45%. Also present are titanomagnetite, phlogopite, apatite, titanite, perovskite and secondary natrolite, cancrinite and calcite, with sometimes up to 10% of K–feldspar in jacupirangite, which thus grades into malignite. The malignite and shonkinite of sub-stage 1 also have specific compositions and contents of pyroxene, which in the malignite is modally 35–45% while in the shonkinite, clinopyroxene and barkevikite comprise 60–70%. Nepheline forms 20–30% of the malignites and 0–7% in shonkinites; they also contain up to 25% of K–feldspar, up to 10–15% titanomagnetite, biotite, apatite, fluorite, albite, zeolites and calcite. The rocks of the first sub-stage form stocks within the dunites and are associated closely with the rocks of the second sub-stage. The rocks of the melanephelinite–alkaline picrite series (second sub-stage) are grouped into three large belts, which are up to 2 km across, and are also represented by dyke-like bodies of alkaline picrite (nepheline and biotite–pyroxene, as well as olivine). In the northeastern part of the complex there are olivine melanephelinites, in which there are shonkinite xenoliths, and to the east there are melanephelinites and olivine melanephelinite with nepheline picrites. In the southern part of the complex melanephelinites and olivine melanephelinites occur which sometimes have an ophitic texture. The jacupirangites and melteigites of the third sub-stage are cut by veins of ijolite and ijolite pegmatite which represent the fourth intrusive stage. The pegmatites are composed mainly of nepheline (55–65%) and clinopyroxene (25–40%) with accessory titanomagnetite, phlogopite, perovskite, titanite and apatite. In the centre of the complex are several bodies of fine- to medium-grained peralkaline syenite (the fifth stage) which are composed of up to 65% K–feldspar, aegirine and small amounts of nepheline.

Rocks of the sixth and final stage comprise essentially carbonatites and phoscorites. The phoscorite rocks (first sub-stage) consist of 20–30% olivine, 30–60% apatite and 25–50% magnetite, and are present as xenoliths in

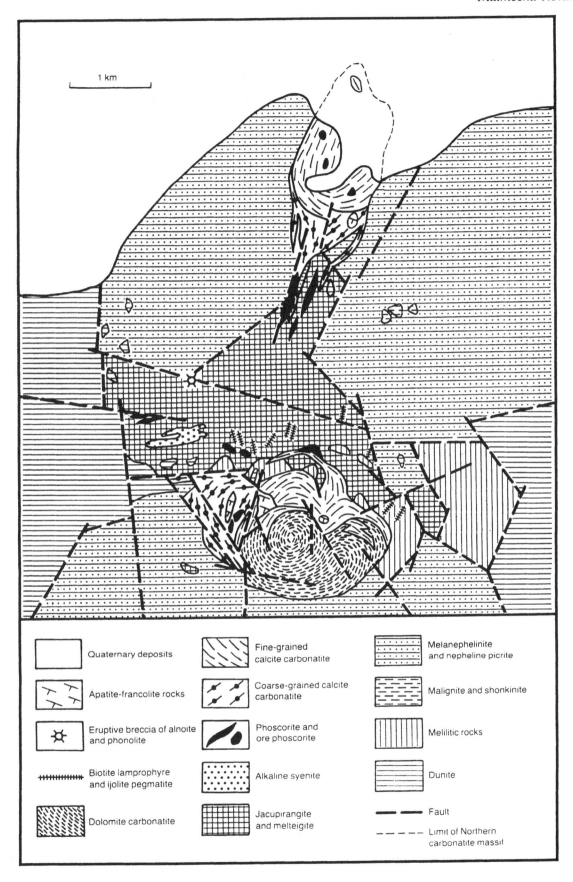

1 km

Quaternary deposits

Apatite-francolite rocks

Eruptive breccia of alnoite and phonolite

Biotite lamprophyre and ijolite pegmatite

Dolomite carbonatite

Fine-grained calcite carbonatite

Coarse-grained calcite carbonatite

Phoscorite and ore phoscorite

Alkaline syenite

Jacupirangite and melteigite

Melanephelinite and nepheline picrite

Malignite and shonkinite

Melilitic rocks

Dunite

Fault

Limit of Northern carbonatite massif

Fig. 113. The central part of the Guli complex (after Egorov, 1991, Fig. 7).

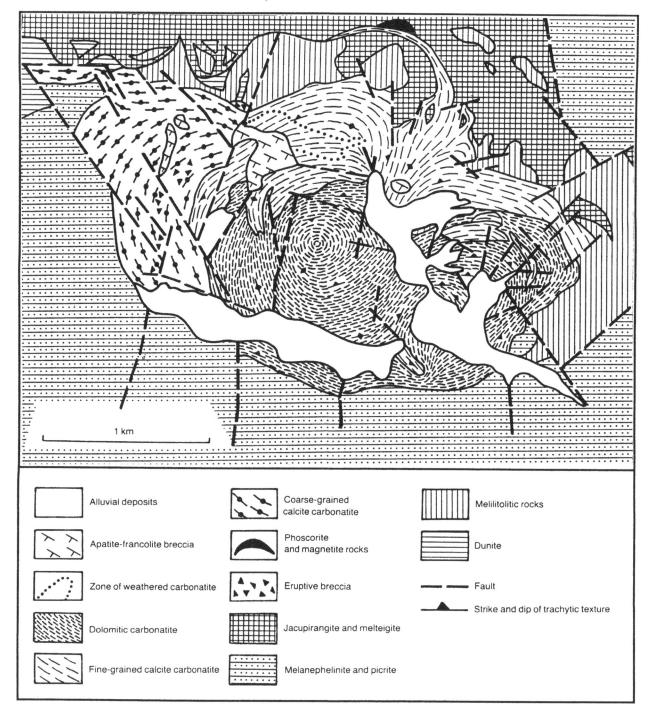

Fig. 114. The southern carbonatite massif of the Guli complex (after Egorov, 1991, Fig. 43).

calcite carbonatites (second and third sub-stages) and veins in melanephelinite and jacupirangite. The rocks of the carbonatite group form two massifs – the Northern and Southern (Fig. 114) and are represented by fine-grained calcite (third sub-stage) as well as dolomite carbonatite (fourth sub-stage). Apart from the dominant carbonates the carbonatites also contain apatite, magnetite, phlogopite, more rarely forsterite, aegirine-diopside, pyrrhotite, pyrochlore, dysanalyte, calzirtite and others.

Economic A deposit of coarse, laminated phlogopite

rock has been discovered (Sheinman, 1947; Prochorova et al., 1966). High concentrations of titanomagnetite in the pyroxenites and chromite in the dunites can also be regarded as potential ore deposits (Egorov et al., 1991).

Age Age determinations by the Rb–Sr and Sm–Nd methods gave 240 Ma (Kogarko et al., 1988). K–Ar determinations gave 206–349 Ma (Egorov, 1991).

References Butakova, 1956; Egorov, *1989 and 1991; Kogarko et al., 1988; Kostyuk, 1974; Prochorova et al.,

1966; Sheinmann, 1947; Vasiliev and Zolotuchin, 1975; Zhabin, 1965.

3 DEBKOGO   70°47′N; 104°19′E

This is a small oval body composed of melteigite and ijolite.

Reference Egorov, 1991.

4 NEMAKIT and SATELLITE PLUGS
71°04′N; 105°10′E
Figs 115, 116 and 117

The Nemakit complex is situated on the right bank of the Medvezhya River, a tributary of the Kotui River, 60 km east of the Odikhincha intrusion and is intruded into

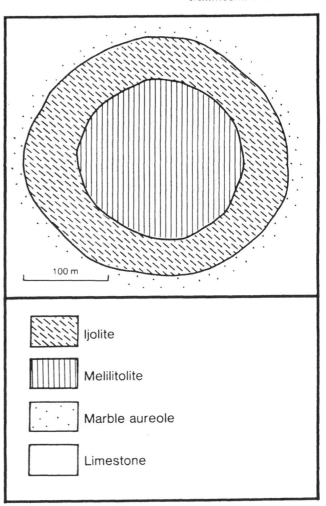

Fig. 116. *Satellite plug northeast of Nemakit (after Egorov, 1991, Fig. 17a).*

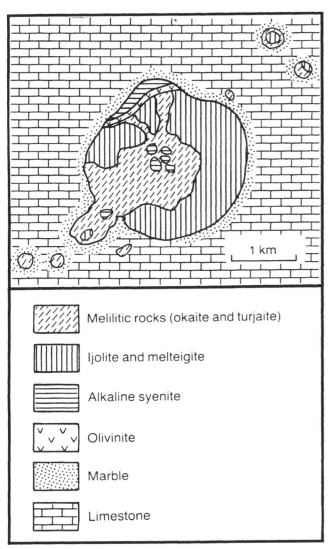

Fig. 115. *Nemakit (after Egorov, 1969, Fig. 7).*

and southwest of Nemakit. The one lying to the northeast (Fig. 116) is circular and has an area of 0.4 km². It comprises a core of melilitolite with peripheral nepheline syenite. A second plug (Fig. 117) contains the same rock types.

Reference Egorov, 1991.

5 ODIKHINCHA   70°58′N; 103°10′E
Fig. 118

This large complex intrusion, which was discovered by G.G. Moor and F.A. Starshinov in 1946, has since been investigated by many geologists. It has an area of 56 km² and is situated 60 km east of the Guli intrusion, on the eastern bank of the Kotui River. The country rocks are Lower and Middle Cambrian dolomites. The massif has a concentric structure controlled by the distribution of the various compositionally contrasting intrusive phases, which, from early to late, are olivinite, melilite rocks, jacupirangite–melteigite and ijolite. They are all cut by dykes and veins of ijolite pegmatite, microijolite, alkaline and nepheline syenites and calcite carbonatite. The olivinites are rocks of predominant olivine with small amounts of titanomagnetite and accessory clinopyroxene, perovskite, phlogopite, calcite, clinohumite, chromite and spinel. They are encountered as xenoliths within the later rocks. The melilite rocks are

lower Cambrian limestones, which are metamorphosed near the contact. According to the data of Moore, Ivanov and Safronov, the principal rocks are ijolite and melilite rocks, including okaite and turjaite, with a rather limited development of olivinite and a little alkaline syenite. Carbonatites also occur as narrow veins. A number of smaller plugs occur to the northeast

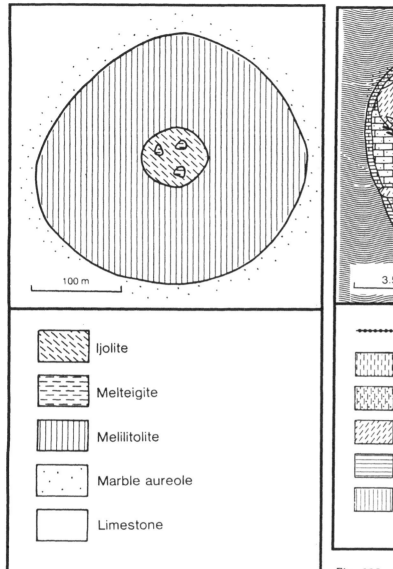

Fig. 117. Satellite plug southwest of Nemakit (after Egorov, 1991, Fig. 17b).

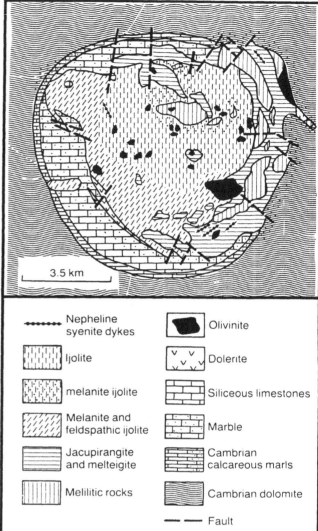

Fig. 118. Odikhincha (after Egorov, 1991, Fig. 13).

cut by later jacupirangite and melteigite and represented by monomineralic melilitolite, pyroxene uncompahgrite and pyroxene–nepheline turjaite. Olivine, titanomagnetite and accessory phlogopite, perovskite, garnet, calcite, apatite and amphibole are present in all rocks. The jacupirangite–melteigite series of rocks is found as a wide strip along all but the western periphery of the pluton and is also present as xenoliths in the later ijolites. The jacupirangites are composed mainly of clinopyroxene (augite and titanaugite); the melteigites also contain 20–30% of nepheline. Biotite, titanomagnetite, titanite, perovskite and accessory olivine, aegirine–augite, K–feldspar and phlogopite are generally present. A central stock having a diameter of 6.5 km and a small satellite west of the main intrusion are formed by ijolites. These rocks are composed of nepheline, pyroxene and accessory titanomagnetite, biotite–phlogopite, melanite, perovskite, titanite, apatite and calcite. Feldspathic ijolites are also developed. Ijolite pegmatites within melilite rocks are attributed to a separate sub-stage. They are phlogopite-bearing and of interest economically.

Dykes and veins of nepheline syenite, in which the main rock-forming minerals are nepheline, K–feldspar and pyroxene, appear to be the last intrusive phase. They include accessory biotite, apatite, amphibole, titanomagnetite, titanite, cancrinite and albite. In some localities these rocks have been recrystallized and enriched in the agpaitic minerals eucolite, lovchorrite (mosandrite), lamprophyllite and others. All the rocks of the complex contain lens-like bodies and dykes of calcite carbonatite, the main minerals of which are calcite, apatite, magnetite and phlogopite with accessory forsterite, aegirine-diopside, pyrrhotite and pyrochlore.

Economic The rocks of the massif are rare metal-bearing and phlogopite occurs in industrial concentrations in some rocks, as does apatite. The complex could be a source of semi-precious stones and chrysolite (Egorov, 1991; Prochorova *et al.*, 1966).

Age Ages of 204–233 Ma have been obtained by K–Ar on biotite (Prochorova *et al.*, 1966).

References Butakova and Egorov, 1962; Egorov, 1969 and 1991; Moor, 1957; Prochorova *et al.*, 1966.

6 CHURBUKA   71°00′N; 104°02′E

This is a small intrusion composed of melteigite and melanephelinite.

Reference Egorov, 1991.

7 SONA   70°43′N; 104°55′E

A small intrusion, oval in shape, Sona extends over some 0.25 km² and consists of melilitolite and melanephelinite.

Reference Egorov, 1991.

8 SONA-WEST   70°43′N; 104°46′E

Sona-west is a small oval-shaped body of about 0.07 km² composed of melteigite and malignite.

Reference Egorov, 1991.

9 KUGDA   70°43′N; 103°28′E
                                       Fig. 119

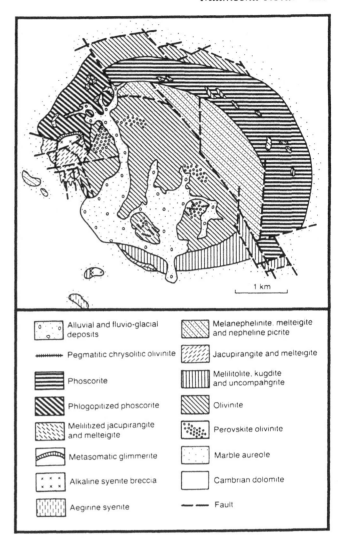

The Kugda complex is situated close to the eastern bank of the Kotui River, 25 km southeast of the Odikhincha complex and is located at the intersection of the Kotui and Kugda faults. The country rocks are Middle Cambrian dolomites which have a sub-horizontal attitude, with no evidence of disruption by emplacement of the intrusion. A metamorphic aureole 200–300 m wide surrounds the complex, which has a circular form and an area of about 16 km². According to magnetic survey data the contacts of the complex dip southeastwards. The morphology of the complex expresses the concentric disposition of the principal rock types, which are olivinite, jacupirangite–melteigite, melilite rocks and phoscorite breccias. Unlike the Odikhincha occurrence, ijolites are developed insignificantly; alkaline syenite and veins of calcite rock are present here. The rocks of the first intrusive stage, medium- to coarse-grained olivinites, compose a stock-like body in the nucleus of the complex, which has an area of 2×3 km. The olivinites consist mainly of weakly serpentinized olivine, titanomagnetite, chromite and spinel; calcite may also be present. A noteworthy feature of the complex is enrichment of the olivinites in ore minerals, the content of which is generally 5–10%, but in some places reaches 20–50%. A subhorizontal layering of the ore-rich olivinites is apparent, which is expressed by rhythmic ore-rich layers 0.2–2 cm thick. Schlieren composed of up to 30–40% titanomagnetite and perovskite are also encountered. Thick, vertical veins of pegmatitic olivinite with phlogopite, clinohumite and gem-quality chrysolite are localized in the middle zone of the olivinite nucleus. A marginal facies, with a width of up to 300–500 m, of fine-grained olivinite impoverished in titanomagnetite can be traced along the outer contact zone.

In the southern part of the complex the olivinite nucleus is separated from the country rock dolomites by an arcuate body of melilite rocks of the second intrusive stage. These rocks are composed of melilite, olivine, clinopyroxene and nepheline and, depending on the ratios of the constituent minerals, the melilite-bearing rocks can be classified as kugdite, melilitolite or uncompahgrite. Titanomagnetite, perovskite, phlogopite, titanium-bearing garnet, monticellite, apatite, wollastonite and calcite are present as minor phases. In the kugdites of the southern part of the body thin veins of turjaite and micro-turjaite, having intrusive contacts, have been found. The melilite rocks in the contact zone

Fig. 119. Kugda (after Egorov, 1991, Fig. 14).

with dolomite acquire a basaltoid appearance. The kugdites and uncompahgrites contain an abundance of olivine xenoliths. Melanocratic alkaline rocks having an area of outcrop of about 7 km² are widespread in the western and eastern parts of the intrusion and trachytic jacupirangite and melteigite, which pass gradually into melanephelinite and olivine melanephelinite, are dominant among these rocks. They are composed of pyroxene, varying in composition from diopside to aegirine–diopside, nepheline and titanomagnetite. Olivine, phlogopite, titanite, perovskite, melanite, apatite, cancrinite, zeolites and calcite are minor constituents. In the contact zones melteigite usually passes into jacupirangite. A post-magmatic process of nephelinization in some places has transformed the rocks into small patches of taxite ijolite. All the rocks of the jacupirangite–melteigite series contain xenoliths of olivinite and melilitic rocks.

Lens-like bodies (100–200 × 500–800 m) of ijolite are located in the southern part of the massif at the contact of the melilite rocks with dolomite. These rocks are composed mainly of pyroxene, varying from diopside to aegirine–diopside, and nepheline. Titanomagnetite, apatite, titanite, biotite and perovskite are the minor constituents. There are no sharp boundaries between the jacupirangite–melteigite and ijolite series within the Kugda complex. However, a sharp boundary does occur

Legend

Alluvial and fluvio-glacial deposits	Melanephelinite, melteigite and nepheline picrite
Pegmatitic chrysolitic olivinite	Jacupirangite and melteigite
Phoscorite	Melilitolite, kugdite and uncompahgrite
Phlogopitized phoscorite	Olivinite
Melilitized jacupirangite and melteigite	Perovskite olivinite
Metasomatic glimmerite	Marble aureole
Alkaline syenite breccia	Cambrian dolomite
Aegirine syenite	Fault

between trachytic melteigite and non-trachytic ijolite in the satellite of the Kugda massif (Fig. 120), which constitutes evidence of two-stage intrusion of a nepheline–pyroxene melt. Thus, the jacupirangite–melteigite suite could be attributed to the third, and those of ijolites to the fourth, intrusive stage.

A small stock of 200 × 500 m and series of dykes of peralkaline aegirine syenite cut the massif in its northwestern part across a direction defined by their trachytic fabric. They are surrounded by a wide contact zone of syenitized rocks. The syenites are composed of K–feldspar and pyroxene (aegirine, aegirine–augite); the minor minerals are nepheline, together with apatite, titanomagnetite, titanite, cancrinite, alkaline amphibole, calcite and zeolites. Direct contacts with ijolites have not been observed, but by comparison with other complexes it may be surmised that the body of alkaline syenites was emplaced later than the ijolites. A large body of phoscorite breccia was developed at the last stages of formation of the complex. This body can be traced as a wide (0.5–9.9 km) belt along the northern and eastern margins of the complex for 7.5 km. The breccia is composed of fragments, with diameters from tens up to hundreds of metres, of all the rocks of the massif, cemented by heterogeneous and at some places calcitized and phlogopitized forsteritites.

Economic Mica-enriched areas are present in the complex. Phlogopite is developed in veins and zones having nest-shaped forms. Gem-quality chrysolite is to be found (Egorov, 1991).

Age K–Ar on phlogopite gave 238 Ma (Egorov, 1991).

References Egorov, 1969 and 1991.

---

10 SOUTHWESTERN SATELLITE OF KUGDA
70°42′N; 103°25′E
Fig. 120

This is an oval body with an area of 0.6 km² located about 1 km from Kugda. It is composed of melteigite, melilitolite and nepheline syenite.

Reference Egorov, 1991.

---

11 ARY-MAS   70°37′N; 103°05′E

This is a circular intrusion of about 200 m diameter, composed of melteigite and melanephelinite.

Reference Egorov, 1991.

---

12 KRESTYAKH   70°31′N; 102°34′E

This is a small, oval body composed of rocks of the melteigite–melanephelinite series.

Reference Egorov, 1991.

---

13 SEDETE   70°35′N; 101°59′E

The Sedete intrusion is situated 25 km south of the Guli complex of which Egorov (1969) considers it to be a satellite. The alkaline rocks outcrop over an area of about 12 km² and the intrusion is sill-like in form. It was discovered in 1959 and first described by K.M. Shichorina (Egorov, 1991) with further investigations carried out by Egorov (1969). The country rocks are Devonian limestones and early Permian sandstones. The intrusion is layered, with the lowest horizon composed

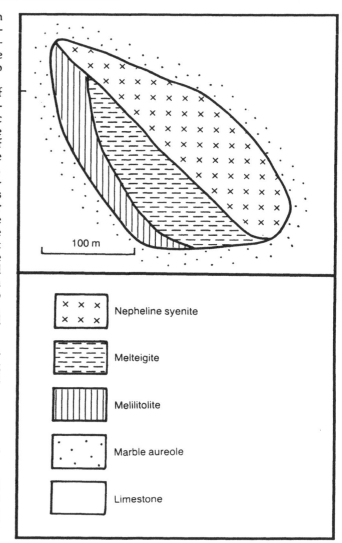

Fig. 120. *Southwest satellite plug of Kugda (after Egorov, 1991, Fig. 17g).*

of olivine melanephelinite (olivine 5–20%, clinopyroxene 30–65%, titanomagnetite 5–15% and K–feldspar 5–10%) and nepheline picrites, which have a higher content of olivine (up to 30–45%). These rocks grade gradually upwards into melteigites that are composed mainly of clinopyroxene (diopside–augite) with nepheline, biotite, titanomagnetite, titanite, perovskite and olivine, minor apatite, K–feldspar, phlogopite and rare aegirine–augite. The melteigites pass upwards through medium-grained malignites enriched in K–feldspar into nepheline syenites. The nepheline-bearing alkaline syenites, which include malignite and shonkinite, have pyroxene contents up to 60–70% by volume and titanomagnetite, biotite, apatite, olivine and albite are present in minor amounts. The layering is considered to have developed *in situ* after the intrusion of the melteigite magma and is illustrated by Egorov (1991, Fig. 8) in a cross-section.

Age An age of 163 Ma was determined by K–Ar on biotite (Prochorova *et al.*, 1966).

References Egorov, 1969, 1991; Egorov *et al.*, 1961; Prochorova *et al.*, 1966.

14 ATYRDYAKH   70°26'N; 100°48'E

Atyrdyakh was discovered in 1958 by Egorov and Surina (1961). It has an oval shape, an area of 0.5–0.6 km² and is emplaced in Upper Cambrian and Ordovician limestones. It is composed, like Changit (No. 17), of jacupirangite, melteigite and olivine melanephelinite of variable texture and mineralogy. The main minerals are augite and titanaugite (30–90%), nepheline (up to 35%), biotite, titanomagnetite, olivine and minor titanite, perovskite, aegirine–augite, apatite, K–feldspar and phlogopite. Towards the margin of the intrusion these rocks grade into a porphyritic facies of melanephelinite, olivine melanephelinite, nepheline picrite and augitite.

References Butakova and Egorov, 1962; Egorov, 1969; Egorov and Surina, 1961.

15 ROMANIKHA   70°20'N; 100°45'E
                                    Fig. 121

The largest intrusion on the western bank of the Meimecha River, Romanikha was discovered by Ya.I. Polkin and studied by Butakova and Egorov (1962). The country rocks are Upper Cambrian and Ordovician limestones. The intrusion has a symmetrical form, occupies an area of 4 km², and has a concentric, zonal structure. The approximately circular nucleus to the intrusion is 1.5 km in diameter and composed of jacupirangite and melteigite containing 50–90% of clinopyroxene (augite and titanaugite), 10–50% of titanomagnetite, nepheline, biotite and accessory titanite, perovskite, olivine, aegirine–augite, apatite, K–feldspar and phlogopite. The nucleus is surrounded by a broad ring-dyke of melanite ijolite, composed of nepheline, pyroxene, melanite, up to 10% titanomagnetite, biotite–phlogopite and accessory perovskite, titanite, apatite and calcite. These rocks contain, and in many cases contaminate, small xenoliths of melilite rocks which comprise melilite, nepheline, pyroxene and olivine. Melilitolite is also found in a small plug a kilometre east of the main intrusion. There are lens-like bodies, up to 100 × 500 m, of coarse-grained diopsidites which grade into typical phoscorites, the main minerals of which are carbonate, apatite and forsterite.

Economic In the diopsidites there are pods of coarse, lamella phlogopite, as well as occurrences of apatite-bearing rocks of the phoscorite series (Butakova and Egorov, 1962).

References Butakova and Egorov, 1962; Egorov, 1991.

16 EASTERN SATELLITE OF ROMANIKHA
                                70°20'N; 100°44'E
                                    Fig. 122

The central part of this circular, 0.9 km² intrusion is composed of nepheline syenite containing xenoliths of melteigite; the peripheral part is composed of melilitolites.

Reference Egorov, 1991.

17 CHANGIT   70°12'N; 100°06'E
                                    Fig. 123

The Changit intrusion is situated 20 km southwest of Romanikha and was discovered by P.S. Fomin in 1943 and studied by Egorov and Surina (1961). The country rocks are limestones of Upper Cambrian and Ordovician age. The intrusion is composed essentially of jacupiran-

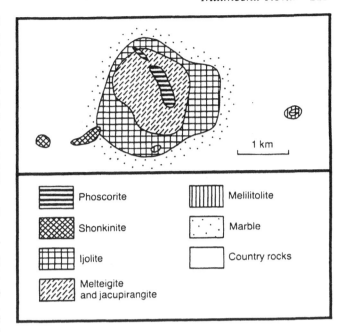

Fig. 121. Romanikha (after Egorov, 1991, Fig. 16a).

gite, melteigite and olivine melanephelinite of variable texture and composition. The main components are augite and titanaugite (30–90%), nepheline (up to 35%), biotite, titanomagnetite, olivine and minor titanite, perovskite, aegirine–augite, apatite, K–feldspar and phlogopite. The feldspathoidal rocks contain schlieren of melilitic rocks including turjaite and nepheline uncompahgrite, which are composed mainly of melilite, pyroxene, nepheline and olivine. Close to Changit are some small (from a few to tens of metres diameter) stock-like bodies and dykes of carbonatite and peralkaline syenites.

Economic The carbonatites contain pyrochlore (Egorov and Surina, 1961).

References Butakova and Egorov, 1962; Egorov, 1969; Egorov and Surina, 1961.

18 SOUTH SATELLITE OF CHANGIT
                                70°05'N; 100°10'E
                                    Fig. 124

This is a small, oval body located 300–600 m from Changit which is composed of calcite carbonatite with an alkaline syenite breccia developed at the periphery.

Reference Egorov, 1991.

19 WEST SATELLITE OF CHANGIT
                                70°12'N; 100°16'E

This small (70–100 m²) intrusion of carbonatite lies 1.5 km to the west of Changit.

Reference Egorov, 1991.

Dalbykha group of intrusions

The Dalbykha group includes 9 intrusive bodies varying from 0.25 to 2 km in diameter, which form a cluster 20 km across, located between the Kotui and Maimecha Rivers (Figs 109, 112 and 125) (Egorov, 1969). Their location is controlled by an intersection of the

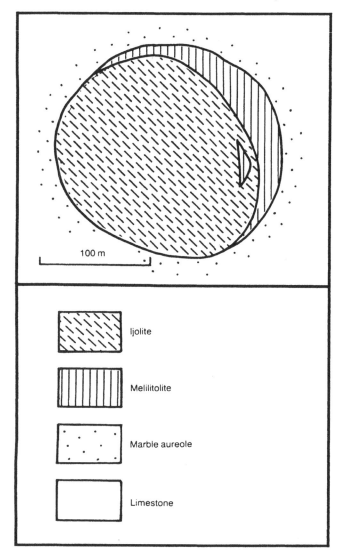

Fig. 122. Eastern satellite plug of Romanikha (after Egorov, 1991, Fig. 17v).

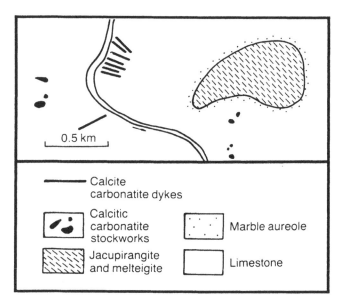

Fig. 123. Changit (after Egorov, 1991, Fig. 16b).

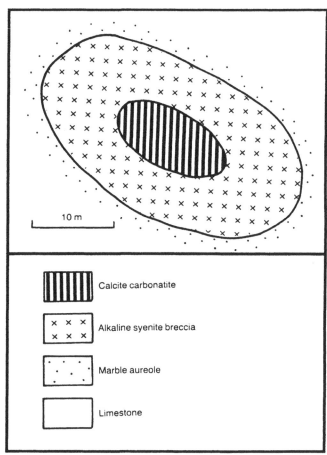

Fig. 124. South satellite of Changit (after Egorov, 1991, Fig. 17d).

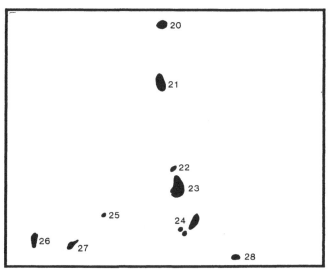

Fig. 125. The distribution of the Dalbykha group of intrusions (after Egorov, 1991, Fig. 9). For general location see Fig. 112.

Maimecha-Kotui and Dalbykha-Maimecha deep faults. The largest intrusion of the group, Dalbykha, is situated at the centre of the intersection. Apart from Dalbykha the group includes the three Bykhyt occurrences, Dalbykha-North, Urukit, Kyndyn, Dalbykha-South and Chara. The country rocks are Middle and Upper

Cambrian dolomites. This group of intrusions was discovered by Yu.M. Sheinman in 1944 (Sheinman, 1947) and studied by Butakova (Butakova and Egorov, 1962) and Egorov (1969).

---

20 KYNDYN

70°23′N; 102°00′E
Fig. 126

Situated 4 km north of the Urukit intrusion, Kyndyn forms a complex stock of 0.5×0.8 km which is oval in plan. It was formed during three intrusive stages. The early melanephelinites (melteigite porphyries) are composed mainly of clinopyroxene (70–80%) and minor olivine (0–5%), titanomagnetite, phlogopite, apatite, titanite and perovskite. These rocks are cut by a plug of ijolite in the centre of the intrusion and by an arc of the same rocks in the west. Veins of alkaline and nepheline syenites were formed during the last stage.

References Butakova and Egorov, 1962; Egorov, 1969.

---

21 URUKIT

70°22′N; 102°00′E
Fig. 127

Urukit is oval in outline and occupies an area of 0.5×1 km (Butakova and Egorov, 1962). It is composed of melteigite and ijolite which are cut by small bodies of

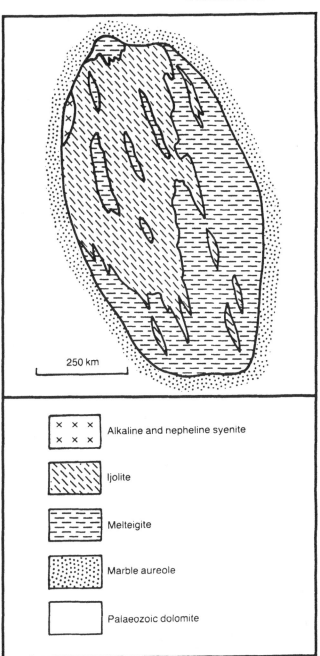

Fig. 127. *Urukit (Egorov, unpublished).*

nepheline syenite. The melteigites have a trachytic texture and are composed essentially of clinopyroxene, with small amounts of nepheline, biotite, titanomagnetite, titanite and perovskite. There followed intrusion of ijolite/nephelinite, which is a massively textured rock containing minor titanomagnetite, phlogopite, perovskite, titanite, apatite and calcite. The country rocks are dolomites, which are metamorphosed at the contacts.

References Butakova and Egorov, 1962; Egorov, 1969.

---

22 DALBYKHA-NORTH

70°20′N; 102°00′E

This is a stock-like body with an area of 0.2×0.25 km which is located 1.5 km north of Dalbykha. It is almost completely composed of melanephelinite. There are cross-cutting veins of ijolite which comprise 70–80%

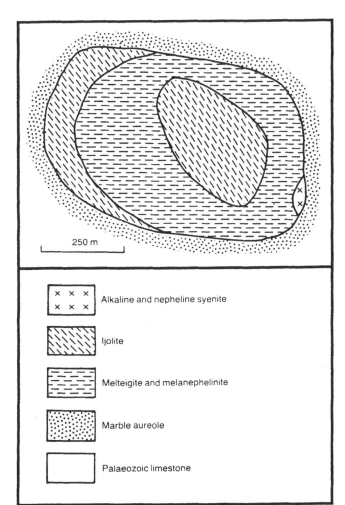

Fig. 126. *Kyndyn (Egorov, unpublished).*

clinopyroxene, up to 5% of olivine and minor titano-magnetite, phlogopite, apatite, titanite and perovskite.

Reference Egorov, 1969.

---

23 DALBYKHA
70°19′N; 102°01′E
Fig. 128

Dalbykha is a complex intrusive body having a zonal ring structure. It occupies an area of about 3 km² (Butakova and Egorov, 1962; Bagdasarov and Danilin, 1982). The central stock of the complex (0.3 × 0.5 km) is composed of melteigite having clinopyroxene as the principal mineral (augite and titanaugite) and with nepheline, biotite, titanomagnetite, titanite and perovskite as well as minor olivine, aegirine–augite, apatite, phlogopite and K–feldspar. Ijolite forms a broad ring around the central stock and consists of nepheline (55–65%), clinopyroxene (25–40%) and titanomagnetite, biotite, phlogopite, melanite, perovskite, titanite and apatite as minor constituents. Later nelsonite and phoscorite containing apatite, magnetite, tetraferriphlogopite and pyrochlore were formed and occur as veins in the older rocks and xenoliths in the younger ones. Younger still are coarse-grained biotite and biotite–pyroxene–calcite carbonatites which form an incomplete outer ring to the complex. These carbonatites are cut by finer grained calcite carbonatites containing tetraferriphlogopite and richterite. A francolite-rich weathered crust extends over Dalbykha and there are numerous radial dykes around it. The dykes comprise phonolite, nephelinite, alnoite, monchiquite and melilite nephelinite.

Economic A large apatite deposit and promising occurrences of rare metals mineralization are known to be present (Butakova and Egorov, 1962; Bagdasarov and Danilin, 1982).

Age Age determinations by the K–Ar method ranged from 167 to 363 Ma (Prochorova *et al.*, 1966).

References Bagdasarov and Danilin, 1982; Butakova and Egorov, 1962; Prochorova *et al.*, 1966.

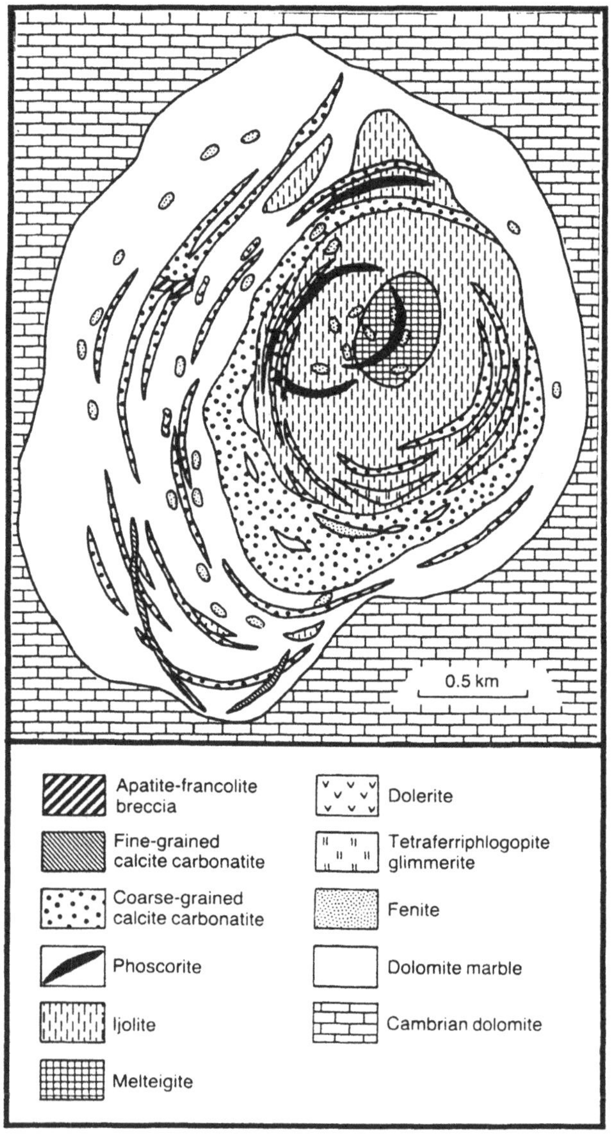

Fine-grained calcite carbonatite
Coarse-grained calcite carbonatite
Phoscorite
Ijolite
Melteigite
Apatite-francolite breccia
Dolerite
Tetraferriphlogopite glimmerite
Fenite
Dolomite marble
Cambrian dolomite

0.5 km

Fig. 128. Dalbykha (after Egorov, 1991, Fig. 10).

24 DALBYKHA-SOUTH
70°18′N; 102°00′E

This includes three distinct bodies, which are situated 1.5 km south-southeast from the Dalbykha complex. They range in size from 0.1 × 0.2 to 0.3 × 1 km. The central part of the largest intrusion is composed of olivine melanephelinite, which contains 30–65% of clinopyroxene, 5–20% of olivine and minor titanomagnetite, phlogopite, apatite, titanite and perovskite. An ijolite dyke forms a ring around the intrusion which, in places, contains xenoliths of melteigite and fine-grained olivinite. The ijolites are cut by dykes of alkaline syenite.

References Butakova and Egorov, 1962; Egorov, 1969.

---

25 BYKHYT-NORTH
70°19′N; 101°56′E

Lying between Dalbykha and the Bykhyt-East occurrence, Bykhyt-North consists of a central stock of melanephelinite which is surrounded by a ring-dyke of nepheline syenite. The melanephelinite comprises 70–80% clinopyroxene with small amounts of olivine (up to 5%), titanomagnetite, phlogopite, apatite, titanite and perovskite.

Reference Egorov, 1969

---

26 BYKHYT-WEST
70°18′N; 101°53′E
Fig. 129

The intrusion is situated 8 km southwest of Dalbykha (Egorov, 1969; Butakova and Egorov, 1962) and is a concentrically zoned stock having an elongate, oval form. The earlier rocks grade into later ones towards the centre of the intrusion. Olivinite was the first phase to be intruded and forms an outer ring-dyke. Melilitolite was emplaced later and these earlier rocks are disrupted and fragmented by younger intrusions of melteigite, jacupirangite and melanephelinite. All these rocks are cut by ijolite in the western part of the intrusion. The formation of the complex was completed by intrusion of peralkaline syenite, which forms a central stock. Numerous veins of peralkaline and nepheline syenites cut all rocks.

References Butakova and Egorov, 1962; Egorov, 1969.

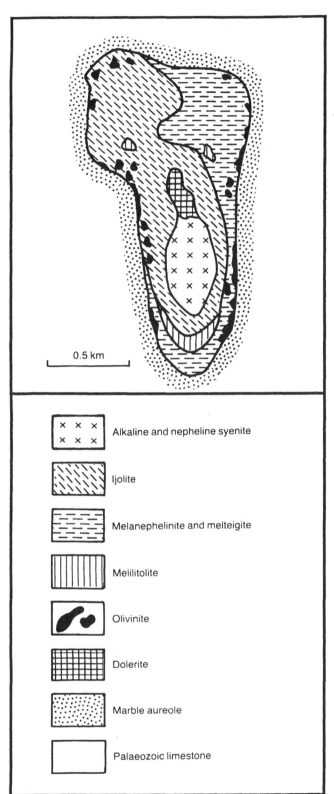

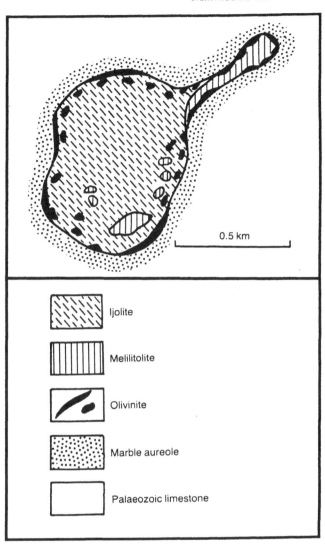

Fig. 130. Bykhyt-East (after Egorov, 1991, Fig. 11b).

Fig. 129. Bykhyt-West (after Egorov, 1991, Fig. 11a).

27 BYKHYT-EAST   70°18′N; 101°56′E
                                                   Fig. 130

This intrusion is situated 2.5 km east of Bykhyt-West (Butakova and Egorov, 1962; Egorov, 1969) and is oval in plan with a narrow northeasterly extension. The succession of intrusions is the same as in Bykhyt-West, that is olivinite, which is concentrated towards the margins, melilitolite, melteigite, ijolite, nepheline and alkaline syenite. The largest part of the massif is occupied by ijolite. The narrow northeasterly extension of the complex is composed of okaite and melilitolite with small amounts of nepheline, pyroxene and olivine. All the other rock types are represented mainly by xenoliths and veins.

References Butakova and Egorov, 1962; Egorov, 1969.

28 CHARA   70°18′N; 102°02′E

This small intrusion of 1.2 km² has a nucleus of melteigite in which there are inclusions of porphyritic melanephelinite and olivine melanephelinite. A ring-dyke of ijolite encircles the nucleus and in the northern part of the intrusion melilitic rocks are developed. There are veins of both ijolite and syenite.

Reference Egorov, 1991.

29 SAGA-CHARA   70°17′N; 102°08′E

This small intrusion is composed of ijolites.

Reference Egorov, 1991.

---

30 BOR-URYAKH   70°00′N; 102°07′E
Fig. 131

Bor-Uryakh is situated 40 km south of the Dalbykha group of intrusions at the intersection of the Maimecha-Kotui and Ambardach faults. It was discovered in 1943–44 by Yu.M. Sheinman and has been studied at different times by E.L. Butakova, L.S.Egorov, A.G. Zhabin, Yu.R. Vasiliev and V.V. Zolotuchin. The complex has a pear-shaped form and occupies an area of 18.5 km². The country rocks are dolomites, sandstones, marls, siltstones and marbles. It is an olivinite stock of simple structure with an outer zone of dunite. Throughout the intrusion there are numerous dykes and veins which are mainly ore olivinite pegmatites and peralkaline syenites, with more rarely jacupirangite pegmatites, melteigites and calcite carbonatites. Mineralogically the rocks comprise combinations of olivine, titanomagnetite and highly variable modal proportions of clinopyroxene, perovskite, chromite, phlogopite, calcite, clinohumite and spinel. In some places the content of ore minerals (titanomagnetite, perovskite and chromite) increases from 5–10% up to 20–25% and even up to 50%. 'Quenched', more fine-grained olivinites, which are enriched in magnetite and chromite, are found in the marginal zones of the complex.

Economic Deposits of phlogopite, rocks enriched in titanomagnetite and apatite, and gem-quality chrysolite are associated with the complex (Egorov, 1991).

Age Ages from 215 to 225 Ma were obtained by K–Ar on biotite (Prochorova *et al.*, 1966).

References Butakova and Egorov, 1962; Egorov, 1969 and 1991; Sheinman *et al.*, 1961; Vasiliev and Zolotuchin, 1975.

---

31 KARA-MENI   69°57′N; 102°10′E
Figs 131 and 132

This is a small intrusion of 0.6 × 1.2 km emplaced in Cambrian dolomites and situated 2 km southeast of the Bor-Uryach complex (see Fig. 131) (Butakova and Egorov, 1962). It consists mainly of melteigite and nepheline picrite, the former comprising 50–90% clinopyroxene, nepheline, biotite and titanomagnetite with accessory titanite, perovskite, olivine, apatite and K–feldspar. About one-third of the intrusion comprises earlier melilitolites (okaites) and olivine turjaites. The okaites consist of melilite and nepheline with accessory olivine, pyroxene, titanomagnetite, biotite, perovskite, apatite and others. Olivinite has been found only as small xenoliths in melteigite.

References Butakova and Egorov, 1962; Egorov, 1969.

---

32 D'OGD'OO   70°27′N; 104°20′E

This small 300 m diameter intrusion consists of melteigite.

Reference Egorov, 1991.

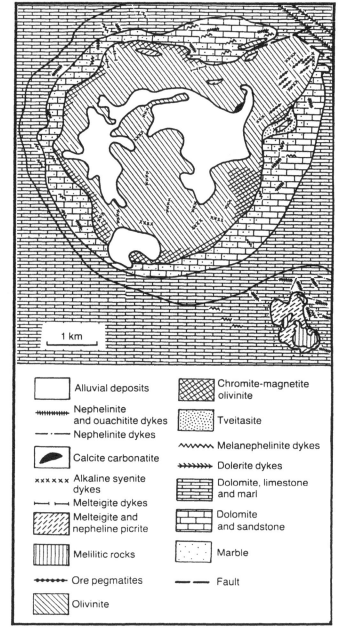

☐ Alluvial deposits	▨ Chromite-magnetite olivinite
┉┉┉ Nephelinite and ouachitite dykes	⬚ Tveitasite
─·─ Nephelinite dykes	
◣ Calcite carbonatite	∿∿∿ Melanephelinite dykes
	⋙ Dolerite dykes
×××× Alkaline syenite dykes	▦ Dolomite, limestone and marl
├─┤ Melteigite dykes	
▨ Melteigite and nepheline picrite	▦ Dolomite and sandstone
▥ Melilitic rocks	⬚ Marble
●●●● Ore pegmatites	── Fault
▨ Olivinite	

Fig. 131. Bor-Uryach, with the small intrusion of Kara-Meni (No. 31 and Fig. 132) to the southeast (after Egorov, 1991, Fig. 12).

33 MAGAN   70°15′N; 105°25′E
Fig. 133

Magan is the largest intrusion in the southern part of the province and is situated on the watershed of the Magan and Diogzho Rivers. It was discovered in 1954 and at different times scientific investigations have been carried out by, amongst others, G.G. Moor, L.S. Egorov, E.L. Butakova, Yu.A. Bagdasarov and S.M. Kravchenko. The country rocks are Riphean carbonate rocks of the Kotuikan suites underlying which are quartz sandstones of the Mukun series. The latter are metamorphosed at contacts with the intrusion and transformed into apatite-bearing rocks of the aegirine fenite series. The Magan intrusion has an irregular oval form, which is elongated in a north–south direction and an area of over 50 km². The massif is composed of a series of

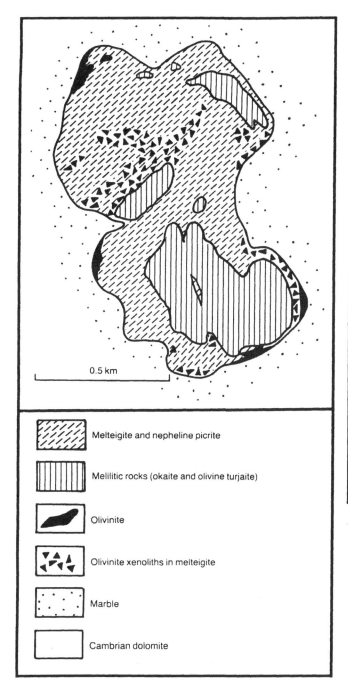

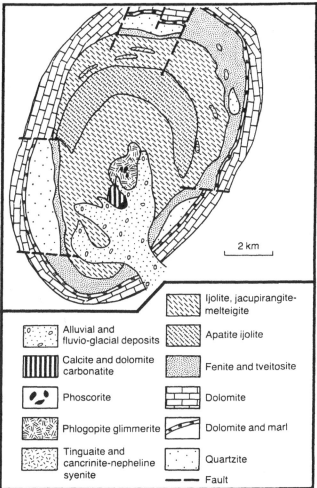

Fig. 132. Kara-Meni (after Egorov, 1991, Fig. 12).

Fig. 133. Magan (after Egorov, 1991, Fig. 18).

intrusive and metamorphic/metasomatic rocks, namely jacupirangite, melteigite, jacupirangite–melteigite metasomatic rocks, ijolite, phlogopite-bearing ijolite pegmatite and micro-ijolite, nepheline and cancrinite syenites, rocks of the phoscorite series, including magnetite rock, ore forsteritite and phoscorite, dolomite–calcite and dolomite carbonatites. The olivinites are found as small altered phlogopitized xenoliths among the nepheline–pyroxene rocks and contain olivine and titanomagnetite as the main minerals with clinopyroxene, perovskite, phlogopite, clinohumite, spinel, chromite and calcite as accessories. The rock-forming minerals of the jacupirangite and melteigite are clinopyroxene (up to 90%), nepheline, biotite, titanomagnetite, titanite and perov-

skite; the minor constituents are olivine, aegirine–augite, apatite, melanite, K–feldspar, phlogopite, zeolites, calcite, cancrinite and tremolite. The ijolites, which comprise more than half of the area of the complex, contain areas of olivinite and jacupirangite. The ijolites contain up to 60% nepheline and up to 50% pyroxene as the main minerals. Titanomagnetite, biotite–phlogopite and melanite are present in minor amounts and accessory minerals are perovskite, titanite, apatite, zeolites and calcite. In many localities the ijolites are cut by veins of phlogopite-bearing ijolite pegmatite and by later dykes of micro- ijolite. Arcuate and lens-like bodies of cancrinite–nepheline syenite are controlled by ring faults in the northern half of the complex. The youngest rocks of the complex are concentrated in the central southern part where the centre is occupied by a 1 km diameter stock-like body of dolomite carbonatite, which on its northern margin borders a body of similar size of magnetite rocks and phoscorites composed of forsterite, apatite and carbonate. South of the stock the jacupirangites, melteigites and ijolites are cut by a series of steeply dipping veins of pegmatitic ore forsteritites. There is no extensive rare element mineralization in the carbonatites but analytical data for rare elements in a range of rock types are given by Bagdasarov (1987).

Economic There are deposits of phlogopite (Goldburt and Landa, 1963) and apatite developed in the outer contact of the complex (Bogaditsa *et al.*, 1983; Egorov,

1960). The apatite resources are discussed by Bagdasarov (1987), and Kravchenko *et al.* (1987) describe the
petrology of the apatite deposits.

Age Ages of 233–245 Ma have been obtained by K–Ar
(Egorov, 1990).

References *Bagdasarov, 1987; Bogaditsa *et al.*, 1983;
Egorov, 1960, 1968, 1980 and 1991; Goldburt and
Landa, 1963; Kravchenko and Bagdasarov, 1987;
*Kravchenko *et al.*, 1987; Moor, 1957.

---

34 YRAAS   70°15′N; 105°10′E
                                              Fig. 134

Yraas is situated 30 km east of the Magan pluton at the
watershed of the Magan and Ilya Rivers and was
discovered in 1961. The country rocks are Riphean
quartz sandstones of the Labaztach Series. The intrusion
has a somewhat complicated outline with many fault-
bounded contacts. Overall dimensions are about
5 × 2.7 km. It is a complex, multistage intrusive comprising two parts: the northwestern part consists sub-
stantially of ijolites and the southeastern part is
represented essentially by intrusive breccias. The ijolites
occupy a lens-like body of 0.5 × 1.7 km and are homogeneous, medium-grained rocks composed mainly of
nepheline and pyroxene (up to 90%) and lesser amounts
of titanomagnetite, biotite–phlogopite and melanite
(up to 10%). The accessory minerals are perovskite,
titanite, apatite and calcite. In some places the ijolite
contains blocks of weakly altered, coarse-grained
melteigite and in places apatite-bearing aegirine fenite.
Veins of pyroxene nelsonite, nepheline syenite and
ore-bearing diopsidite are widespread. The intrusive
breccia in the southeastern part of the complex occupies
an area of 6.3 km², contains numerous fragments of
basement rocks, including two pyroxene and garnet

plagio-gneisses, granulite and schist in a heterogeneous
nelsonite–carbonatite cement (apatite, magnetite and
carbonate). Large basement blocks, up to hundreds of
metres across, which are only partly displaced, have
been found in the western part of the intrusive breccia.
In the northern and eastern parts of the breccia body
more than 50% of the gneisses have been transformed
into carbonate–mica metasoma tites. The Riphean quartz
sandstone country rocks at the margin of the northwestern ijolite body have been metamorphosed, with a
150–200 m wide zone of these rocks transformed into
apatite- bearing aegirine fenites. Thus, the generation of
the Yraas complex took place in five stages. The first
three involved the successive emplacement of melteigite,
ijolite and nepheline syenite, with fenitization of quartz
sandstone. Finally, the intrusion of a fluid saturated in
iron and phosphate (nelsonite) and carbonatite magmas
triggered violent explosions within basement rocks and
the formation of an intrusive explosion breccia.

Economic Apatite deposits have been discovered in
rocks of the phoscorite series (Egorov, 1991).

References Egorov, 1969 and 1991; Prochorova *et al.*,
1966.

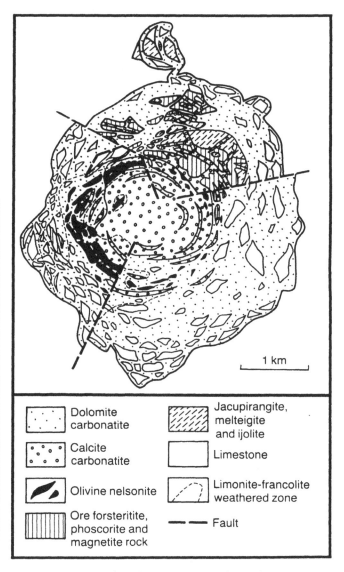

⣿ Dolomite carbonatite	▨ Jacupirangite, melteigite and ijolite
⣿ Calcite carbonatite	☐ Limestone
⬬ Olivine nelsonite	⌇ Limonite-francolite weathered zone
⦀ Ore forsteritite, phoscorite and magnetite rock	— — Fault

Fig. 135. *Essei (after Egorov, 1991, Fig. 20).*

☰ Calcite carbonatite	⣿ Sandstone
△△△ Intrusive breccia	▦ Gneiss
⫽ Ijolite	⣶ Brecciated gneiss
⣿ Aegirine fenite	— — Fault

Fig. 134. *Yraas (after Egorov, 1991, Fig. 19).*

35 ESSEI   68°42′N; 102°08′E
                                        Fig. 135

The Essei complex is located 30 km north of Essei lake and is the most southerly occurrence of the Maimecha-Kotui Province. The complex and an apatite deposit associated with it have been investigated by Porshnev and Suvalova (1970), Egorov (1991) and Kravchenko and Bagdasarov (1987). According to geophysical and exploration data the complex has an oval form and dimensions of 3.5 × 4.2 km. The country rocks are mainly limestones, including dolomites, which are cut by dolerite dykes. The complex has a symmetrical, concentrically zoned structure and is composed principally of carbonatite and rocks of the phoscorite series with jacupirangite, melteigite and ijolite occupying less than 0.7 km². Relicts of olivinite have been encountered in jacupirangite and melteigite. The jacupirangite and melteigite, which appear to form a discontinuous, steeply dipping body, are cut by veins of phoscorite and carbonatite and are also known from xenoliths in the younger rocks. The main minerals of the jacupirangite and melteigite are clinopyroxene (up to 90%), nepheline, biotite, titanomagnetite, titanite and perovskite with minor olivine, aegirine–augite, apatite, melanite, K–feldspar, phlogopite, zeolites and cancrinite. The ijolites are composed of 50–60% nepheline, 40–50% pyroxene with titanomagnetite, biotite–phlogopite, melanite and accessory perovskite, titanite, apatite, cancrinite, zeolite and calcite. The rocks of the phoscorite series form two discontinuous belts in the northern central area of the complex. The more northerly belt comprises magnetite phoscorite rich in forsterite, whereas the more southerly belt contains nelsonites. The principal minerals of the phoscorites are calcite, apatite, magnetite, forsterite and biotite with accessory baddeleyite and zircon. Panina and Shatskiy (1974a) studied primary melt and liquid-gas inclusions in apatite and forsterite from these rocks and inclusions in nepheline and pyroxene from silicate rocks (Panina and Shatskiy, 1974b). Dolomite carbonatite forms a central circular stock and a series of arcuate dykes and veins. It is composed of dolomite, apatite, magnetite and minor forsterite, aegirine–diopside, pyrrhotite and pyrochlore. In the very centre of this body the carbonatite is almost free of none-carbonate minerals. The dolomite stock is surrounded by a 0.5–1.5 km wide calcite carbonatite, in which calcite is the sole carbonate mineral; olivine, magnetite, apatite and phlogopite are also present in these carbonatites.

Economic A large apatite–magnetite deposit has been located in the phoscorites (Egorov, 1991). There is also a phosphate-rich weathering zone (Danilin *et al.*, 1982).

References Bogaditsa *et al.*, 1983; Danilin *et al.*, 1982; Egorov, 1969, 1980 and 1991; Kravchenko and Bagdasarov, 1987; *Panina and Shatskiy, 1974a and 1974b; Porshnev and Shuvalova, 1970.

Maimecha-Kotui references

*BAGDASAROV, Yu.A. 1987. Geologic and geochemical features of the apatite–iron-ore rocks and the carbonatites of the Magan pluton. *International Geology Review*, **29**: 264–80.

BAGDASAROV, Yu.A. and DANILIN E.L. 1982. The carbonatite massif of Dalbycha. *Doklady Akademii Nauk SSSR*, **267**: 1440–4.

BOGADITSA, V.P., BRAGINA, V.I., GERDT, A.A., DANILIN, K.L., EGOROV, L.S., KAZITSKY, M.L., KOROLEV, E.M., MALISHEV, A.A., MATUKHINA, V.G., MKRTCHYAN, A.K., RUSAKOV,

D.K., SERGEEV, V.L., SHALMINA, G.G. and SHERMAN, M.L. 1983. In Y.N. Zanin (ed.) *The apatite deposits of the Maimecha-Kotui province and their geological and economic assessment*. Siberian Branch of the Academy of Sciences, Novosibirsk. 83 pp.

BUTAKOVA, E.L. 1956. On the petrology of the Maimecha-Kotui complex of ultrabasic and alkaline rocks. *Trudy Nauchno-Issledovatel'skogo Instituta Geologii Arktiki, Moscow*, **89**: 201–49.

*BUTAKOVA, E.L. 1974. Regional distribution and tectonic relations of the alkaline rocks of Siberia. *In* H. Sorensen (ed.), *The alkaline rocks*, 172–89. John Wiley, London.

BUTAKOVA, E.L. and EGOROV, L.S. 1962. The Meimecha-Kotui complex of formations of alkaline and ultrabasic rocks. *Petrography of Eastern Siberia*. **1**: 417–589. Izd-vo AN SSSR, Moscow.

DANILIN, E.L., ZANIN, Y.N., VAHRAMEEV, A.M., GILINSKAYA, L.G., KRIVOPUTSKAYA, L.M. STOLKOVSKAYA, V.N. and ZAMIRAILOVA, A.G. 1982. *Phosphate-bearing weathering ores and phosphorites of the Maimecha-Kotui province of ultrabasic-alkaline rocks*. Nauka, Moscow. 73 pp.

EGOROV, L.S. 1960. On the problem of nephelinization and iron magnesium-calcium metasomatism in intrusive alkaline and ultrabasic rocks. *Trudy Nauchno-Issledovatel'skogo Instituta Geologii Arktiki, Leningrad*, **114**: 1102–8.

EGOROV, L.S. 1968. Apatite of the Maimecha-Kotui complex of ultrabasic alkaline rocks. In D. Vorobyeva and V. Petrov (eds), *Apatite*, 227–32. Nauka, Moscow.

EGOROV, L.S. 1969. *The melilite rocks of the Maimecha-Kotui province*. Nedra, Leningrad. 247 pp.

EGOROV, L.S. (ed.). 1976. Apatite-bearing rocks of the north of Siberia. *Sbornik Nauchnykh Trudov, Nauchno-Issledovatel'skogo Instituta Geologii Arktiki, Leningrad*, 116 pp.

EGOROV, L.S. 1980. The rocks of the phoscorite series (apatite–magnetite ores) of the Essei massif and some general problems of petrology, classification and nomenclature of apatite–olivine–magnetite rocks of ijolite-carbonatite complexes. *In* L.S. Egorov (ed.) *Alkaline magmatism and apatite-bearing rocks of the north of Siberia*. 39–60. Trudy Nauchno-Issledova tel'skogo Instituta Geologii Arktiki, Leningrad.

*EGOROV [YEGEROV], L.S. 1989. Form, structure and development of the Guli ultramafic-alkalic and carbonatite pluton. *International Geology Review*, **31**: 1226–39.

EGOROV, L.S. 1991. *Ijolite carbonatite plutonism (case history of the Maimecha-Kotui complexes northern Siberia)*. Nedra, Leningrad. 260 pp.

EGOROV, L.S. and SURINA, N.P. 1961. The carbonatites of the Changit intrusive at the north of the Siberian platform. *Trudy Nauchno-Issledovatel'skogo Instituta Geologii Arktiki, Leningrad*, **125**: 160–78.

EGOROV, L.S., GOLDBURT, T.L., SHIKORINA, K.M., EPSTEIN, E.M., ANIKEEVA, L.I. and MIKHAOLOVA, A.F. 1961. *Geology and petrology of magmatic rocks of the Guli intrusion*. State Publishing House of Literature for Mining (Gosgortekhisdat), Moscow. 272 pp.

*GLADKIKH, V.S. 1991. Geochemistry of melilitic volcanic rocks in the Maymecha-Kotuy province. *Geochemistry International*, **28**(11): 119–28.

GOLDBURT, T.L. and LANDA E.A. 1963. A new phlogopite deposit Magan at the north of the Siberian platform. *Uchenye Zapiski Nauchno-Issledovatel's kogo Instituta Geologii Arktiki, Leningrad*, **1**: 35–43.

KOGARKO, L.N., KARPENKO, S.F., LYALIKOV,

A.V. and TEPTELEV, M.P. 1988. Isotopic criteria for the origin of melteigite magmatism (Polar Siberia). *Doklady Akademii Nauk SSSR*, **301**: 939–42.

KONONOVA, V.A. 1976. *The jacupirangite–urtite series of alkaline rocks*. Nauka, Moscow. 212 pp.

KOSTYUK, V.P. 1974. *Mineralogy and problems of origin of alkaline rocks of Siberia*. Nauka, Siberian Branch of the Academy of Sciences, Novosibirsk.

KRAVCHENKO, S.M. and BAGDASAROV, Yu.A. 1987. *Geochemistry, mineralogy and origin of the apatite-bearing massifs (Meimecha-Kotui carbonatite province)*. Nauka, Moscow. 128 pp.

*KRAVCHENKO, S.M., KOLENKO, Yu.A. and RASS, I.T. 1987. The petrology of the Magan apatite deposit in the Maymecha-Kotuy province. *International Geology Review*, **29**: 281–94.

MOOR, G.G. 1957. The differentiated alkaline intrusives of the northern margin of the Siberian Platform (the right bank of the lower stream of the Kotui river). *Izvestiya Akademii Nauk SSSR, Seriya Geologicheskaya*, **8**: 400–52.

*PANINA, L.I. and SHATSKIY, V.S. 1974a. Inclusions of melt in magnetite–apatite rock of the Yessey carbonatite intrusion. *Doklady Earth Science Sections, American Geological Institute*, **209**: 149–51.

*PANINA, L.I. and SHATSKIY, V.S. 1974b. Traprock and ultramafic alkalic rocks of the Yessey carbonatite intrusion. *Doklady Earth Science Sections, American Geological Institute*, **209**: 144–7.

PORSHNEV, G.J. and SHUVALOVA, V.Z. 1970. The Essei alkaline-ultrabasic massif and a deposit of magnetite–apatite ores associated with it. Geology and mineral deposits of the western part of the Siberian platform. *Transactions of the Krasnoyarsk Geological Office*, 238–44.

PROCHOROVA, S.M., EVZIKOVA, N.Z. and MICHAILOVA, A.F. 1966. Phlogopite-bearing rocks of the Maimecha-Kotui province of ultrabasic alkaline rocks. *Trudy Nauchno-Issledovatel'skogo Instituta Geologii Arktiki, Leningrad*, **140**: 1–196.

SHEINMAN, Yu.M. 1947. On a new petrographic province of the north of the Siberian Platform. *Izvestiya Akademii Nauk SSSR, Seriya Geologicheskaya*, 1: 123–4.

SHEINMAN, Yu.M., APELTSYN, F.R. and NECHEVA, E.A. 1961. *Alkaline intrusives, their localization and mineralization associated with them. Geology of deposits of rare elements*. Institute of Mineral Resources, Moscow. 12–13: 213 pp.

VASILIEV, Yu.R. and ZOLOTUCHIN, V.V. 1975. *Petrology of ultrabasites of the north of the Siberian platform and some problems of their origin*. Nauka, Novosibirsk. 269 pp.

ZHABIN, A.G. 1965. On the structure and succession of formation of the Guli dunite complex, ultrabasic and ultrabasic-alkaline lavas, alkaline rocks and carbonatites. *In* L. Borodin (ed.), *Petrology and geochemical features of complexes of ultrabasites, alkaline rocks and carbonatites*. 160–92. Nauka, Moscow.

ZHABIN, A.G. and OTTEMAM, J. 1976. Ontogeny of chromite–olivine eutectic in dunites. *In* A.I. Ginsburg (ed.), *Essays on genetic mineralogy*. Nauka, Moscow.

* In English

ANABAR

Of the two alkaline provinces lying in the northern part of the Siberian Platform, the Maimecha-Kotui Province is situated west of the Anabar Shield and the Anabar Province (Fig. 136) immediately east of the shield (Fig. 110). The province as a whole contains numerous kimberlite occurrences as well as dykes and stocks of picritic and basaltic rocks, with alkaline rocks relatively minor. Tomtor is a very large complex of carbonatite, nepheline and other syenites. The occurrences of Bogdo and Tomtor comprise the Olenek Province of Borodin (1974).

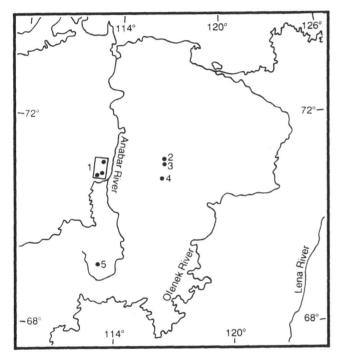

Fig. 136. Distribution of alkaline igneous rocks and carbonatites in the Anabar province.

1 Orto-Yrigakhskoe, Tundrovoye and Nomotookhskoe	3 Tomtor
	4 Chimaara
2 Bogdo	5 Mald'zhangarskii

1 ORTO-YRIGAKHSKOE, TUNDROVOYE and NOMOTOOKHSKOE

71°06'N; 113°10'E
70°58'N; 113°09'E
70°53'N; 112°52'E

In 1964–5 numerous carbonatitic dykes and eruptive pipes were discovered on the eastern slope of the Anabar shield in three distinct areas. They are emplaced in Proterozoic and Cambrian sedimentary rocks. A great number of kimberlite pipes are also developed in the area. The carbonatite pipes have diameters between 75 and 275 m and comprise tuffaceous rocks with a breccia texture of xenoliths in a fine-grained cement. The xenoliths are angular and consist of sedimentary limestones, dolomitized limestones, sandstones and granite gneisses. The cement, which forms about 80% of the rock, consists of dolomite, apatite, magnetite, phlogopite and

fine-grained calcite with accessory pyrochlore, baddeleyite, zircon, columbite, fersmite, pyrite, chalcopyrite, sphalerite, alkali amphibole and aegirine. Lapin et al. (1976) give chemical compositions of magnetite, including trace elements, and show that they are chemically similar to magnetite in typical plutonic carbonatites. There are nine dykes of carbonatite varying from 0.1 to 1 m in thickness. They comprise massive textured rocks of calcite, phlogopite, apatite and subordinate serpentine, magnetite, perovskite and rutile.

Age Geological evidence indicates an Upper Triassic age (Marshintsev, 1974).

References *Lapin et al., 1976; Marshintsev, 1974; *Marshintsev and Balakshin, 1969.

2 BOGDO

71°17'N; 116°32'E
Fig. 137

The Bogdo complex has a diameter of about 2.5 km. The country rocks are Proterozoic to lower Cambrian dolomites and lower Jurassic sandstones and argillites. The complex consists principally of nepheline syenite and juvite. The latter forms several stock-like bodies which are composed of nepheline and K–feldspar with lesser amounts of aegirine, schorlomite, apatite, biotite, titanite and titanomagnetite. The northern part of the complex is composed principally of nepheline syenites, the main rock-forming minerals of which are nepheline, K-feldspar and aegirine with subordinate titanite, apatite and sodalite. In the central part of the intrusion there is a ring zone 300 to 400 m wide of xenoliths consisting of fenites and marbles. To the south of the intrusion there

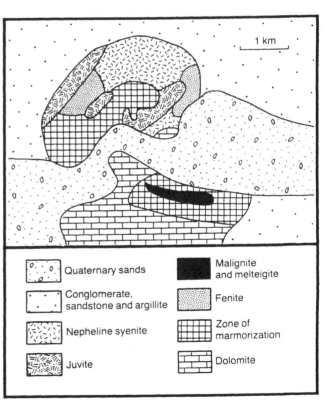

Fig. 137. Bogdo (after Erlikh, 1964, Fig. 3).

125

is an isolated dyke up to 250 m wide of malignite and melteigite, around which is a broad aureole of marbles.

Reference Erlikh, 1964.

---

3 TOMTOR
71°00′N; 116°35′E
Fig. 138

The Tomtor complex is located on the Udzhinsky rise, that is composed of tuffaceous and carbonate deposits of Upper Proterozoic age. The complex is covered by Quaternary sediments with a thickness of 10–140 m, and was discovered as the result of geophysical work; it has been investigated by drilling. It is circular in form with a diameter of 16 km; the outer contacts are vertical. The principal rock types are jacupirangite–ijolite, nepheline and peralkaline syenites and a carbonatite series, which includes apatite–magnetite rocks – phoscorites. In the central part there is a stock-like body of calcite carbo-

natites towards the periphery of which the carbonatites change into a zone of breccia with a carbonatitic cement. Dykes of alnoite, limburgite, augitite and picrite are concentrated in this zone. To the west of the carbonatites there is a block of Upper Proterozoic rocks which have been intensively metasomatized, and to the east and north of the carbonatites there is a zone of alkaline-ultrabasic rocks which includes nepheline pyroxenites, melteigites, ijolites and associated magnetite ores. All these rocks are characterized by heterogeneous textures and structures and are intensively metasomatized and phlogopitized. The peripheral zone of the Tomtor complex consists of nepheline and peralkaline syenites which are cut by numerous dykes, necks and pipes of alkaline ultrabasic rocks (alnoite, picrite and alvikite). Overlying the massif there is a mantle of weathered rocks with a thickness of 50–200 m which comprises kaolinitic, francolitic and carbonatitic facies which are rich in iron oxides. The geochemistry and origin of the complex are discussed by Kravchenko et al. (1993).

Economic In the core of the overlying weathered rocks there are concentrations of vermiculite, francolite and kaolinite. The francolite-rich rocks occur in the form of lenses up to 50 m thick and have an average grade of 12.5% P_2O_5. Kravchenko et al. (1990) describe Sc, REE and Nb ores in the weathered mantle overlying altered

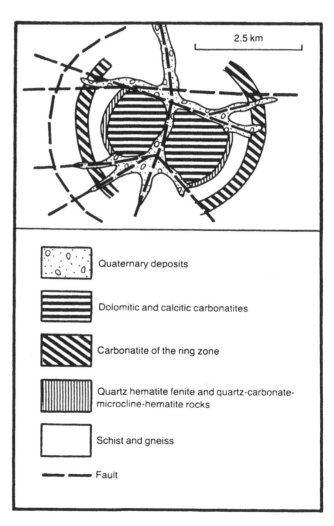

Legend (Fig. 138):
- Sandstone, shale and conglomerate
- Crust of weathering
- Alvikite, limburgite, augitite and alkaline picrite pipes and necks
- Alvikite, limburgite and augitite sills and dykes
- Eruptive breccia
- Siderite and dolomite carbonatites
- Calcite carbonatite
- Phoscorite
- Limit of intrusion from geophysical data
- Phlogopite-magnetite ores
- Phlogopite-pyroxene-magnetite-apatite-carbonate metasomatites
- Nepheline and alkaline syenites
- Jacupirangite-melteigite-ijolite
- Carbonate-mica-apatite-chlorite-zeolite metasomatites
- Marble aureole
- Proterozoic dolomite, limestone, sandstone and shale
- Fault

Legend (Fig. 139):
- Quaternary deposits
- Dolomitic and calcitic carbonatites
- Carbonatite of the ring zone
- Quartz hematite fenite and quartz-carbonate-microcline-hematite rocks
- Schist and gneiss
- Fault

Fig. 138. Tomtor (after Porshnev and Stepanov, 1981, Fig. 2).

Fig. 139. Mald'zhangarskii (after Shakhotko and Bagdasarov, 1983, Fig. 1).

carbonatite in some parts of which pyrochlore concentrations give up to 12% Nb_2O_5 and there are up to 37% RE oxides. Apatite–magnetite and phlogopite–magnetite ores are associated with the alkaline ultrabasic rocks (Porshnev and Stepanov, 1981).

Age According to geological evidence the complex must be early-middle Palaeozoic in age (Porshnev and Stepanov, 1981; Erlikh and Zagrusina, 1981).

References Entin *et al.*, 1990; Erlikh and Zagrusina, 1981; *Kravchenko *et al.*, 1990 and 1993; Porshnev and Stepanov, 1980 and 1981.

---

4 CHIMAARA   70°51′N; 116°30′E

In this area to the south of the Tomtor complex (No. 3) and, like Tomtor, located on the Udzhinsky rise, there are a number of small dykes, veins and stock-like bodies of alkaline ultrabasic rocks, nepheline syenites and carbonatites. The last are represented by calcite, dolomite and ankerite–siderite varieties.

References Porshnev and Stepanov, 1980; Shakhotko and Bagdasarov, 1983.

---

5 MALD'ZHANGARSKII   69°23′N; 113°10′E
Fig. 139

Mald'zhangarskii is situated along the upper reaches of the River Mald'zhangarka and forms a steep highland area with a diameter of about 4.5 km. The country rocks are Archaean gneisses and schists. The central stock of the intrusion is composed of medium- and fine-grained calcitic and dolomitic carbonatites which contain xenoliths of quartz–K–feldspar fenites. Apart from carbonates, the principal minerals of the carbonatite stock are amphibole, aegirine, phlogopite, apatite and subordinate magnetite, ilmenite, leucoxene, pyrochlore and zircon. Peripheral to the carbonatite stock there is a fenite zone about 40 m wide of quartz–hematite and quartz–carbonate–microcline–hematite rocks. There is an incomplete outer ring of carbonatite cutting the Archaean basement rocks.

Reference Shakhotko and Bagdasarov, 1983.

Anabar references

BORODIN, L.S. (ed.) 1974. *Principal provinces and formations of alkaline rocks.* Nauka, Moscow. 376 pp.
ENTIN, A., ZAITSEV, A.I., NENASHEV, N.I., VASILENKO, V.B., ORLOV, A.N., TIAN, O.A.,

OLKHOVOK, Yu.A., OLSHTYNSKII, S.P. and TOLSTOV, A.V. 1990. On the sequence of geological events involved in the intrusion of the Tomtor massif of ultrabasic alkaline rocks and carbonatites (Northwestern Yakutia). *Geologiya i Geofizika.* Novosibirsk, Part **12**: 42–51.
ERLIKH, E.N. 1964. A new province of alkaline rocks on the north of the Siberian Platform. *Zapiski Vsesoyuznogo Mineralogicheskogo Obshchestva.* Moskva, **93**: 682–93.
ERLIKH, E.N. and ZAGRUSINA, I.A. 1981. Geological aspects of the geochronology of the north-eastern part of the Siberian Platform. *Izvestiya Akademia Nauk SSSR, Seriya Geologiya,* N.G. **8**: 5–13.
*KRAVCHENKO, S.M., BELYAKOV, A.Yu. and POKROVSKIY, B.G. 1993. Geochemistry and origin of the Tomtor massif in the North Siberian Platform. *Geochemistry International,* **30**: 20–36.
*KRAVCHENKO, S.M., BELYAKOV, A.Yu., KUBYSHEV, A.I. and TOLSTOV, A.V. 1990. Scandium-rare-earth-yttrium-niobium ores – a new economic resource. *International Geology Review,* **32**: 280–4.
*LAPIN, A.V., GARANIN, V.K., MARSHINTSEV, V.K. and KUDRYAVTSEVA,G.P. 1976. Composition and internal structure of magnetite from explosion carbonatite breccia of the northeastern margin of the Siberian Platform. *Doklady Earth Science Sections, American Geological Institute,* **228**: 129–32.
MARSHINTSEV, V.K. 1974. *Carbonatitic formations of the eastern slope of the Anabar Arc* Yakut Branch of the Academy of Sciences, Yakutsk, 199 pp.
*MARSHINTSEV, V.K. and BALAKSHIN, G.G. 1969. Nature of carbonatite at the east flank of the Anabar Arch. *Doklady Earth Science Sections, American Geological Institute,* **188**: 70–3.
PORSHNEV, G.I. and STEPANOV, L.L. 1980. Geological structure and phosphate-bearing rocks of the Tomtor massif (north-western Yakutia). *Alkaline magmatism and apatite-bearing rocks of north Siberia.* 84–100. Nauchno-Issledovatel'skogo Instituta Geologii Arktiki, Leningrad.
PORSHNEV, G.I. and STEPANOV, L.L. 1981. Geology and mineralogy of the Ud'in Province (north-western Yakut ASSR). *Sovetskaya Geologiya.* Moskva, **12**: 103–6.
SHAKHOTKO, L.I. and BAGDASAROV, Yu.A. 1983. A new carbonatite massif of the Anabar Shield. *Doklady Akademii Nauk SSSR,* **273**: 186–9.

*In English

CHADOBETSKAYA

Chadobetskaya is located east of the Enisei Province, as shown on Fig. 140.

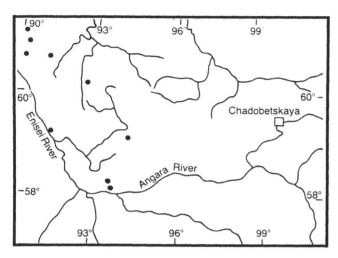

Fig. 140. The location of Chadobetskaya east of the Enisei Province. For comparison see Fig. 142.

---

1 Chadobetskaya

---

1 CHADOBETSKAYA

59°20'N; 99°17'E
Fig. 141

The Chadobetskaya occurrences (Fig. 141) are located in the southwestern part of the Siberian platform. They include alkaline ultrabasic rocks which are closely associated with kimberlites (but see below) and belong to the Chadobetsky upland, which comprises two small domes formed by Riphean terrigenous and carbonate strata. The largest outcrops of the alkaline ultrabasic rocks, which are mostly as pipes, sheet-like bodies and dykes, are disposed in the central parts of the domes. According to geophysical evidence the igneous intrusions that have sheet-like forms have areas of up to 100 × 300 m and thicknesses of 50–70 m. The intrusions are formed of alnoites, micaceous melilite–nepheline peridotites and pyroxene peridotites. Sometimes they are intersected by veins of melteigite. Zoned veins are also found in which the central parts are formed of porphyritic picrites and the margins have phlogopite–apatite–pyroxene compositions. All the rocks are rather uniform in mineralogy, the rock-forming minerals being forsterite, diopside-augite, phlogopite, ilmenite, apatite and perovskite. The quantitative relationships between these minerals is somewhat variable. In the porphyritic picrites, forsterite comprises 22–45% of the volume of the rock; in phlogopite–olivine porphyries forsterite comprises 20–40% and phlogopite up to 15%, although usually the forsterite is totally replaced by secondary minerals. The groundmass of these rocks consists of diopside-augite together with phlogopite, apatite, magnetite, ilmenite, perovskite and other minerals. Secondary minerals, which are present in large quantities, include carbonates, generally calcite with some dolomite and ankerite, serpentine, chlorite, actinolite, limonite and others. Melilite–nepheline peridotites are fine- and medium-grained almost black rocks with a massive, and more

rarely, irregular texture; they are sometimes porphyritic. The principal minerals comprising these rocks are forsterite, monticellite, nepheline, melilite, perovskite and, in some places, phlogopite. Amongst the accessory minerals are titanomagnetite and apatite. Carbonatites have been found in four occurrences. They are represented by series of veins, the thicknesses of which vary from several centimetres to 0.2–0.3 m, as well as irregularly shaped bodies. Both calcite and dolomite occur amongst the carbonates with the calcite carbonatites coarse-grained rocks which are essentially monomineralic. Albite, quartz, mica and pyroxene are present in insignificant amounts. Dolomitic carbonatites are more widespread and they contain small quantities of pyrite, iron hydroxides, apatite and accessory pyrochlore and sometimes sphalerite. Bagdasarov et al. (1969) give carbon isotope data on a range of carbonatites, carbonatite-like rocks and carbonate country rocks. Vasilenko (1989), based on the application of statistical methods and about 70 chemical analyses, considers that the rocks described as kimberlites are not so, but are in fact melilitites.

Age Judging from the presence of fragments of Permian sedimentary rocks and traps in pipes, they are Triassic in age.

References Bagdasarov et al., *1969 and 1972; Osokin et al., 1974; Polunina, 1966; Vasilenko et al., 1989.

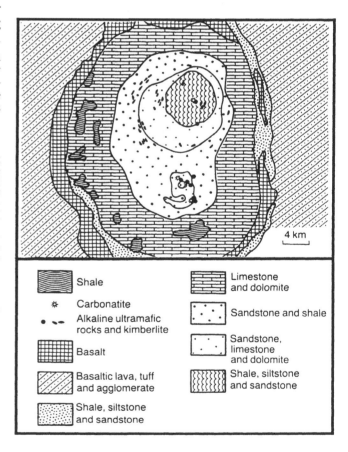

Fig. 141. The distribution of the alkaline rocks of Chadobetskaya (after Bagdasarov et al., 1972, Fig. 13).

128

Chadobetskaya references

*BAGDASAROV, Yu.V., GALIMOV, E.M. and PRO-KHOROV, V.S. 1969. Isotopic composition of carbon in ankerite carbonatite and the source of carbonate material present in sedimentary rocks. *Doklady Earth Science Sections, American Geological Institute*, **188**: 201–4.

BAGDASAROV, Yu.V., NECHAEVA, E.A. and FROLOV, A.A. 1972. Chadobetskaya province of ultramafic alkaline rocks and carbonatites. *In* A.I. Ginzburg (ed.) *Geology of the rare elements deposits. 35. Geology, mineralogy and genesis of the carbonatites.* 79–91. Nedra, Moscow.

OSOKIN, E.D., LAPIN, A.V., KAPUSTIN, YU.L.,

POHVISNEVA, E.A. and ALTUHOV, E.N. 1974. Alkaline provinces of Asia. Siberia-Pacific group. *In* L.S. Borodin (ed.) *Principal provinces and formations of alkaline rocks.* 91–166. Nauka, Moscow.

POLUNINA, L.A. 1966. The Chadobetsky complex of ultrabasic-alkaline rocks. *Geology of the Siberian Platform.* 318–20. Nauka, Moscow.

VASILENKO, V.B., KRYUKOV, L.G. and KUZNET-SOVA, L.G. 1989. Petrochemical types of alkaline-ultrabasic rocks of the Chadobetsky uplift. *Geologiya i Geofizika. Novosibirsk,* **8**: 46–54.

* In English

ENISEI

The Enisei Province occurs close to the western margin of the Siberian platform and is located within the Riphean fold system. The alkaline occurrences define a north-northwest–south-southeast-trending zone to the east of the Enisei River (Fig. 142).

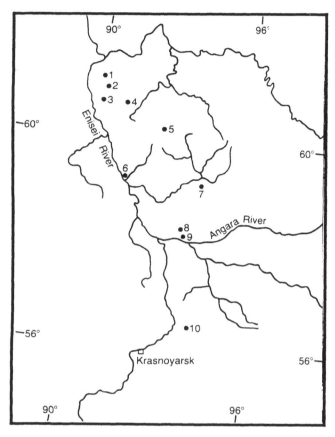

Fig. 142. *Distribution of occurrences of alkaline rocks in the Enisei Province (after Osokin et al., 1974, Fig. 33) with additions of the authors.*

1 Srednevorogovskii	6 Kiiskii
2 Zakhrebetninskii	7 Penchenginskii
3 Kutukas	8 Srednetatarskii
4 Chapinskii	9 Pogromninskii
5 Noibinskii	10 Porozhninskii

1 SREDNEVOROGOVSKII   61°15′N; 90°02′E

This occurrence comprises one body of 2.2 × 0.8 km and about 70 smaller ones each occupying only hundreds of square metres and concentrated in an area of 10–12 km². The alkaline intrusions, which appear to represent the upper parts of eroded intrusive cupolas, are small stocks, dykes and veins. The larger body consists mainly of leucocratic quartz syenite (nordmarkite), which gradually grades into peralkaline syenite, grano-syenite and peralkaline granite, the last situated close to the contact zone. Small bodies (from 1 to 100 m across) and dykes of granite porphyry and syenite porphyry belong to the final stage of the magmatic history. Nord-

markite is the most widespread rock type and consists of microcline perthite (70–80%), antiperthite (10–15%), quartz (5–10%), biotite (5–8%) and some riebeckite and arfvedsonite. Accessories are zircon, apatite, titanite, monazite, fluorite, rutile, Fe–Ti oxides and sulphides, including pyrite, pyrrhotite, molybdenite and chalcopyrite. Peralkaline syenites are compositionally similar to the nordmarkite except for less quartz (up to 3%) and some additional accessories including tourmaline and orthite. Peralkaline granite consists of microcline perthite (30–40%), quartz (30–35%), albite (20–25%), biotite (1–3%) and sometimes riebeckite. Accessories are apatite, garnet, rutile, titanite, zircon, ilmenorutile, pyrochlore, euxenite, tourmaline, Fe oxides, cassiterite and sulphides. Postmagmatic alteration has enriched some rocks in albite and microcline. There are quartz-fluorite-feldspar-carbonate and quartz veins enriched in rare minerals.

Age K–Ar dating of nordmarkite gave 567 ± 9 Ma, of quartz syenite 526 ± 6 Ma and of syenite porphyry 536 ± 8 Ma (Nozhkin and Trofimov, 1982).

Reference Nozhkin and Trofimov, 1982.

2 ZAKHREBETNINSKII   61°02′N; 90°15′E

This occurrence occupies about 20 km² and is composed principally of peralkaline syenites; calc-alkaline and nepheline syenites are also present. The last forms several small lenses and consists of albitized alkali feldspar (67%), nepheline replaced by cancrinite (18%), analcime (2%), aegirine (13%) and biotite (1%) with accessory orthite, zircon, apatite and magnetite. Nepheline-bearing syenites contain 3–5% nepheline. The peralkaline syenites contain up to 82% alkali feldspar with albite–oligoclase 14% and biotite plus alkali amphibole 4%.

Age K–Ar on peralkaline syenite gave 475 and 486 ± 6 Ma and on nepheline syenite 478 ± 8 Ma (Nozhkin and Trofimov, 1982).

References Kornev *et al.*, 1974; Nozhkin and Trofimov, 1982.

3 KUTUKAS   60°46′N; 90°05′E

Peralkaline syenites occupy an area of 0.5 × 0.7 km and are surrounded by alaskite granite. The syenite consists of microcline perthite (85%), aegirine replaced by amphibole (3–12%) and quartz (0.5%) with minor zircon, monazite, apatite, magnetite and fluorite.

Age K–Ar determinations on whole rocks gave 600–635 Ma (Datsenko, 1984).

References Datsenko, 1984; Nozhkin and Trofimov, 1982.

4 CHAPINSKII   60°44′N; 91°06′E

About 50 small bodies of alkaline ultramafic rocks occur as pipes, dykes and small volcanic fields. The dykes vary from 0.3 to 30 m thick and up to 600 m in length; the pipes have diameters of 25 to 460 m. The volcanic fields extend over 1 × 2 and 1 × 3 km with the volcanics up to 70 m thick. Both flows and tuffs occur but all are much altered. The dominant rock type is low alkali basalt but micaceous picrites are also present. The latter rock consists of altered olivine, which comprises as much as 40% of some flows, altered clinopyroxene with relicts of titanaugite, biotite and phlogopite. Volcanic glass is replaced by carbonate, serpentine, chlorite and iron hydroxides. A suite of heavy minerals consists of ilmenite, perovskite, apatite, zircon and, from lavas, garnet, baddeleyite and chromite.

Age K–Ar dating of whole rocks gave 550–660 Ma and these rocks occur as fragments in Middle Cambrian conglomerates (Kornev *et al.*, 1974).

References Karpinsky and Kachevskova, 1973; Kornev *et al.*, 1974.

---

5 NOIBINSKII   60°20′N; 92°38′E

This occurrence is represented by three approximately circular bodies of quartz syenite which gradually grade into syenite. The rocks are composed of microcline, albite, quartz, biotite and barkevikite. Nepheline-bearing rocks, of which nepheline syenite porphyries are the most widely developed, are found in talus deposits. Phenocrysts (15%) are represented by feldspar, nepheline and hastingsite. The groundmass has a trachytic texture and comprises microcline (70%), nepheline (17%), aegirine (6%) and hastingsite (4%); accessories are apatite, titanite and titanomagnetite.

Age K–Ar dating of nepheline syenite porphyry gave 543 Ma, peralkaline syenite porphyry 497–503 Ma and quartz syenite 453–553 Ma (Nozhkin and Trofimov, 1982).

References Nozhkin and Trofimov, 1982; Osokin *et al.*, 1974.

---

6 KIISKII   59°14′N; 91°30′E
                                 Fig. 143

The Kiiskii occurrence has a circular form and occupies 12 km². The country rocks are granite gneisses and crystalline schists of Lower Proterozoic age. Four stages can be distinguished in the formation of the complex, the first of which involved emplacement of urtite, ijolite, melteigite and jacupirangite in the central and northern parts of the occurrence. These rocks are characterized by highly variable compositions and textures. Ijolites are the most abundant and comprise nepheline and aegirine-diopside. During the second stage nepheline syenites were developed over most of the western part of the occurrence. They consist of nepheline, orthoclase-perthite, aegirine–diopside and hastingsite. Rocks of the third stage are leucocratic syenite porphyries which occupy the central and eastern parts of the complex. They contain orthoclase–cryptoperthite as phenocrysts up to 1 cm in diameter in a groundmass composed of orthoclase, albite, biotite and calcite. The fourth stage consists of veins of tinguaite and camptonite–monchiquite˙ which are found throughout the intrusion, as are small bodies of trachytic porphyry. All the rocks were intensively altered and replaced by K- and Na-feldspar and carbonate. This resulted in the production of amphibole–apatite, biotite–apatite and feldspar-carbonate rocks. The central part of the massif is occupied by calcite carbonatite with dolomite, ankerite, siderite and calcite veinlets with rare earth mineralization. What stage is represented by the carbonatites has not been determined.

Economic An iron ore deposit of ankerite-siderite type with 20–30% iron is present with reserves of some hundreds of tons (Frolov, 1984).

Age K–Ar determinations on whole rocks gave 550–660 Ma (Kornev *et al.*, 1974) but Plyusnin *et al.* (1990) obtained 251 ± 12 Ma from a Rb–Sr whole rock isochron.

References Frolov, 1984; Kornev *et al.*, 1974; Osokin *et al.*, 1974; *Plyusnin *et al.*, 1990; Samoilova, 1962; Sheinman *et al.*, 1961.

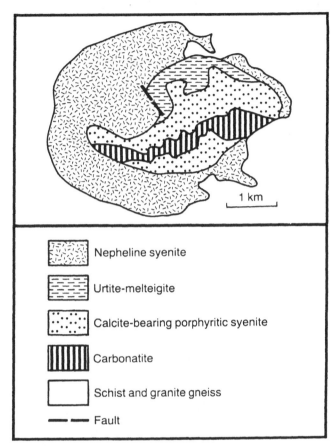

Nepheline syenite

Urtite-melteigite

Calcite-bearing porphyritic syenite

Carbonatite

Schist and granite gneiss

Fault

Fig. 143. Kiiskii (after Sheinman, et al., 1961, Fig. 10).

---

7 PENCHENGINSKII   59°17′N; 94°12′E
                                    Fig. 144

Dykes and veins of alkali gabbro, lusitanite, pedrosite, camptonite, solvsbergite and carbonatite occupy a zone 1.5 km wide and more than 18 km in length. The lusitanite dykes are up to 200 m thick and 2 km long and fenitize the surrounding rocks. They consist of 60–95% magnesioriebeckite, up to 25% alkali feldspar, aegirine, biotite and apatite. Accessories include pyrochlore, columbite, magnetite, ilmenite, titanite and chalcopyrite, with secondary quartz and calcite also present. The pedrosites are essentially magnesioriebeckite (95–99%) rocks which form vein-like bodies up to 2 m wide within lusitanites. The solvsbergites are porphyritic aegirine–feldspar rocks with phenocrysts up to 1.5 cm diameter of aegirine, occupying about 25% of the rock, and albite; the aegirine has inclusions of magnesioriebeckite, albite and microcline. Carbonatite forms dykes up to 200 m wide and 3 km in length. They are compositionally variable with calcite, dolomite–calcite and ankerite–calcite carbonatites represented. The calcite carbonatites consist of up to 95% calcite as crystals up to 10 cm in diameter, needles of magnesioriebeckite (5%)

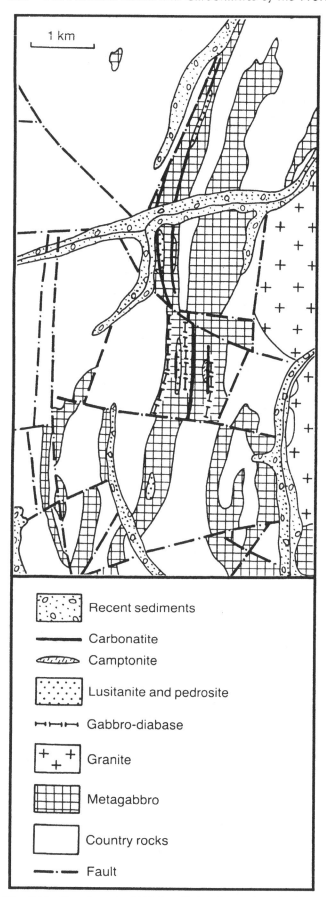

Recent sediments

—— Carbonatite

◁▭▷ Camptonite

Lusitanite and pedrosite

⊢▬⊣▬⊣ Gabbro-diabase

Granite

Metagabbro

Country rocks

—·—·— Fault

*Fig. 144. Penchenginskii (after Zabrodin and Malyshev,
1975, Fig. 1).*

up to 4 cm in length, apatite (2%) and iron oxides. The dolomite–calcite carbonatite consists of 40–85% calcite, 10–55% dolomite, 5–7% amphibole, apatite and iron oxides. Among the ankerite–calcite carbonatites are lenses, up to 2 m across, enriched in pyrochlore (15–35%). There is in places enrichment of the carbonatites in sulphides (up to 15–20%), including pyrrhotite, pyrite and chalcopyrite.

Age Geological evidence indicates a late Proterozoic age.

References *Zabrodin and Malyshev, 1975 and 1978.

---

8 SREDNETATARSKII   58°24′N 93°35′E
(Zaangarskii)   Fig. 145

The Srednetatarskii intrusion cuts limestones in which are beds of coal and clay. In plan it is approximately circular and stock-like and predominantly comprises foyaite and ijolite with an area of muscovite syenite in the north, which is unlikely to be consanguineous with the Srednetatarskii alkaline occurrence. Country rock xenoliths are widespread within the intrusion. The foyaites are leucocratic medium-grained nepheline syenites comprising nepheline (26–28%), microperthite and other feldspars (50–65%), aegirine (2–10%), titanite (0.4%), fluorite (0.1–0.6%) and lavenite (0.1–2.2%). In a narrow zone near the contact the proportion of dark coloured minerals increases up to 10–15%. Lying

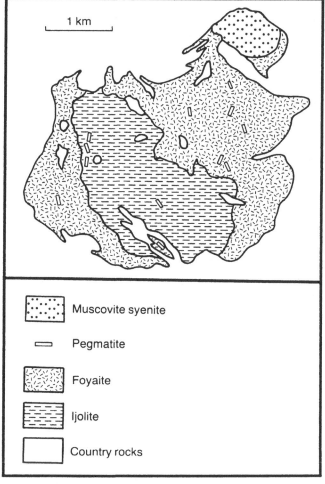

Muscovite syenite

▱ Pegmatite

Foyaite

Ijolite

Country rocks

*Fig. 145. Srednetatarskii (after Sveshnikova et al.,
1976, Fig. 1 and Yashina, 1984, Fig. 43).*

beneath the ijolites in the central part of the occurrence foyaites have been encountered at a depth of 50–200 m. The ijolites are likely to form a subhorizontal body overlying the foyaites. As well as the ijolites, urtites and melanocratic juvites are also present. Their mineral composition is variable. Feldspar-free varieties include nepheline (65–85%), aegirine (13–32%), titanite (2–3%) and lavenite (0.5%) and the juvites nepheline (10–20%), microcline and microperthite (up to 26%), aegirine (14–23%), titanite (1–2%), fluorite (0.2–0.3%), apatite (0.1–0.2%) and lavenite (0.5%). Various types of pegmatite and postmagmatically altered rocks, which have undergone microclinization, albitization or calcite-cancrinite hydrothermal alteration, are present. In the pegmatites astrophyllite, eudialyte, zircon and other rare minerals are found.

Age K–Ar on nepheline from ijolites gave 675 ± 25 Ma and from foyaite 662 ± 25 Ma; on lepidomelane 660 ± 26 Ma and microcline 660 ± 26 Ma, the last from microlinized nepheline syenite, according to Sveshnikova *et al.* (1976, Table 1).

References Sveshnikova *et al.*, 1976; Yashina, 1984.

---

9 POGROMNINSKII   58°13′N; 93°40′E

This occurrence is composed predominantly of quartz-bearing rocks but nepheline syenites are also found in a talus deposit.

Reference Osokin *et al.*, 1974.

---

10 POROZHNINSKII   56°28′N; 94°07′E

Porozhninskii consists of five small syenitic massifs of from 0.5 to 5 km². They are surrounded by trachybasalts, andesites and rhyolites. Peralkaline and mildly alkaline syenites are dominant and occupy the central parts of the intrusions. They consist of microcline perthite, orthoclase, riebeckite, albite–oligoclase, aegirine, hastingsite, arfvedsonite, biotite and Fe–Ti oxides. From the centres to the outer contacts the peralkaline syenites grade into mildly alkaline granites and granosyenites.

Age Lower Devonian.

Reference Datsenko, 1984.

Enisei references

DATSENKO, V.M. 1984. *Granitoid magmatism of the south-west margin of the Siberian Platform.* Nauka, Novosibirsk. 120 pp.
FROLOV, A.A. 1984. Iron ore deposits of carbonatite-alkaline ultrabasic massifs with ring structures. *Geology of Ore Deposits*, 1: 9–21.
KARPINSKY, R.B. and KACHEVSKOVA, G.I. 1973. Alkaline ultrabasic magmatism of the northern part of the Eniseisk mountain ridge. *In* V.S. Sobolev (ed.) *Problems of magmatic geology*, 143–58. Nauka, Novosibirsk.
KORNEV, T.Yu, DATSENKO, A.V. and BOZIN, A.V. 1974. *Riphean magmatism and ores from the Enisei mountains.* Nedra, Moscow. 132 pp.
NOZHKIN, A.D. and TROFIMOV, Yu.P. 1982. The peralkaline granite–syenite association of Srednevogovsky massif (Enisei ridge). In V.S. Sobolev (ed.) *Geology of coloured metal deposits from the folded margin of the Siberian Platform.* 61–9. Nauka, Novosibirsk.
OSOKIN, E.D., LAPIN A.V., KAPUSTIN, Yu.L., POHVISNEVA, E.A. and ALTUHOV, E.N. 1974. Alkaline provinces of Asia. Siberian-Pacific group. *In* L.S. Borodin (ed.) *Principal provinces and formations of alkaline rocks*, 91–166. Nauka, Moscow.
*PLYUSNIN, G.S., KOLYAGO, Ye.K., PAKHOL'CHENKO, Yu.A., ALMYCHKOVA, T.N. and SANDIMIROVA, G.P. 1990. Rubidium–strontium age and genesis of the Kiya alkalic pluton, Enisei Ridge. *Doklady Earth Science Sections, American Geological Institute*, 305: 207–10.
SAMOILOVA, N.V. 1962. Petrochemical specific features of the association of the ijolite–melteigite rocks and nepheline syenites (an example of alkaline intrusions of the Enisei Ridge). *In* O.A. Vorob'eva (ed.) *Alkaline rocks of Siberia*, 143–68. Akademii Nauk SSSR, Moskva.
SHEINMAN, U.M., APELTSIN, F.R. and NECHAEVA, E.A. 1961. Alkaline intrusions, their locations and mineralization. *In* A.I. Ginzburg (ed.) *Geology of rare element deposits*, 12–13: Gosgeoltexisdat, Moscow. 177 pp.
SVESHNIKOVA, E.V. SEMENOV, E.I. and HOMYKOV, A.P. 1976. *The Transangarsky alkaline massif, its rocks and minerals.* Nauka, Moscow. 80 pp.
YASHINA, R.M. 1984. Magmatic series of the alkaline intermediate (syenoids) rocks from the shield and the consolidated folded areas. *In* V.A. Kononova (ed.) *Magmatic rocks: Alkaline rocks* 138–84. Nauka, Moscow.
*ZABRODIN, V.Yu. and MALYSHEV, A.A. 1975. New alkalic ultramafic and carbonatite complex on the Yenisei Ridge. *Doklady Earth Science Sections, American Geological Institute*, 223: 195–8.
*ZABRODIN, V.Yu. and MALYSHEV, A.A. 1978. A new association of basic-alkaline rocks and carbonatites on the Yenisey ridge. *International Geology Review*, 20: 517–24.

*In English

EAST SAYAN

The occurrences of alkaline rocks of the East Sayan province are divided into two subprovinces. The first subprovince comprises occurrences located within the Siberian Platform to the northeast of the East Sayan mountain ridge, and the second subprovince is located along and to the southwest of the ridge (Fig. 146). The ridge is located at the margin of the platform and separates folded rocks to the southwest from the Archaean platform. The Sayan subprovince is thus related to the uplifted and fractured margin of the Siberian Plateau. The basement blocks of the platform are divided by grabens (the Urik graben and the Onot graben) composed of Riphean green schists. The Bolshetagninskii, Nizhnesayanskii and Verkhnesayanskii occurrences are related to the Urik graben and Zhidoy to the Onot graben. The second subprovince, i.e. lying along and to the southwest of the ridge, is related to the Okin depression within the Proterosayan anticlinorium where alkaline rocks are associated with gabbroids, for example Botogol, or granitoids, for example N'urganskii.

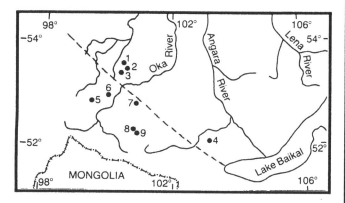

Fig. 146. Distribution of alkaline rocks and carbonatites in the East Sayan province (after Osokin et al., 1974, Fig. 30 and Konev, 1982, Fig. 6). The broken line indicates the location of the East Sayan mountain ridge, which subdivides the alkaline province.

1 Bolshetagninskii	6 Sorok
2 Nizhnesayanskii	7 Khaitanskii
3 Verkhnesayanskii	8 Botogol
4 Zhidoy	9 Khushagol
5 N'urganskii	

1 BOLSHETAGNINSKII   53°38'N; 100°28'E
(Tagninskii, Tagna)   Fig. 147

This complex has a circular outline with a diameter of about 3.6 km. It is a stock-like or a pipe-like body cutting Riphean phyllitic schists and was emplaced in several phases. The earliest rocks are ijolites and melteigites which have been preserved as large blocks, up to 1 km², within peralkaline and nepheline syenites. The ijolite occupies about 20% of the total area of the complex. The ijolites are composed mainly of nepheline and diopside-augite, with lesser amounts of biotite, feldspar, schorlomite, titanite and apatite. The second intrusive phase is represented by aegirine/hedenbergite–nepheline syenites, cancrinite syenites and aegirine syenites. These

were followed by the intrusion of leucocratic syenites, which cover about 30% of the complex. The syenites consist predominantly of microcline and biotite; nepheline, cancrinite, calcite and pyroxene are present in small amounts. The syenites are cut by arcuate dykes of porphyritic picrite, alnoite and damkjernite. The nucleus of the massif is composed of carbonatites, more than

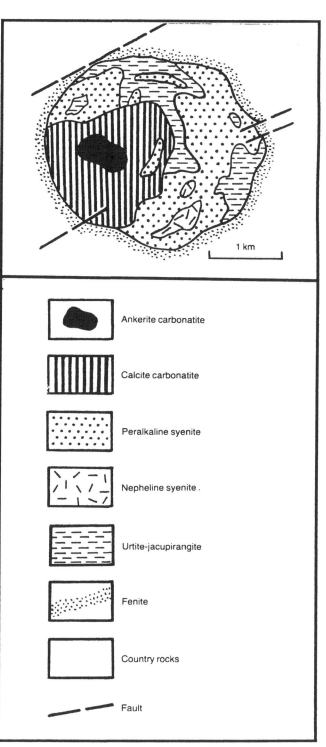

Fig. 147. Bolshetagninskii (after Frolov, 1975, Fig. 12).

134

80% of which are calcite carbonatites. At a later stage fluorite–ankerite–calcite carbonatites were emplaced. The calcite carbonatites are massive, occasionally banded, medium- and fine-grained calcite rocks containing biotite, pyroxene, apatite and accessory pyrite, pyrochlore and zircon. Sr, C and O isotopic data for primary calcite in foyaite are available in Konev *et al.* (1984).

Age 700–650 Ma, which is similar to other carbonatite occurrences of the East Sayan Province (Kononova, 1976). A whole-rock Rb–Sr isochron age of 628 ± 21 Ma was obtained by Chernysheva *et al.* (1992).

References *Chernysheva *et al.*, 1992; Frolov, 1975; Konev, 1982; *Konev *et al.*, 1984; Kononova, 1976.

---

2 NIZHNESAYANSKII   53°31'N; 100°31'E
(Lower Sayan, Beloziminskii)   Fig. 148

The Nizhnesayanskii intrusion takes the form of an oval, vertical, tube-shaped body in which the rocks are concentrically zoned. The alkaline rocks cut schists, diabases and conglomerates of Riphean age. Within a radius of 500–800 m from the contact the country rocks are fenitized, and near the contact they are transformed into banded biotite–pyroxene–feldspar and biotite–albite rocks. The central part of the complex consists of carbonatite while around the periphery are areas of nepheline-pyroxene rocks and nepheline syenites. The oldest rocks are pyroxenites which are preserved as small, altered blocks among ijolites. They are composed of clinopyroxene with accessory titanomagnetite, perovskite and phlogopite. The blocks of pyroxenite have been nephelinized and are transformed around their margins to nepheline–pyroxene rocks of highly variable texture. The second phase forms a stock of ijolite in which the rocks vary from fine- to coarse-grained. They consist of aegirine–hedenbergite, nepheline and small amounts of schorlomite, biotite and calcite. The third phase of intrusion involved the emplacement of nepheline and cancrinite syenites over the whole area of the complex as numerous veins and dykes which vary in thickness from several centimetres to 10–20 m, and can be up to 300–400 m in length. The nepheline syenites are leucocratic, trachytic rocks of nepheline, microcline and biotite. The next phase is represented by dykes and pipes of picrite porphyry. The intrusive events closed with the emplacement of a complex stock of carbonatites. Three stages of carbonatite have been identified. The first stage comprises calcite carbonatites with phlogopite, diopside, forsterite and dysanalyte. The second stage also involves calcite carbonatites but with apatite, magnetite, phlogopite and pyrochlore. The third stage carbonatites are of dolomite-ankerite with amphibole, molybdenite, pyrite, sphalerite, galenite, parisite, bastnaesite, monazite with a little quartz, fluorite, strontianite and apatite. Bagdasarov (1981) describes, with illustrations, complex vermicular intergrowths involving forsterite, magnetite, andradite and calcite in early calcite carbonatites. Calcite from carbonatite with phlogopite, of the second carbonatite stage, gave $d^{18}O^0/_{00}$ values of +6.9 and $d^{13}C^0/_{00}$ −6.3, and calcite from third-stage amphibole carbonatite gave $d^{18}O^0/_{00}$ values of +7.3 (Kononova and Yashina, 1984). $d^{34}S^0/_{00}$ values for sulphides from urtite were +1.9 to 0.6, from syenite −0.8 to −1.2 and from carbonatite +2.9 to −4.2 (Grinenko *et al.*, 1970). Compositions of rocks and minerals are available in monographs by Kononova (1976) and Pozharitskaya and Samoilov (1972).

Economic There are ores of the rare elements associated with carbonatites including apatite-pyrochlore,

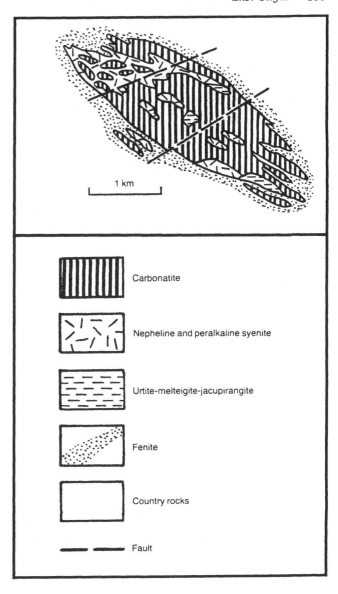

Fig. 148. Nizhnesayanskii (after Frolov, 1975, Fig. 11)

Legend:
- Carbonatite
- Nepheline and peralkaline syenite
- Urtite-melteigite-jacupirangite
- Fenite
- Country rocks
- Fault

1 km

pyrochlore–hatchettolite, pyrochlore and thorite-monazite types. Apatite deposits from the weathering of the carbonatite occupy an area of 3.4 km² and have apatite concentrations of about 25%. The mode of apatite in the central stock is 8–10% (Frolov, 1975). Apatite–magnetite ores contain 5–15% iron (Frolov, 1984).

Age K–Ar on amphibole gave 720 ± 22 Ma, on phlogopite 715 ± 25Ma and on tetraferriphlogopite 675 ± 20 Ma (Kononova, 1976), and a Rb–Sr whole-rock isochron age was obtained by Chernysheva *et al.* (1992).

References *Bagdasarov, 1981; *Chernysheva *et al.*, 1992; Frolov, 1975 and 1984; Grinenko *et al.*, 1970; Kononova, 1976; *Kononova and Yashina, 1984; Osokin *et al.*, 1974; Pozharitskaya and Samoilov, 1972.

---

3 VERKHNESAYANSKII   53°27'N; 100°25'E
(Sredneziminskii)   Fig. 149

The massif, oval in outline, cuts Upper Proterozoic conglomerates, schists and sandstones. The country rocks

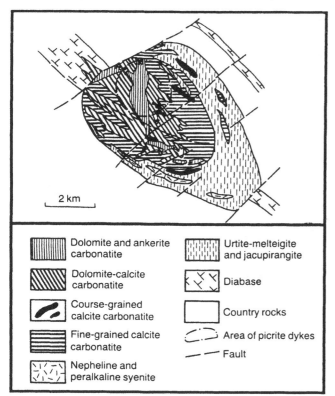

Fig. 149. *Verkhnesayanskii (after Samoilov, 1977, Fig. 1 and Frolov, 1975, Fig. 13).*

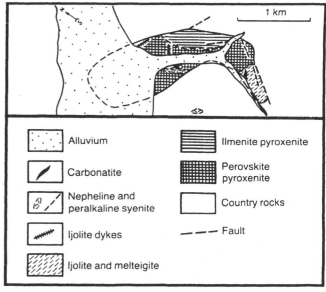

Fig. 150. *Zhidoy (after Konev, 1970, Fig. 4).*

are intensively fenitized up to 300–400 m from the contact zone. In the formation of the massif, four series have been distinguished: (1) pyroxenite, ijolite-melteigite, (2) nepheline and cancrinite syenites, (3) picrite porphyrite and alnoite and (4) carbonatites. Pyroxenites have been preserved as xenoliths among nepheline–pyroxene rocks and carbonatites. They are fine-grained rocks composed of clinopyroxene and Ti–Fe oxide minerals. Ijolites comprise nepheline and aegirine-hedenbergite; among minor minerals are biotite, schorlomite, microcline and calcite. Nepheline syenite forms dykes 1–30 m thick and 10–400 m long which are situated predominantly in the southern part of the complex cutting ijolites. They are coarse-grained massive rocks, which are occasionally porphyritic, with phenocrysts including K-feldspar, nepheline and pyroxene, while in some cancrinite syenites the cancrinite also forms phenocrysts. In the syenites minor and accessory minerals include biotite, apatite, titanite and zircon. Porphyritic picrite and alnoite are found as blocks among the carbonatites and also occur as small dykes in ijolite-melteigite. The porphyritic picrites consist of olivine and diopside in serpentine, and the alnoites comprise olivine, melilite, phlogopite and serpentine. Carbonatites make up the core of the complex as well as numerous veins and dykes. Within the main carbonatite body are many fragments of silicate rocks amongst which ijolite-melteigites are dominant. There are many types of early carbonatites including varieties with biotite, diopside, apatite, hatchettolite, pyrochlore, zircon, magnetite and pyrite; the later ankerite carbonatites include chlorite, quartz, baryte, strontianite, burbankite, ancylite, parisite, thorite, monazite, sulphides and zeolites. Calcite from phlogopite-bearing calcite carbonatite (second-stage) gave $d^{18}O^0/_{00}$ values of $+6.6$ and $d^{13}C^0/_{00}$ -6.3, while amphibole-bearing calcite carbonatite (third-

stage) gave $d^{18}O^0/_{00}$ values of $+6.4$ and $d^{13}C^0/_{00}$ -6.3; ankerite carbonatite of the fourth stage gave $d^{18}O^0/_{00}$ values from $+13$ to $+14$ (Kononova and Yashina, 1984). $d^{34}S^0/_{00}$ for sulphides from peralkaline syenites were $+2.9$ and from carbonatites $+1.6$ to -5.1 (Grinenko et al., 1970).

Economic There are economic deposits of rare metals associated with the carbonatites.

Age K–Ar on arfvedsonite from carbonatite gave 725 ± 25 Ma and on phlogopite 660 ± 20 Ma (Kononova, 1976).

References Frolov, 1975; Grinenko et al., 1970; Kononova, 1976; *Kononova and Yashina, 1984; Pozharitskaya and Samoilov, 1972.

---

4 ZHIDOY

52°03′N; 103°02′E
Fig. 150

This arcuate massif has an area of only 0.85 km² and is situated in the Onot graben, which is formed of Riphean schists and sandstones. It is composed predominantly (90%) of perovskite and ilmenite pyroxenites. Ijolite dykes, feldspar ijolites, feldspathoid (nepheline, cancrinite) syenites and carbonatites cut the pyroxenites, among which perovskite pyroxenites dominate. These are medium-grained rocks with distinct trachytic textures in which pyroxene phenocrysts may comprise 10–50% of the rock. They consist of clinopyroxene (66%), a small amount of hornblende (4%), biotite, titanite, Ti–Fe oxide minerals (15%), apatite (7%), perovskite (6.5%) and ilmenite (1%). Pyrrhotite, pentlandite, chalcopyrite and pyrite are accessories. Ilmenite pyroxenites are fine-grained rocks with massive and sometimes layered structures. They contain clinopyroxene (72%), a small amount of hornblende, biotite and titanite, Ti–Fe oxide minerals (10%), ilmenite (7.5%) and apatite (6%). Accessory minerals include sulphides and occasionally perovskite. Ijolite–melteigites form dykes up to 20 m thick. They are medium-grained rocks comprising clinopyroxene (50%), nepheline (7–25%), Ti–Fe oxide minerals, perovskite, apatite, biotite and calcite. Feldspathic ijolite is a black, fine-grained rock which forms a dyke about 0.5 m thick consisting of aegirine, nepheline and anorthoclase. Feldspathoidal syenites occur as dykes and veins and are composed of anorthoclase (54%), nepheline (16%), cancrinite (16%) and aegirine (12%), a little biotite and albite and accessory zircon, titanite, fluorite, clinozoisite, calcite and baddeleyite; columbite also occurs. Carbonatites are found as veins 0.2–5 m thick. They are calcite carbonatites with biotite, apatite, sulphides and pyrochlore. Further information on rock and mineral compositions is available in Konev (1970).

Age K–Ar on phlogopite from calcite carbonatite gave 680 ± 20 Ma Kononova, 1976).

References Konev, 1970 and 1982; Kononova, 1976.

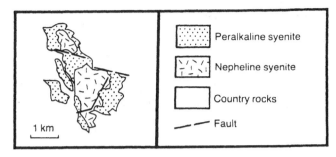

Fig. 151. *Botogol (after Lobzova, 1975, Fig. 2).*

---

5 N'URGANSKII   52°55'N; 99°30'E

This stock-like intrusion has a zonal structure and cuts Proterozoic gneisses, schists and limestones. In the peripheral areas the most abundant rock is granosyenite which is replaced inwards by peralkaline syenites; the central part consists of nepheline syenites. The peralkaline syenites consist of dominant microcline, but with orthoclase also present, aegirine–hedenbergite and minor biotite. The nepheline syenites contain microcline–perthite, nepheline, biotite, hastingside and minor aegirine–hedenbergite with accessory titanite, apatite and fluorite. Data on rock and mineral compositions are given by Kost'uk and Bazarova (1966).

References Kost'uk and Bazarova, 1966. Osokin *et al.*, 1974.

---

6 SOROK   53°01'N; 99°59'E

Peralkaline syenites and granosyenites are dominant in this intrusion. The syenites are coarse-grained, massive rocks comprising alkali feldspar (60–70%) and femic minerals (25–30%) including dominant aegirine–hedenbergite, but also amphibole and biotite.

Reference Kost'uk and Bazarova, 1966.

---

7 KHAITANSKII   52°50'N; 100°50'E

This occurrence is mainly composed of melanocratic nepheline syenites. It cuts a Proterozoic limestone sequence.

Reference Kost'uk and Bazarova, 1966.

---

8 BOTOGOL   52°20'N; 100°45'E
                            Fig. 151

Botogol occupies an area of 10 km². The intrusion comprises nepheline and peralkaline syenites containing a variety of xenoliths including limestone, quartzite and schist. The country rock limestones near the contact with the alkaline massif are metamorphosed with the formation of andradite, diopside and orthoclase while nepheline, apatite, zircon and cancrinite are present in some rocks. The massif has a zonal structure with nepheline syenites, together with small juvite and ijolite bodies, towards the centre and nepheline-free syenites at the margins. The nepheline syenites have a trachytic texture and comprise microcline (40–60%), nepheline (20–40%) and aegirine–hedenbergite (15–30%). The syenites are predominantly leucocratic rocks, with modal aegirine–hedenbergite varying from 5 to 30%, but in the vicinity of the outer contact the proportion of dark minerals increases and they become more melano-

cratic. Within the massif large concentrations of graphite have been discovered, the majority of these occurrences being related to xenoliths of carbonate rocks and localized in nepheline syenites and occasionally in peralkaline pyroxene syenites. For a detailed account of rock and mineral compositions see Lobzova (1975) and Kost'uk and Bazarova (1966).

Economic About 30 graphite bodies have been located in the complex, mining having been confined to the northern area.

Age K–Ar determinations on biotite gave 324 ± 12 Ma and on nepheline 378 ± 12 Ma (Lobzova, 1975).

References Kost'uk and Bazarova, 1966; Lobzova, 1975.

---

9 KHUSHAGOL   52°18'N; 100°47'E

This massif covers an area of about 7 km² and cuts Proterozoic schists, granites and limestones, the last of which contain graphite. The intrusion consists of syenites, in which gabbro xenoliths occur, peralkaline syenites, which are the dominant rocks of the complex and include riebeckite and riebeckite plus pyroxene types, granosyenites and nepheline syenites. Gabbros, gabbro–diorites and syenite–diorites were the first rocks to be emplaced and occur as minor bodies of 0.2 × 0.2 km up to 0.4–2.8 km across. They are fine- and medium-grained rocks of plagioclase (45–50%), hedenbergite (15–20%), biotite (10–20%), amphibole (10–15%) and olivine (0.6%) with accessory apatite, carbonate, ilmenite, magnetite and zircon. These were followed by hedenbergite syenites then by the intrusion of nepheline syenites, which are essentially foyaites. Igneous activity was completed by the injection of aegirine–hedenbergite–riebeckite syenites. Peralkaline granites are also present, but their place in the igneous sequence is not clear, although they are likely to be a younger part of the complex. Details on rock and mineral chemistry and temperatures of crystallization will be found in Romanov (1974 and 1976).

References Konev, 1982; Kost'uk and Bazarova, 1966; Lobzova, 1975; Romanov, 1974 and 1976.

East Sayan references

*BAGDASAROV, Yu.A. 1981. Original features of composition and structures of early carbonatites. *International Geology Review*, **23**: 753–60.
*CHERNYSHEVA, E.A., SANDIMIROVA, G.P., PAHOL'CHENKO, U.A. and KUZNETSOVA, S.V. 1992. Rb–Sr age and some specific features of the genesis of the Bolshetagninskii carbonatite complex (East Sayan). *Transactions (Doklady) of the USSR Academy of Sciences. Earth Science Sections*, **323**: 942–7.

FROLOV, A.A. 1975. *Structure and metallogeny of the carbonatite massives.* Nedra, Moscow. 161 pp.

FROLOV, A.A. 1984. Iron ore deposits at the carbonatite-alkaline ultrabasic massifs with ring structure. *Geology of Ore Deposits,* **1**: 9–21.

GRINENKO, L.N., KONONOVA, V.A. and GRINENKO, V.A. 1970. Isotopic composition of sulphur of sulphides from carbonatites. *Geokhimiya. Akademiya Nauk SSSR, Moskva,* **1**: 66–75.

KONEV, A.A. 1970. *Zhidoy alkaline-ultrabasic pluton.* Nauka, Moscow, 84 pp.

KONEV, A.A. 1982. *Nepheline rocks of the Sayan-Baikal mountain region.* Nauka, USSR Academy of Sciences, Novosibirsk. 201 pp.

*KONEV, A.A., VOROB'YEV, Ye.I., PAVLOVA, L.V. and BRANDT, G.S. 1984. Geochemical and isotopic data on the origin of calcite in the Baykal region nepheline rocks. *Geochemistry International,* **21**: 131–7.

KONONOVA, V.A. 1976. *The jacupirangite–urtite series of alkaline rocks,* Nauka, Moscow, 214 pp.

*KONONOVA, V.A. and YASHINA, R.M. 1984. Geochemical criteria for differentiating between rare metal carbonatites and barren carbonatite-like rocks. *Indian Mineralogist (Sukheswala Volume),* 136–50.

KOST'UK, V.P. and BAZAROVA, T.U. 1966. *Petrology of the alkaline rocks from the eastern part of the East Sayan.* Nauka, Moscow. 168 pp.

LOBZOVA, R.V. 1975. *Graphite and alkaline rocks of the Botogol massive area.* Nauka, Moscow. 123 pp.

OSOKIN, E.D., LAPIN, A.V., KAPUSTIN Yu.L. POHVISNEVA, E.A. and ALTUHOV, E.N. 1974. Alkaline provinces of Asia. Siberian-Pacific group. *In* L.S. Borodin (ed.) *Principal provinces and formations of alkaline rocks.* 91–166. Nauka, Moscow.

POZHARITSKAYA, L.K. and SAMOILOV, V.S. 1972. *Petrology, mineralogy and geochemistry of carbonatite from East Siberia.* Nauka, Moscow. 266 pp.

ROMANOV, I.A. 1974. Mineralogy and petrochemistry of the gabbro diorite from the Khyshagol massive (East Sayan). *Problems of the petrography and mineralogy of the basic and ultrabasic rocks from Eastern Siberia.* Siberian Branch of the USSR Academy of Sciences, Irkutsk, 79–88.

ROMANOV, I.A. 1976. Specific features of tin distribution in the rocks and minerals of the multistage Khyshagol massive (East Sayan). *Problems of the mineralogy and geochemistry of the igneous rocks from East Siberia.* Siberian Branch of the USSR Academy of Sciences, Irkutsk, 90–7.

SAMOILOV, V.S. 1977. *Carbonatites.* Nauka, Moscow. 291 pp.

*In English

KUZNETSK-MINUSINSK

The Kuznetsk-Minusinsk Province occupies the territory of the Salairids (the Salairian phase is part of the Caledonian Orogeny) in the Kuznetsk Alatai (Fig. 152) and an approximately north–south-trending system of post-Caledonian troughs, including the Minusinsk Depression, and others, which separate the province from the Eastern Sayan Ripheids (Upper Proterozoic). The province contains numerous alkaline massifs which extend over an area of about 120,000 km² (Fig. 152). The alkaline rocks were emplaced in Lower to Middle Devonian times and are represented by effusives (basanite, tephrite, phonolite), as well as genetically related intrusive complexes, including alkaline gabbro, nepheline syenite and sometimes urtite, ijolite and alkaline syenite. According to the data of Andreeva (1968) and Kononova (1976) the compositions of these massifs greatly depend upon specific features of their basement structure. In particular, the alkaline massifs which include urtite, ijolite and juvite (Kiya-Shaltyr, Kurgusul and Belogorsk) are confined to the earliest consolidated structures of intra-geosynclinal troughs, which have the most complete successions of Sinian (Upper Proterozoic) and Cambrian rocks. These basement rocks are composed predominantly of green schists, terrigenous limestones and volcanic rocks, while pure carbonate rocks are subordinate. Considerable thicknesses of these rocks (up to 6 km) accumulated in the parts of the intra-geosynclinal troughs which underwent the most subsidence. Nepheline syenites of foyaite type are widespread in occurrences located within stable blocks, for example the Patyn and Balankul massifs, which are characterized by having undergone less subsidence at the geosynclinal stage of development, so that there are notably reduced thicknesses (up to 3 km) of sediments, predominantly limestones and dolomites. Nepheline syenites of miaskitic type, found in, for example, the Pestraya and Bericul' massifs, are widespread within intermediate zones, which separate the very large structures mentioned above. The structures of the intermediate zones are characterized by the widespread development of variously orientated faults and fracture zones as well as by unusually sharp variations in the facies and lithology of the rocks.

1 Bericul'	16 Dedova
2 Batanaueskii	17 Telyashkin Ulus
3 Semenovskii	18 Burovskii
4 Goryachegorskii	19 Vysokaya
5 Gavrilovskii	20 Saybar
6 Cheremushinskii	21 Tyrdanov
7 Kurgusul'	22 Bulankul'
8 Andryushkina Rechka	23 Patyn
9 Pestraya	24 Matyr
10 Belogorskii	25 Synzas
11 Nichkuryup	26 Kultaiginskii
12 Tuluyul'	27 Karatag
13 Kiyashaltyr'	28 Kobarzinskii
14 Petropavlovskii	29 Sokol
15 Kiiskii	

1 BERICUL' 55°30'N; 88°33'E
Fig. 153

This stock-like body, which has an area of about 6 km², is mainly composed of gabbro, pyroxenite and rocks intermediate between gabbro and syenite. These rocks are cut by peralkaline syenite and dykes of nepheline

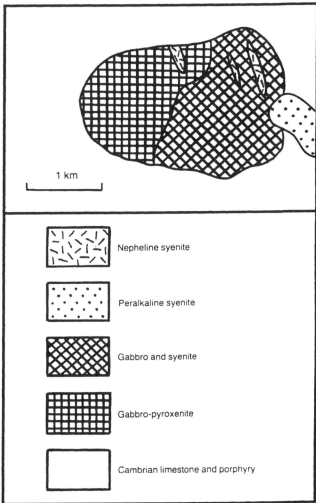

Nepheline syenite

Peralkaline syenite

Gabbro and syenite

Gabbro-pyroxenite

Cambrian limestone and porphyry

Fig. 153. Bericul' (after Vrublevsky, 1963, Fig. 1).

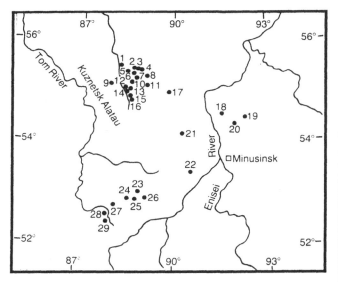

Fig. 152. The distribution of alkaline rocks in the Kuznetsk-Minusinsk Province (after Kononova, 1976, Fig. 13b and Luchitsky, 1960. Fig. 21).

syenite. The nepheline syenite dykes are 3–12 m thick and vary in structure and composition with foyaite, nepheline syenite aplite, miaskitic nepheline syenite, nepheline monzonite and pulaskite represented. Some dykes have complex structures, for example, foyaite with central aegirine zones and marginal miaskitic aplite, or central foyaite zones with marginal melanocratic nepheline monzonite. The foyaites are composed of nepheline (20–45%), microcline (up to 40%), aegirine–augite, aegirine, hastingsite and biotite. Secondary minerals are cancrinite, liebenerite, zeolite and accessories include Fe–Ti oxide minerals, titanite, apatite and fluorite. The miaskitic nepheline syenites have modal compositions of albite (32%), nepheline (23%), other feldspar (25%), hastingsite, biotite and muscovite. The secondary and accessory minerals are the same as in the foyaites. Nepheline monzonites comprise hastingsite (28%), nepheline (25%), plagioclase (20%), K-feldspar (12%) and andradite (13%) and sometimes contain biotite and clinopyroxene. Peralkaline syenites, which include pulaskites and nordmarkites, have porphyritic textures; they are mainly composed of microcline–perthite and aegirine–augite and sometimes Fe–Ti oxide minerals; accessories are apatite, titanite and zircon.

Age K–Ar on nepheline syenite gave 414 Ma (Skobelev, 1963).

References Skobelev, 1963; Vrublevsky, 1963.

---

2 BATANAULSKII   55°28′N; 88°38′E

This sheet-like body covers an area of 5 × 0.6–1.5 km. It is surrounded by Lower and Middle Devonian volcanic rocks and is composed of theralite porphyry, nepheline monzonite and bereshite (a monzonite or monzogabbro rich in nepheline phenocrysts and with analcime and augite rimmed by aegirine). These rocks are cut by dykes of peralkaline and mildly alkaline syenite and dolerite. According to Skobelev (1963) nine lava flows of bereshite have been distinguished.

References Dovgal' and Shirokih, 1980; Skobelev, 1963.

---

3 SEMENOVSKII   55°28′N; 88°49′E

Semenovskii occupies an area of about 6 km² and is surrounded by volcanic rocks of Devonian age. It comprises several stocks, sills and dykes composed of melteigite, theralite, nepheline monzonite, nepheline syenite, dolerite and calc-alkaline and peralkaline syenites.

Reference Dovgal' and Shirokih, 1980.

---

4 GORYACHEGORSKII   55°27′N; 88°55′E
                                                          Fig 154

The stock-like Goryachegorskii intrusion of 0.9 km² cuts volcanic rocks of Lower Devonian age. It is composed, from west to east, of theralite and leucotheralite, which comprise the first phase, feldspathic ijolite and urtite, the second phase, and nepheline syenite, the third phase. Theralite is the most abundant rock type and consists of nepheline (56–60%), plagioclase of basic and intermediate composition (20–25%), alkali feldspar (2–3%), clinopyroxene, mainly fassaite, (15%), olivine, alkali amphibole, biotite and Fe–Ti oxide minerals. Feldspathic urtites (nepheline-rich nepheline syenite) are characterized by dominant nepheline (61%) with sodic plagioclase (21%), clinopyroxene (13%), olivine, alkali amphibole and Fe–Ti oxide minerals. The nepheline syenites comprise alkali feldspar and sodic and sodic-

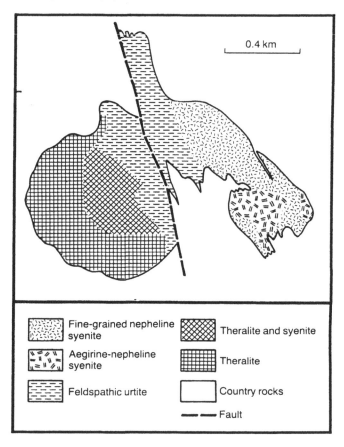

Fig. 154. Goryachegorskii (after Kononova, 1976. Fig. 14, II and including data of V.G. Mikhalev).

calcic plagioclase (55–65%), nepheline (20–25%) and mafic minerals, including aegirine.

Economic This intrusion has been prospected for the recovery of nepheline for alumina production.

References Bazhenov, 1963; Kononova, 1976.

---

5 GAVRILOVSKII   55°18′N; 88°35′E

Nepheline syenite forms small bodies (0.25 × 0.125 km) which cut a large intrusion of gabbro, pyroxenite and rocks intermediate between gabbro and syenite. The nepheline syenite comprises microcline (55%), cancrinite and nepheline (20%), albite (10%), clinopyroxene (9%) and biotite as well as titanite, zircon, apatite, orthite, fluorite, calcite and liebenerite.

Reference Vrublevsky and Kortusov, 1963.

---

6 CHEREMUSHINSKII   55°17′N; 88°43′E

The Cheremushinskii intrusion has an area of 1.25 km² and cuts metamorphic rocks of Upper Proterozoic age and Cambrian volcanics. It is composed of gabbro, melteigite and nepheline syenite. The melteigite has phenocrysts of nepheline (10%) and clinopyroxene (10%) in a groundmass of nepheline (55%) and pyroxene (40%) varying from titanaugite to aegirine–augite. The nepheline syenite consists of nepheline (40%), feldspar (20%), andesine (8%), augite and aegirine–augite (8%), arfvedsonite (14%), olivine (3%) and cancrinite.

Reference Dovgal' and Shirokih, 1980.

7 KURGUSUL' 55°16'N; 88°45'E
Fig. 155

The principal massif has an oval shape and occupies an area of about 1 km². It cuts marbles of Proterozoic age and consists mainly of nepheline syenites, of which those towards the centre are distinguished by their trachytic texture, containing nepheline (26–48%), anorthoclase (34–38%), aegirine–augite (6–13%) and hastingsite (4–7%), as well as apatite and Fe–Ti oxide minerals. The northwestern part of the massif is composed of theralite and nepheline-bearing gabbro (essexite), which occupy 23% and 2% of the area of the massif respectively. The theralite consists of andesine (56%), liebeneritized nepheline (25%), aegirine–augite and amphibole (15%). The essexite is characterized by a lower nepheline content. Dykes of tinguaite and teschenite are associated with the complex.

References Ivashkina, 1963; Osokin *et al.*, 1974.

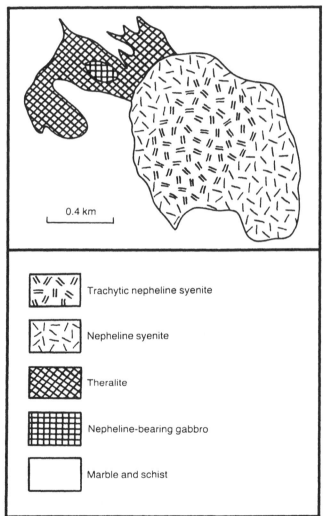

Trachytic nepheline syenite

Nepheline syenite

Theralite

Nepheline-bearing gabbro

Marble and schist

Fig. 155. Kurgusul' (after Ivashkina, 1963, Fig. 22 with additions according to A.V. Bozin and V.G. Mikhalev).

8 ANDRYUSHKINA RECHKA 55°15'N; 89°05'E

Bereshite (see locality 2) sheets and cross-cutting dykes have been observed among nappes of diabase porphyry and basalt in the northwestern part of the Minusinsk trough. Bereshite is notable for containing segregations of idiomorphic nepheline phenocrysts having diameters from 2–3 mm to 3–5 cm. Their abundance is widely variable and they are usually partially or completely replaced by secondary minerals such as cancrinite, zeolite and hydromica. Small euhedral phenocrysts of augite also occur. These rocks, besides nepheline, also contain clinopyroxene of varying composition (titanaugite, aegirine–augite), plagioclase (labradorite to oligoclase), and sometimes olivine, as well as accessory apatite and Fe–Ti oxide minerals. The chemical composition of these rocks can be found in Luchitsky (1960).

Age K–Ar on nepheline gave 444 ± 20 Ma (Kononova and Shanin, 1973).

References Kononova and Shanin, 1973; Luchitsky, 1960.

9 PESTRAYA 55°08'N; 87°38'E
Fig. 156

The alkaline complex of Pestraya forms a steep-sided body that has an area of about 10 km² and is elongated in an approximately north–south direction. The intrusion cuts granite and granodiorite of Proterozoic and Upper Cambrian age, as well as Upper Cambrian and Devonian pyroxenite, peridotite, gabbro-pyroxenite and diorite. The intrusion consists of various types of nepheline syenite which are distributed in a distinct banded structure. Various types of miaskitic nepheline syenite and medium-grained foyaite alternate with each other from east to west. The composition of the nepheline syenites varies slightly. The foyaites contain microcline–perthite, nepheline, albite and hastingsite with lesser amounts of lepidomelane, garnet, fluorite, titanite, apatite, Fe–Ti oxide minerals and sulphides. In the miaskitic nepheline syenites the prevailing dark-coloured mineral is lepidomelane and there is more albite than in the foyaites. The banded structure of the miaskites is mainly determined by variations of grain sizes and sometimes of composition. The widest bands (2–10 m) are composed of coarse-grained nepheline syenite containing the hackmanite variety of sodalite.

Age K–Ar on foyaite gave 358 ± 30 Ma and on miaskite 390 ± 16 Ma (Andreeva, 1968).

References Andreeva, 1968; Kortusov, 1963a.

10 BELOGORSKII 55°12'N; 88°38'E
Fig. 157

This occurrence, which has an area of about 1 km², cuts limestones of Cambrian age. Some 82% of the area of the complex is composed of theralite with nepheline syenites present in the northern part. Leucotheralite or theralite-syenite surrounds the nepheline syenite. These rocks grade gradually into each other. Close to the southwestern contact is a zone of melteigite.

References Dovgal' and Shirokih, 1980; Kononova, 1976; Osokin *et al.*, 1974.

11 NICHKURYUP 55°06'N; 89°07'E

Intrusive sheets of nephelinite and phonolite are encountered within diabase nappes. The nephelinites contain abundant small nepheline phenocrysts in a groundmass composed of titanaugite, alkali amphibole, magnetite

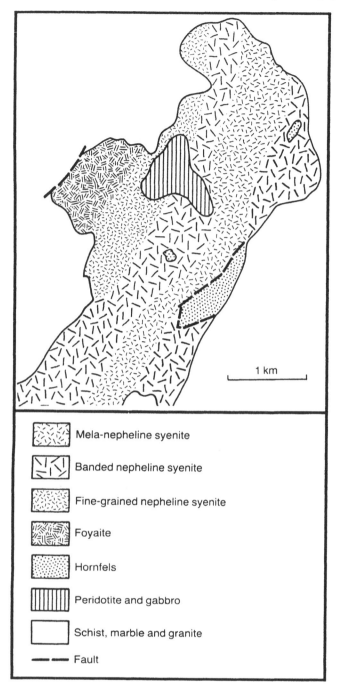

Mela-nepheline syenite

Banded nepheline syenite

Fine-grained nepheline syenite

Foyaite

Hornfels

Peridotite and gabbro

Schist, marble and granite

━ ━ ━ Fault

Fig. 156. Pestraya (after Andreeva, 1968, Fig. 4).

and apatite. In the phonolites the phenocrysts are alkali feldspar and albite. The groundmass has a trachytic texture and consists of thin laths of alkali feldspar, pseudomorphs after nepheline, aegirine and Fe–Ti oxide minerals. The chemical composition of the rocks can be found in Luchitsky (1960).

Reference Luchitsky, 1960.

─────────────

12 TULUYUL'

55°05′N; 88°11′E
Fig. 158

This small alkaline intrusion is confined to the contact zone of Lower Cambrian limestones and Upper Cam-

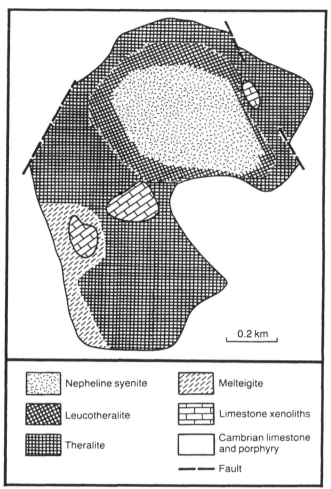

Nepheline syenite

Leucotheralite

Theralite

Melteigite

Limestone xenoliths

Cambrian limestone and porphyry

━ ━ ━ Fault

Fig. 157. Belogorskii (after Kononova, 1976, Fig. 14, IY and including data of V.G. Mikhalev).

brian–Lower Ordovician granitoids. The intrusion is predominantly composed of alkaline gabbro, theralite and feldspathic ijolite but nepheline syenite also outcrops in the eastern and northern parts of the intrusion and is partly replaced by liebenerite and altered to liebenerite syenite of feldspar, liebenerite (30%), albite and relicts of coloured minerals. Dykes of various compositions are known including diabase, lamprophyre, nepheline syenite and microsyenite.

Age K–Ar on nepheline syenite gave 384 Ma (Andreeva, 1968).

References Andreeva, 1968; Dovgal' and Shirokih, 1980; Skobelev, 1963.

─────────────

13 KIYA-SHALTYR'

55°04′N; 88°33′E
Fig. 159

The complex, which has an area of 2.1 km² and an irregular form, is located along the contact of limestone and volcanic-sedimentary sequences of Lower Cambrian age. The major part, some 75% of the area, is composed of gabbro and leucocratic gabbro with a trachytic texture, which were the first phases to be emplaced. In the southwestern part of the complex an elongate urtite

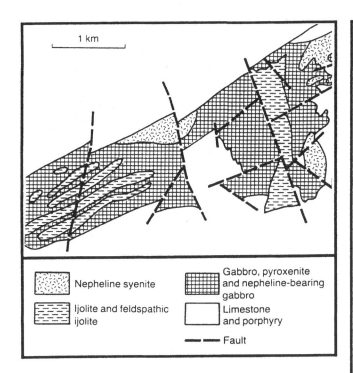

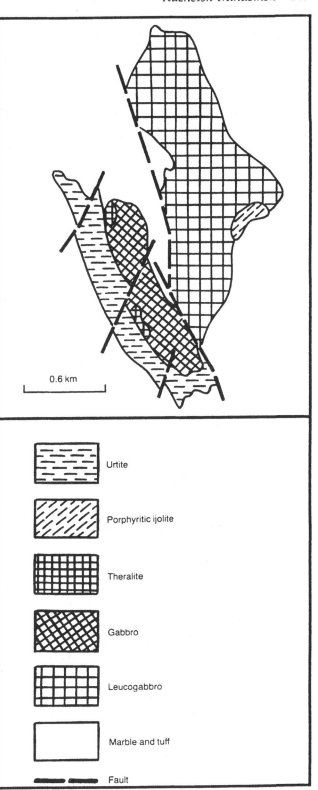

Fig. 158. Tuluyul' (after Skobelev, 1963, Fig. 1 including data of P.V. Osipov and N.A. Makarenko).

body occurs. Theralite and porphyritic ijolite bodies are also present and the intrusion of urtite was accompanied by metasomatic alteration of the wall rocks, including nephelinization of gabbro and volcanics. Dykes of various compositions including ijolite porphyry, nepheline syenite, micro-ijolite and alkaline syenite, having thicknesses up to 3–4 m, are present in the complex. Also found are dykes of nepheline–pyroxene–pyrrhotite and pyroxene–pyrrhotite rocks, which have been described by Rodygina and Grinev (1989). The urtites have been studied in detail, because of their economic potential, and consist of nepheline (75–90%), fassaite (10–25%) and accessory apatite, pyrrhotite and Fe–Ti oxides. The theralites have equigranular or porphyritic textures amd consist mainly of plagioclase (54%) and fassaite (30%) as well as small quantities of nepheline (8%), olivine (4%), biotite, barkevikite and minor apatite, pyrrhotite and Fe–Ti oxide minerals. The nepheline syenite (foyaite) dykes are composed of microcline–perthite, nepheline and aegirine–augite; accessory minerals are represented by lavenite, zircon, titanite, apatite, Fe–Ti oxide and sulphide minerals. The compositions of rocks and minerals are given in publications of Kononova (1976), Andreeva (1968) and Klyushkina *et al.* (1963).

Age K–Ar on urtite gave 388 Ma (Skobelev, 1963). K–Ar on nepheline from urtite gave 426 ± 12 Ma (Kononova and Shanin, 1973).

Economic Nepheline is recovered from urtites for aluminium production.

References Andreeva, 1968; Klyushkina *et al.*, 1963; Kononova, 1976; Kononova and Shanin, 1973; Mostovsky, 1978; *Rodygina and Grinev, 1989; Skobelev, 1963.

Fig. 159. Kiya-Shaltyr' (after Mostovsky, 1978, Fig. 1).

---

14 PETROPAVLOVSKII   55°00′N; 88°12′E
                            Fig. 160

This stock of subalkaline gabbro is injected by intrusions of theralite, feldspathic ijolite-urtite and foyaite. The theralites are porphyritic rocks with nepheline pheno-crysts. The groundmass is composed of nepheline (15%), plagioclase (41%), clinopyroxene (7%), hornblende and biotite; apatite, titanite and magnetite occur as accessory phases. The feldspathic ijolites, apart from nepheline (42–60%) and clinopyroxene (30%), contain a little microcline and andesine. In nepheline syenites the pro-portions of the minerals vary greatly, notably nepheline (22–42%), feldspar (46–52%) and aegirine–augite (2–18%); hornblende and biotite are also present and the accessory minerals are titanite, apatite and Fe–Ti oxide minerals. In the central and northern parts of the com-plex small bodies and veins of carbonatite are present which, as a rule, have sharp contacts. In some places the adjacent rocks are altered with the development of garnet, idocrase, apatite, magnetite and calcite. The car-bonatite mineralogy is 20–70% calcite, 1–5% clino-pyroxene, 7–60% monticellite, 5–20% apatite, 1–10% magnetite, 0–50% phlogopite. Homogenization tem-peratures have been determined for micro-inclusions in carbonatite minerals and have up to 890°C for monti-cellite, 890–700°C for clinopyroxene, 700°C for phlogo-pite, 650–550°C for apatite and 590–400°C for calcite.

References Metshanskaya, 1963; Vrublevsky *et al.*, 1989.

15 KIISKII 54°55′N; 88°32′E

Kiiskii comprises numerous dykes and dyke-like bodies of nepheline syenite which cut syenites of the Udarninsk pluton. The nepheline syenite usually has a gneissose structure and consists of nepheline (18–41%), orthoclase-perthite (32–60%), albite (2–20%), hastingsite and biotite (2–30%); sometimes aegirine–augite (5–9%) is present. The usual accessory minerals are titanite, apatite and Fe–Ti oxide minerals.

Reference Kortusov, 1963b.

16 DEDOVA 54°55′N; 88°35′E
Fig. 161

The Dedova alkaline massif, which forms a small moun-tain, cuts plagioclase porphyries of Lower Cambrian age. It is composed of leucocratic trachytic textured

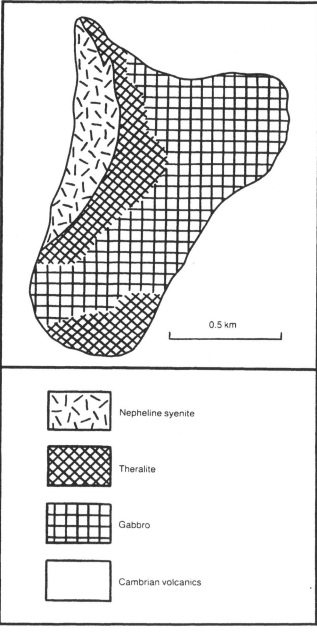

Fig. 161. Dedova (after Andreeva, 1968, Fig. 2).

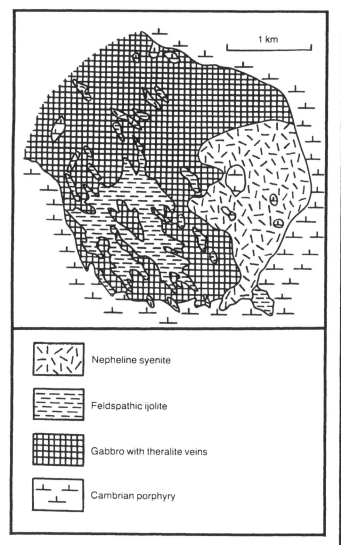

Fig. 160. Petropavlovskii (after Vrublevsky et al., 1989, Fig. 1).

gabbro, theralite and nepheline syenite. The last forms two bodies the larger of which forms an elongate lens-like intrusion situated along the western boundary of the gabbro while the smaller is a dyke. Theralite lies between the nepheline syenite and gabbro. The nepheline syenite is a foyaite consisting of alkali feldspar, nepheline, tita-naugite and accessory titanite, apatite and Ti–Fe oxides. The nepheline syenite of the dyke is an albitized ferro-hastingsite–nepheline syenite in which a varied suite of accessory minerals has been found including yttrium-bearing garnet, lavenite, zircon, eudialyte, orthite, Fe–Ti oxide minerals and sulphides.

Age K–Ar on nepheline from the nepheline syenite dyke gave 316 ± 17 Ma (Andreeva, 1968; Osokin *et al.*, 1974).

References Andreeva, 1968; Osokin *et al.*, 1974.

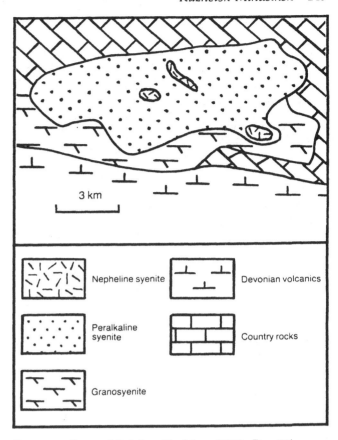

Fig. 162. *Burovskii (after Yashina, 1982, Fig. 19).*

17 TELYASHKIN ULUS 54°57′N; 89°50′E

This is a small stock having a diameter of up to 100 m and composed predominantly of teschenite. The stock is confined to a dome formed by volcanogenic strata of Devonian age and is intruded into lower Devonian rocks. The teschenites contain needle-like barkevikite crystals and segregations of labradorite laths with the interstices occupied by aggregates of zeolite and anal-cime. Clinopyroxenes, which may be zoned from fas-saite in the centre to aegirine–augite at the margins, occur together with small quantities of alkali feldspar and nepheline; a little apatite and Fe–Ti oxide minerals are also present. The teschenite is characterized by the presence of occasional fine veinlets of urtite comprising nepheline, aegirine–augite, hornblende, apatite and analcime. Chemical analyses of the rocks are given by Luchitsky (1960).

Reference Luchitsky, 1960.

18 BUROVSKII 54°32′N; 92°01′E
Fig. 162

The Burovskii occurrence is located in a contact zone of Cambrian rocks, including limestone, with Lower Devonian volcanics. In plan the intrusion is oval in shape and occupies an area of about 70 km². It is composed of coarse-grained, often trachytic peralkaline syenites, which contain 85–90% perthitic alkali feldspar as well as aegirine–augite, and rarely aegirine. Quartz-bearing peralkaline syenites are also present. Nepheline syenite forms a number of small bodies among the peralkaline syenites. They are porphyritic with alkali feldspar phenocrysts up to 5–6 cm in diameter. The nepheline syenites also contain 10–25% nepheline, in some varieties 4–8% sodalite and 6–8% mafic minerals including aegirine–augite, amphibole and biotite. The nepheline syenites are albitized, with the formation in some places of albitite. The last intrusions are aphyric and porphyritic basalt dykes and syenite porphyry veins.

References Luchitsky, 1960; Yashina, 1982.

19 VYSOKAYA 54°25′N; 92°40′E

This intrusion occurs amongst limestones and volcanic rocks of Cambrian age. Pink, coarse-grained, peralkaline syenites are the dominant rocks and consist predominantly of alkali feldspar, together with a little nepheline

and aegirine. Small bodies of aegirine–nepheline syenite and dykes of fine-grained syenite are present in the intrusion. Nepheline is replaced, as a rule, by an aggregate consisting of secondary hydromica, cancrinite and zeolite.

Reference Luchitsky, 1960

20 SAYBAR 54°15′N; 92°25′E

The alkaline rocks of this occurrence, which are predominantly peralkaline syenites, form intrusive sheets. The country rocks are siliceous shales and volcanics of variable composition; there are limestones which are presumed to be of Cambrian age. The peralkaline syenites are mainly represented by two varieties: coarse-grained trachytic syenite and fine-grained, massive granosyenite. The latter forms dyke-like bodies among the coarse-grained syenite. Within the coarse-grained peralkaline syenite intrusive sheets of nepheline syenite are also found; their size varies from 10 × 10 to 200 × 500 m. The nepheline syenites contain nepheline, aegirine, alkali feldspar, albite–oligoclase, alkali amphibole and such accessory minerals as apatite, fluorite, titanite and Fe–Ti oxide minerals. The texture is usually trachytic with idiomorphic nepheline and aegirine, the former usually being replaced by natrolite and hydromica. The nepheline syenites have a banded structure reflecting quantitative variations of the aegirine content. Chemical analyses of the rocks are given by Luchitsky (1960).

Reference Luchitsky, 1960.

21 TYRDANOV

54°10′N; 90°25′E
Fig. 163

This small peralkaline body cuts Cambrian limestones. The alkaline rocks have a banded structure produced by alternating leucocratic and melanocratic varieties. Nepheline syenite–diorite, nepheline syenite and syenite are the most abundant rock types and they contain xenoliths of gabbro and marble derived from the country rocks. The nepheline-bearing rocks consist of up to 35% nepheline, up to 40% fassaite, plagioclase of An_{17-18} or An_{23-27} in nepheline syenites but andesine in nepheline syenite–diorite. Microcline is poikilitic and includes plagioclase, clinopyroxene and amphibole. Secondary minerals are liebenerite, sericite and muscovite; accessories include andradite, calcite, titanite and apatite.

References Bognibov, 1979; Dovgal' and Shirokih, 1980; Luchitsky, 1960.

---

22 BULANKUL'

53°25′N; 90°35′E
Fig. 164

The Bulankul' occurrence, which has an area of about 3 × 0.5 km, lies at the contact of Lower Cambrian limestones with intrusions of gabbro–diorite and syenite of the same age. The alkaline complex was emplaced in the order essexite, nepheline-bearing diorite–syenite, nephe-

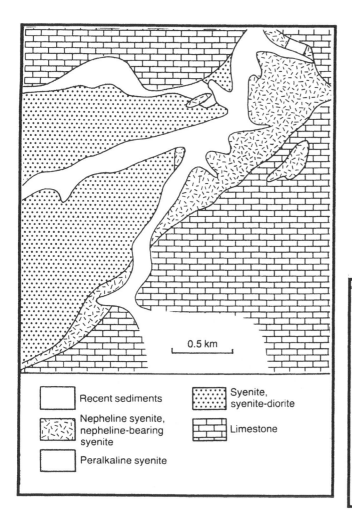

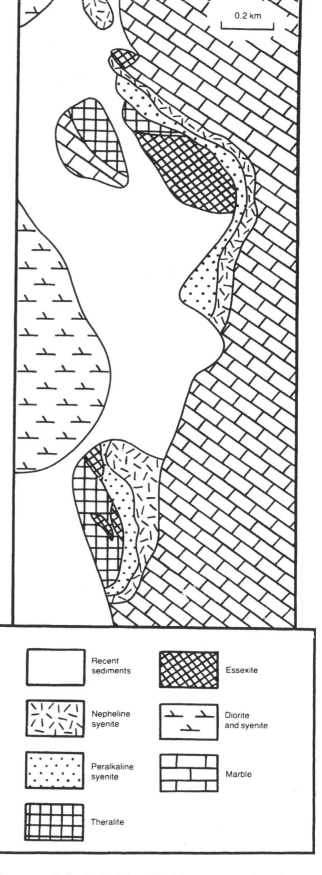

Fig. 163. Tyrdanov (after Dovgal' and Shirokih, 1980, Fig. 12).

Fig. 164. Bulankul' (after Shokhina, 1961, Fig. 1).

line and nepheline-bearing syenite followed by a vein series of tinguaite and pegmatitic foyaite. The major part of the massif is composed of essexite which has the composition plagioclase (An$_{55-45}$) (47–65%), olivine (1–3%), augite (7–18%), barkevikite (6–16%), nepheline (3–12%), biotite (5–10%), alkali feldspar (3–7%), apatite and Fe–Ti oxide minerals. Nepheline syenite consists essentially of microcline perthite (35–70%) and nepheline; albite and oligoclase (1–20%), augite and aegirine-augite (0–6%), hastingsite and barkevikite (3–20%) and biotite (1–8%) are also present; accessories include titanite, apatite and Fe–Ti oxide minerals. The nepheline diorite–syenite contains 20–65% andesine (An$_{26-37}$), nepheline, alkali feldspar (10–50%), augite, diopside (8–18%), barkevikite and arfvedsonite (14–22%), apatite, titanite and Fe–Ti oxides.

References Dovgol' and Shirokih, 1980; Osokin *et al.*, 1974; Shokhina, 1959 and 1961.

---

23 PATYN   53°03′N; 88°50′E

This massif, of an area of some 50 km², cuts Cambrian limestones, including dolomite, with layers of carbonaceous shale and sandstone. The intrusion is composed predominantly of gabbro, which contains small bodies and dykes of alkaline rocks, including dykes of melilite-nepheline and nepheline–sodalite rocks and peralkaline syenite. In the southwestern part of the massif a body of feldspathic melteigite–jacupirangite is present within recrystallized limestone and nepheline–melilite dykes are met with. The peralkaline syenites are coarse-grained rocks consisting mainly of orthoclase–perthite (65%), albite (30%), aegirine–augite (14%), biotite (1%) and accessory titanite, apatite and Fe–Ti oxides. The melilite-nepheline rocks usually contain garnet (25–60%) and admixtures of titanite, ilmenite, apatite, calcite, titanaugite and sulphides (up to 15–20%). Feldspathic melteigite consists of titanaugite (75%), plagioclase (12%) and nepheline (8%), as well as apatite, garnet and Fe–Ti oxide minerals. Pyrrhotite from a melilite rock gave d^{34}S values of +3.5 to −0.5 and melilite rocks gave δ^{18}O values of +6.0 to +8.1 and δ^{13}C of +12.5 to +20.4 (Pokrovskiy and Andreeva, 1991). The same authors obtained ^{87}Sr/^{86}Sr ratios on gabbro of 0.7043 and 0.7047, on melilite rocks of 0.7065 and 0.7077, and on marble of 0.7078–0.7087. They concluded that 60–85% of the Sr in the melilite rocks was derived from the country rocks. Chemical analyses of these rocks are given by Bogatikov (1966).

References Bogatikov, 1966; Ilyenok, 1963b; Osokin *et al.*, 1974; *Pokrovskiy and Andreeva, 1991.

---

24 MATYR   52°55′N; 88°12′E
                            Fig. 165

A narrow strip of peralkaline and nepheline syenite occurs along the contact of a Devonian volcanogenic sequence and Cambrian limestone, with the peralkaline syenite the predominant rock type. It consists of microcline–perthite and albite (up to 93%), aegirine–augite (9%), lepidomelane (6%) and minor apatite and magnetite. Nepheline syenite (foyaite) forms two elongate bodies and consists of microcline–perthite and albite (56–69%), nepheline (23–27%), aegirine–augite, barkevikite and lepidomelane, as well as accessory apatite and magnetite. The alkaline massif is crossed by a large number of dykes, up to 1–1.5 m thick, consisting of tinguaite, syenite porphyry, fine-grained foyaite and kersantite.

Reference Yanishevskaya, 1963.

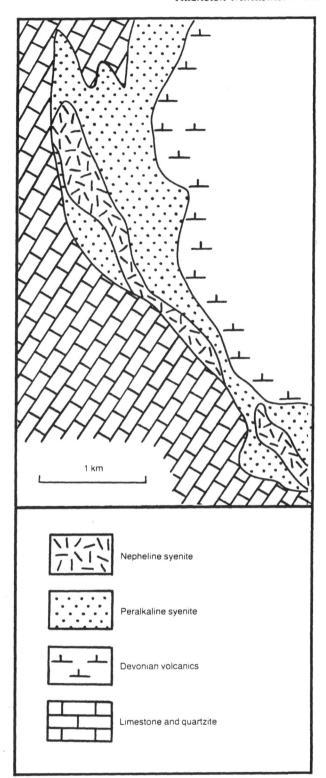

Nepheline syenite

Peralkaline syenite

Devonian volcanics

Limestone and quartzite

Fig. 165. Matyr (after Yanishevskaya, 1963, Fig. 2).

---

25 SYNZAS   52°53′N; 88°30′E

This small (0.25 km²) stock of peralkaline syenite cuts lower Cambrian limestones. The syenites are medium-grained, massive rocks which consist of alkali feldspar (60%), albite (20%), biotite and alkali amphibole (1–2%), apatite, diopside, magnetite and probable analcime. Boulders of cancrinitized foyaite have been found in recent sedimentary deposits.

Reference Zabolotnikova and Khvatov, 1963a.

---

26 KULTAIGINSKII
52°54′N; 89°00′E
Fig. 166

This oval-shaped intrusion, having an area of about 40 km², cuts Lower Cambrian sequences of sedimentary and metamorphic rocks which consist mainly of limestones. The southern part of the complex is composed of gabbro and the central and northern parts of peralkaline syenite and nordmarkite. The gabbro formed a funnel-shaped body, the northern part of which was destroyed by emplacement of the syenite and nordmarkite intrusion which takes the form of a stock. Altered xenoliths of gabbro are preserved in the syenite. In the central parts of the stock the peralkaline syenites change to nordmarkite and quartz syenite and in some places into granosyenite and granite. At the southern outer contact of the gabbro a small, 100–200 m long, occurrence of miaskitic nepheline syenite has been found. The nepheline syenite cuts both the gabbro and country rock marble and quartzite. The peralkaline syenites are coarse-grained, sometimes porphyritic, rocks consisting of orthoclase perthite (70–80%), albite (8%), aegirine-augite (7–10%), biotite, amphibole and accessory apatite, magnetite and sparse zircon.

References Dovgal' and Shirokih, 1980; Ilyenok, 1963a; Yashina, 1982.

---

27 KARATAG
52°50′N; 87°55′E

Three intrusions, with a total area of about 45 km², cut Upper Proterozoic and Lower Cambrian limestones and Devonian sandstones. The occurrence of Bol'shoi (Big) Karatag Mountain is composed of peralkaline syenite, nepheline syenite, diorite and syenite–diorite. Maly

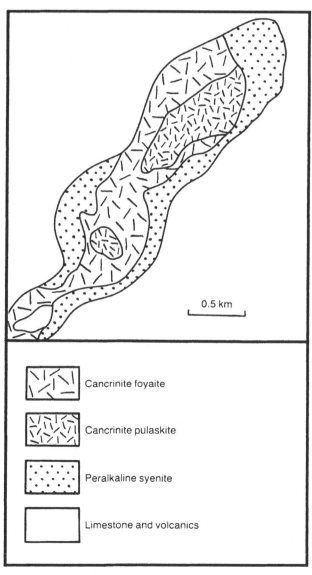

Fig. 166. Kultaiginskii (after Yashina, 1982, Fig. 15).

Fig. 167. Kobarzinskii (after Zabolotnikova and Khvatov, 1963b, Fig. 1).

(Small) Karatag Mountain, however, consists of grano-syenite, granite, mildly alkaline syenite, syenite, diorite and gabbro. Two stages are distinguished, with earlier gabbros followed by granite and syenite. Nepheline syenite from the Bol'shoi Karatag Mountain occurrence consists of alkali feldspar and albite (50–70%), nepheline and liebenerite (20–40%), aegirine–augite, barkevikite and biotite (4–11% total coloured minerals). The peralkaline syenites are rich in alkali feldspar (80–90%), with aegirine–augite (0–5%), barkevikite (0–7%) and biotite (2–12%).

Reference Dovgal' and Shirokih, 1980.

---

28 KOBARZINSKII
52°35'N; 87°33'E
Fig. 167

At Kobarzinskii peralkaline and feldspathoidal syenites cut volcanic and sedimentary sequences including lime-stones of Cambrian age. The peralkaline syenites consist of alkali feldspar (57–69%) with biotite (up to 6%) and magnetite; there has been extensive replacement by muscovite which may form as much as 32% of the rock. Nepheline syenite has been observed in alluvial blocks and consists of microcline perthite (44–59%), nepheline (17–43%), aegirine–augite (up to 7%), biotite (up to 8%), sodalite, cancrinite, muscovite and minor apatite and titanite. Feldspathoid syenite contains very little nepheline and clinopyroxene, but is enriched in cancrinite and sodalite.

Reference Zabolotnikova and Khvatov, 1963b.

---

29 SOKOL
52°29'N; 87°33'E

Covering 3.6 × 1 km, Sokol is composed of foyaite, juvite and peralkaline syenite. The juvite and foyaite consist of alkali feldspar, nepheline (15–45%), cancrinite (10–40%) and sodalite (up to 30%). The syenites include biotite, pyroxene–biotite, cancrinite and cancrinite-sodalite varieties. Dykes of various compositions occur including pseudoleucite porphyry, nepheline syenite porphyry, spessartite, grorudite and solvsbergite.

Reference Dovgal' and Shirokih, 1980.

Kuznetsk-Minusinsk references

ANDREEVA, E.D. 1968. *Alkaline magmatism of Kuznetsk Alatau.* Nauka, Moscow. 169 pp.

BAZHENOV, I.K. 1963. Nephelinitic rocks of the Goryachaia Mt. *In* I.K. Bazhenov and Y.D. Skobelev (eds) *The geology and petrography of nepheline rocks of Kuznets Alatau.* 122–6. Gosgeoltekhizdat, Moscow.

BOGATIKOV, O.A. 1966. *Petrology and metallogeny of gabbrosyenite complexes of the Altay-Sayan region.* Nauka, Moscow. 240 pp.

BOGNIBOV, V.I. 1979. The occurrences of nepheline-bearing rocks in connection with the lower Palaeozoic granitoids of the Kuznetsk Alatau. *In* V.A. Kuznetzov (ed.) *Granitoid complexes from Siberia.* 49–57. Nauka, Siberian Branch of the Academy of Sciences, Novosibirsk.

DOVGAL' V.N. and SHIROKIH, V.A. 1980. *History of the development of the high alkaline magmatism of the Kuznetsk Alatau.* Proceedings of the Institute of Geology and Geophysics, Nauka, Siberian Branch of the Academy of Sciences, Novosibirsk. 457. 216 pp.

ILYENOK, S.S. 1963a. Alkaline rocks of the Kul-Taiga locality. *In* K. Bazenov and Yu.D. Skobelev (eds) *The geology and petrography of nepheline rocks of Kuznetsk Alatau.* 216–26. Gosgeoltekhizdat, Moscow.

ILYENOK, S.S. 1963b. Alkaline rocks of the Patyn Mountain. *In* I.K. Bazenov and Yu.D. Skobelev (eds) *The geology and petrography of nepheline rocks of Kuznetsk Alatau.* 226–42. Gosgeoltekhizdat, Moscow.

IVASHKINA, R.N. 1963. Nephelinitic rocks of the Kurgusul'-Listvenny Massif. *In* I.K. Bazenov and Yu.D. Skobelev (eds) *The geology and petrography of nepheline rocks of Kuznetsk Alatau.* 78–100. Gosgeoltekhizdat. Moscow.

KLYUSHKINA, A.V., PRUSEVICH, A.M. and SKOBE-LEV, Yu.D. 1963. The Kiya-Shaltyr massif of alkaline gabbroids. *In* I.K. Bazenov and U.D. Skobelev (eds) *The geology and petrography of nepheline rocks of Kuznetsk Alatau.* 46–77. Gosgeoltekhizdat, Moscow.

KONONOVA, V.A. 1976. *The Jacupirangite-urtite series of alkaline rocks.* Nauka, Moscow. 213 pp.

KONONOVA, V.A. and SHANIN, L.L. 1973. Application of the radiogenic argon method on nepheline for age determinations of alkaline complexes. *In* G.D. Afanas'ev (ed.) *Geologic-radiologic interpretation of contradictory age determinations.* 40–52. Nauka, Moscow.

KORTUSOV, M.P. 1963a. Nepheline syenite of the upper reaches of the Toydona River (the Pestraya Mt.) *In* I.K. Bazenov and Yu.D. Skobelev (eds) *The geology and petrography of nepheline rocks of Kuznetsk Alatau.* 151–60. Gosgeoltekhizdat, Moscow.

KORTUSOV, M.P. 1963b. The Kiisk gabbro-syenite intrusive complex of the Maryinskaya Taiga (Kuznetsk Alatau). *In* Yu.A. Kuznetsov (ed.) *Magmatic complexes of the Altay-Sayan fold area.* 78–91. USSR Academy of Sciences, Novosibirsk.

LUCHITSKY, I.V. 1960. *Volcanism and tectonics of Devonian depressions of the Minusinsk intermontane trough.* Publishing House of the USSR Academy of Sciences, Moscow. 275 pp.

METSHANSKAYA, L.B. 1963. Nepheline rocks of the Petropavlovsk locality. *The geology and petrography of nepheline rocks of Kuznetsk Alatau.* 100–22. Gosgeoltekhizdat, Moscow.

MOSTOVSKY, A.I. 1978. Formation conditions of alkaline massifs and associated nepheline ores in Kuznetsk Alatau. *In* V.P. Petrov (ed.) *Nepheline ore deposits.* 66–70. Nauka, Moscow.

OSOKIN, E.D., LAPIN, A.V., KAPUSTIN, Yu.L., POHVISNEVA, E.A. ALTUHOV, E.N. 1974. Alkaline provinces of Asia. Siberian-Pacific roup. *In* L.S. Borodin (ed.) *Principal provinces and formations of alkaline rocks.* 91–166. Nauka, Moscow.

*POKROVSKIY, B.G. and ANDREEVA, Ye.D. 1991. Petrography and isotope geochemistry of melilite rocks associated with the Patyn pluton. *International Geology Review*, 33: 689–703.

*RODYGINA, V.G. and GRINEV, O.M. 1989. Nepheline-pyroxene-pyrrhotite and pyroxene-pyrrhotite rocks of the Kiya-Shaltyr pluton. *International Geology Review*, 31: 72–8.

SHOKHINA, O.I. 1959. Specific features of geology of the Bulankul' alkaline massif (south of the Krasnoyarsk Region). *Proceedings of Higher Educational Institutions. Geology and Exploration, Moscow,* 3: 54–64.

SHOKHINA, O.I. 1961. *Alkaline rocks of the Bulancul' massif (the Krasnoyarsk Region).* Publishing House of the Siberian Branch of the USSR Academy of Sciences, Novosibirsk. 70 pp.

SKOBELEV, Yu.D. 1963. The Tuluyul massif of alkaline rocks. In I.K. Bazenov and Yu.D. Skobelev (eds) *The geology and petrography of nepheline rocks of Kuznetsk Alatau.* 126–34. Gosgeoltekhizdat, Moscow.

VRUBLEVSKY, V.A. 1963. The geology and petrography of nepheline syenite in the Staryi Bericul' area in Maryinskaya Taiga. *In* I.K. Bazenov and Yu.D. Skobelev (eds) *The geology and petrography of nepheline rocks of Kuznetsk Alatau.* 135–50. Gosgeoltekhizdat. Moscow.

VRUBLEVSKY, V.A. and KORTUSOV, M.P. 1963. Nepheline syenite of the right side of the Kiya River in the vicinity of the Gavrilovka Settlement. *In* I.K. Bazenov and Yu.D. Skobelev (eds) *The geology and petrography of nepheline rocks of Kuznetsk Alatau.* 193–201. Gosgeoltekhizdat, Moscow.

VRUBLEVSKY, V.V., BABANSKY, A.D., TRONEVA, N.V. and ELISAFENKO, V.N. 1989. Conditions of formation of carbonatite minerals in Kuznetsk Alatau. *Izvestiya Akademii Nauk SSSR, Seriya Geologiya,* **12:** 65–81.

YANISHEVSKAYA, I.A. 1963. Alkaline and nepheline syenite of the Matyr locality. *In* I.K. Bazenov and Yu.D. Skobelev (eds) *The geology and petrography of nepheline rocks of Kuznetsk Alatau.* 160–73. Gosgeoltekhizdat, Moscow.

YASHINA, R.M. 1982. *Alkaline magmatism of fold-block areas.* Nauka, Moscow. 275 pp.

ZABOLOTNIKOVA, I.I. and KHVATOV, V.V. 1963a. Synzas locality. *In* I.K. Bazenov and Yu.D. Skobelev (eds) *The geology and petrography of nepheline rocks of Kuznetsk Alatau.* 173–7. Gosgeoltekhnizdat. Moscow.

ZABOLOTNIKOVA, I.I. and KHVATOV, V.V. 1963b. Nepheline and sodalite–cancrinite rocks of the Kobarzinsk locality. *In* I.K. Bazenov and Yu.D. Skobelev (eds) *The geology and petrography of nepheline rocks of Kuznetsk Alatau.* 177–93. Gosgeoltekhizdat, Moscow.

* In English

EAST TUVA

Numerous alkaline complexes are known in eastern Tuva, where they form an approximately north–south arc in the eastern part of the Tuva region adjacent to the border with Mongolia. Three subprovinces can be identified (Fig. 168). The first is located on the southern slopes of an anticlinorium of East Sayan and extends into the northeastern part of Tuva (Fig. 168, area A). Although these occurrences are adjacent to those of East Sayan they are traditionally included in the East Tuva province, principally because they are located on the same Caledonian structures. The second subprovince lies in eastern Tuva and is part of the Okinskii block (Fig. 168, area B), and the third subprovince is in the Sangilen Upland (Fig. 168, area C). The occurrences vary in area from 1–2 to 55–65 km² but the largest, the Kadyross complex, covers approximately 150 km². The first two subprovinces have much in common in so far as the structures and rock types are concerned. Two or sometimes three magmatic phases are distinguished. During the first, and principal, intrusive phase peralkaline syenites are predominant with less frequently miaskitic nepheline syenites, juvites and feldspathic urtites. The second intrusive phase is generally represented by peralkaline and mildly alkaline granites, granosyenites and quartz syenites but, according to some authors,

peralkaline granites are likely to be found as a third phase also. The alkaline complexes of southeastern Tuva, the third subprovince, are rich in nepheline syenites with urtites, ijolites and juvites also common.

The province of East Tuva extends across the international border into Mongolia and the province as a whole is reviewed, in English, by Pavlenko (1974) under the name of the 'Mongol-Tuva province'.

1 Katun	19 Bayankol
2 Chamdzhyak	20 Vostochno-
3 Aksug	Honchulskii
4 Irelig	21 Ust'-Haigasskii
5 Dotot	22 Terekhol
6 Chongai	23 Khaltrik
7 Sorug	24 Chartis
8 Ulagarchin	25 Pichekhol
9 Karakhol	26 Solbelder
10 Dugda	27 Dakhunur
11 Kadyross and Kyshtag	28 Toskul
12 Surkhai	29 Ust'-Kunduss
13 Kadrauss	30 Khorygtag
14 Chavach	31 Chekbek
15 Verkhnebrenskii	32 Agash
16 Ulugtanzek	33 Korgeredaba
17 Arukty	34 Ulanerge
18 Kharly	35 Chakhyrtoi
	36 Chik

1 KATUN

53°32'N; 96°30'E
Fig. 169

This intrusion, which has an area of 75 km², is located in the most northerly part of Tuva on the southern slope of the East Sayan ridge. The intrusion cuts Proterozoic crystalline schists and dolomitic marbles, Lower Palaeozoic gabbroids and plagiogranites and Cambrian and Devonian volcanic and sedimentary rocks. The intrusion has a complex form and three intrusive phases can be distinguished. The first, and principal, phase comprises coarse- and medium-grained, massive biotite or hornblende–biotite granites, leucocratic and mesocratic pyroxene (augite) or biotite–pyroxene syenites, and occasionally amphibole peralkaline syenites and mesocratic, sometimes trachytic, nepheline syenites with Na-hedenbergite, aegirine–diopside and biotite. The rocks of the first phase comprise about 80% of the outcropping intrusive rocks. Rocks of the second intrusive phase are represented by fine- or medium-grained or porphyritic granosyenites and granites in which the dark minerals may be biotite, hornblende or alkaline amphibole. Small, fine-grained dykes, the third intrusive phase, of riebeckite granite sometimes have a gneissose structure. Petrographic and chemical studies by Kotina et al. (1971) led them to suggest that pre-magmatic metasomatic processes were responsible for differences between the various rock types.

References Dmitre'ev and Kotina, 1966; Kotina et al., 1971.

2 CHAMDZHYAK

53°31'N; 96°48'E

The Chamdzhyak massif includes peralkaline and calc-alkaline syenites and granosyenites. Small dykes and veins of riebeckite granite are present. The more alkaline varieties of rocks are most frequent in those parts of the

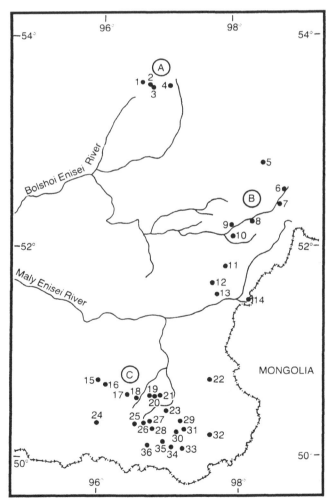

Fig. 168. Distribution of alkaline intrusive rocks in East Tuva. (after Makin and Pavlenko, 1966, Fig. 55). For discussion of subprovinces A–C, see text.

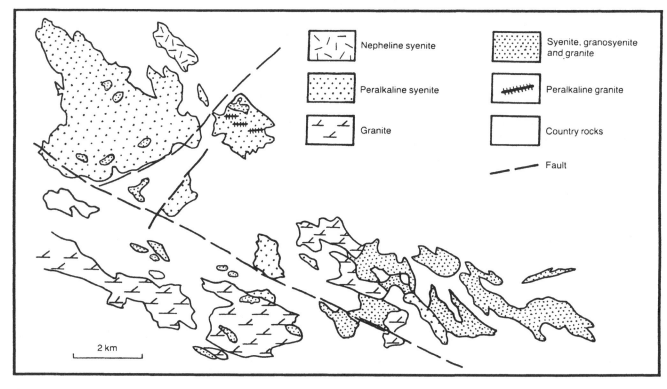

Fig. 169. Katun (after Dmitre'ev and Kotina, 1966, Fig.1).

intrusion where limestone country rocks are more abundant. The composition of the syenites is variable, with alkali feldspar 32–84%, katophorite 1–9%, biotite 0–8%, pyroxene 0–8%; and plagioclase 0–54%; magnetite is minor.

Age K–Ar determinations on syenite gave 408 and 430 Ma (Kovalenko and Popolitov, 1970).

Reference Kovalenko and Popolitov, 1970.

3 AKSUG   53°19′N; 96°51′E

This small stock of only 0.2 km² cuts Lower Palaeozoic gabbro–diorites. The rocks belong to two intrusive phases, the younger of which consists of amphibole syenites, the older of aegirine–riebeckite granites. The rocks are intensively altered by post-magmatic processes, involving the development of microcline and albite, which are associated with various minerals enriched in Zr and Ta–Nb as well as fluorite, cryolite and bastnaesite.

Age The Pb method on euxenite gave 430±60 Ma (Zykov *et al.*, 1961).

References Kovalenko and Popolitov, 1970; Zykov *et al.*, 1961.

4 IRELIG   53°19′N; 97°16′E

The Irelig intrusion comprises predominantly peralkaline plagioclase-free granosyenites and granites as well as small dykes and veins of peralkaline riebeckite granite. Accessories include titanite, apatite and zircon.

Reference Kovalenko and Popolitov, 1970.

5 DOTOT   52°50′N; 98°30′E

Peralkaline syenite is the dominant rock type in this occurrence.

Reference Makhin and Pavlenko, 1966.

6 CHONGAI   52°29′N; 98°44′E

No details of this occurrence have been published.

Reference Makhin and Pavlenko, 1966.

7 SORUG   52°23′N; 98°35′E
                                                  Fig. 170

The Sorug alkaline intrusive complex cuts Precambrian marbles and gabbroic rocks of Lower Palaeozoic age. It includes peralkaline syenites, nepheline-bearing syenites and plagioclase-bearing nepheline syenites; nepheline syenite pegmatites are abundant. There are small cross-cutting granosyenite bodies. The nepheline syenites are composed of nepheline (9–35%), alkali feldspar (40–79%), aegirine–augite (5–17%), biotite (1–18%), amphibole, plagioclase, garnet and not very plentiful Fe–Ti oxide minerals. Mineral compositions are given by Kovalenko and Popolitov (1970).

Reference Kovalenko and Popolitov, 1970.

8 ULAGARCHIN   52°12′N; 98°25′E

No details of this occurrence have been published.

Reference Makhin and Pavlenko, 1966.

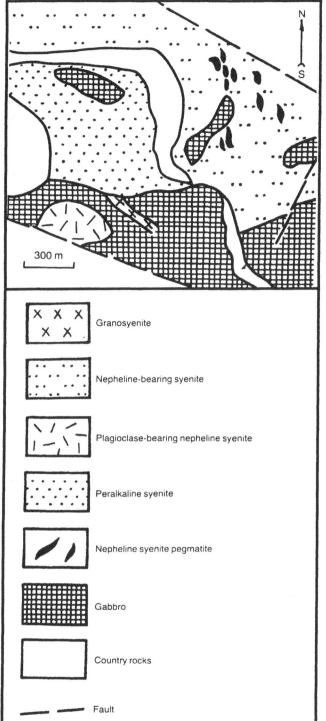

Fig. 170. Sorug (after Kovalenko and Popolitov, 1970, Fig. 15).

9 KARAKHOL   52°10′N; 97°55′E

No details of this occurrence have been published.

Reference Makhin and Pavlenko, 1966.

10 DUGDA   52°07′N; 97°58′E
Fig. 171

This 40 km² complex cuts Upper Proterozoic marbles and schists as well as gabbroic rocks and granites of Palaeozoic age. The first intrusive phase comprises amphibole–biotite syenites and the second trachytic amphibole- and amphibole–biotite–nepheline syenites; there is also a series of veins which includes feldspathoidal syenite porphyries, phonolite porphyries and nepheline–feldspar pegmatites. The pegmatites are associated with aegirine–nepheline–albite rocks. The compositions of the alkaline rocks are variable and apparently related to the composition of the country rocks. At the contact with fenites, which were formed from schists, in the northern part of the complex the intrusive rocks are amphibole–biotite pulaskites and syenites. At the contact with gabbroids, on the other hand, nepheline syenites are usually rich in pyroxene, plagioclase and nepheline. Trachytic hastingsite–

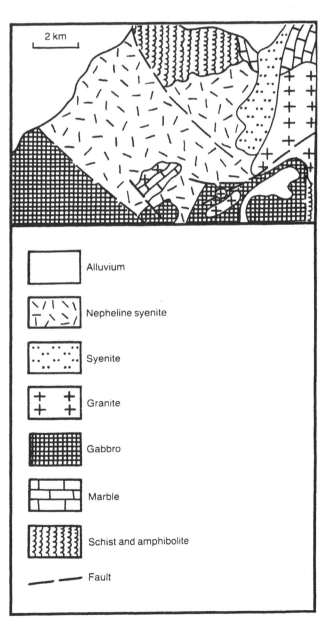

Fig. 171. Dugda (after Yashina, 1982, Fig. 21).

nepheline syenites comprise 50–70% microcline, 20–30% hastingsite, 15–25% nepheline and secondary biotite, cancrinite, sodalite and zeolite.

Age K–Ar on biotite from syenite–diorite gave 278 ± 11 Ma (Yashina and Borisevich, 1966); Pb on thorianite and pyrochlore gave 290 ± 10 Ma (Zykov *et al.*, 1961).

References Osokin *et al.*, 1974; Yashina, 1982; Yashina and Borisevich, 1966; Zykov *et al.*, 1961.

11 KADYROSS and KYSHTAG   51°50′N; 97°50′E

Kadyross is the largest massif of East Tuva covering about 150 km² and, according to Makhin and Pavlenko (1966), the combined occurrences of Kadyross and Kyshtag cover more than 250 km². Kyshtag lies several kilometres to the northeast of Kadyross. Peralkaline syenites are extremely abundant and peralkaline and calc-alkaline granites are common in the eastern part of the complex. There are granite dykes and veins, syenite porphyry, peralkaline syenite pegmatite and aplite.

Age K–Ar on alkaline granite gave 212 ± 10 Ma (Yashina and Borisevich, 1966).

References Kovalenko *et al.*, 1965; Kudrin, 1962; Makhin and Pavlenko, 1966; Yashina and Borisevich. 1966.

12 SURKHAI   51°47′N; 97°29′E

No details of this occurrence have been published.

Reference Makhin and Pavlenko, 1966.

13 KADRAUSS   51°35′N; 97°37′E

This intrusion contains both peralkaline and calc-alkaline granites and the relationship between them is complex. In some places the peralkaline granite cuts the calc-alkaline type but in others a gradual change from amphibole granite to amphibole–biotite and biotite granite has been noted. The massif has a zonal structure with the outer part composed of riebeckite granite which changes inwards to biotite, biotite–amphibole and diopside–biotite granite. Aegirine–amphibole granites can be found within the riebeckite granites. Veins of granite porphyry, peralkaline syenite and quartz–albite rocks are present.

References Kovalenko *et al.*, 1965; Makhin and Pavlenko, 1966.

14 CHAVACH   51°31′N; 98°12′E

Approximately 100 dykes of nepheline syenite 0.1–350 m thick and up to 2 km long are developed over an area of 6 km². They are emplaced within schists. The majority of the dykes are albitized, a process that has destroyed the nepheline. The albite rocks contain biotite, aegirine–augite, fluorite, magnetite, and microcline–perthite relicts; accessories include zircon, britholite, thorite, molybdenite and betafite.

Age Pb determinations on thorite gave 420 ± 40 Ma (Zykov *et al.*, 1961).

References Osokin *et al.*, 1974; Zykov *et al.*, 1961.

15 VERKHNEBRENSKII   50°41′N; 96°00′E

No details of this occurrence have been traced.

Reference Makhin and Pavlenko, 1966.

16 ULUGTANZEK   50°40′N; 96°16′E

This intrusion extends over 0.02 km² only and cuts Lower Palaeozoic diorites. Earlier rocks are coarse-grained peralkaline riebeckite syenite and younger ones fine-grained aegirine–riebeckite granite. Metasomatic quartz–microcline–albite rocks replace the peralkaline syenites and granites and are enriched in rare minerals including pyrochlore, columbite and lavenite.

References Kudrin *et al.*, 1965; Makhin and Pavlenko, 1966.

17 ARUKTY   50°37′N; 96°29′E

This intrusion, covering approximately 3 km², cuts Upper Proterozoic dolomitic marbles and Lower Palaeozoic gabbro–diorites and granites. The massif includes nepheline syenites containing xenoliths of country rock gabbro–diorites and dolomitic marble.

Age K–Ar on K-feldspar gave 244 ± 10 Ma, on biotite 226 ± 9 Ma and on nepheline 232 ± 9 Ma (Yashina and Borisevich, 1966). Pb from zircon gave 400 ± 40 Ma (Zykov *et al.*, 1961).

References Yashina, 1963; Yashina and Borisevich, 1966; Zykov *et al.*, 1961.

18 KHARLY   50°34′N; 96°32′E
Fig. 172

The Kharly complex is a stock-like intrusion of 9.2 km². It cuts Proterozoic marbles and in the north is terminated by a large fault. In the southern part a primary zonal structure is preserved. The following zones, going from the country rock dolomitic marbles towards the central nepheline syenites, have been distinguished: (1) An outer contact zone, 0.5–2.5 m thick, of recrystallized marble with apatite, nepheline, pyroxene and injections of ijolite and foyaite. (2) A central contact zone, varying from 15–110 m to 400–550 m wide, of fine- and medium-grained ijolites with numerous xenoliths of altered country rock limestones. Commonly the xenoliths are present as banded, stratified bodies which range in size from 0.8 × 1.5 to 40 × 175 m. (3) The inner contact zone is from several metres to 60 m wide; it is composed of juvites. There is a sharp boundary between the juvites and the ijolites of the outer contact zone, but there is a gradational boundary between the juvites and nepheline syenites of the central·part of the massif. At the contact with juvites the ijolites are recrystallized and porphyritic and poikiloblasts of Na-orthoclase appear while pyroxene is replaced by ferro-hornblende and garnet develops. (4) The central part of the Kharly zoned ring structure consists of trachytic aegirine–hedenbergite foyaites. They comprise 50–70% microcline, 20–30% aegirine–augite, 20–30% nepheline and albite, calcite, hastingsite, biotite, titanite and accessories including zircon, ilmenite and magnetite. A series of veins consists of nepheline-feldspar pegmatites with apatite, calcite and lepidomelane; calcite veins contain clinopyroxene, mica, apatite and titanomagnetite and look like carbonatite. Calcite from ijolites, foyaite pegmatite and carbonatite-like veins gave $\delta^{18}O°/_{oo}$ SMOW from 15.4 to 17.2 and $\delta^{13}C°/_{oo}$ PDB from −2.8 to +1.5, apparently indicating a mixed source. Dykes of aegirine–diopside syenite are also present.

Age K–Ar on nepheline gave 404 ± 20 Ma (Yashina and Borisevich, 1966).

References Kononova and Yashina, 1984; Yashina, 1982; Yashina and Borisevich, 1966.

19 BAYANKOL
50°36′N; 96°40′E
Fig. 173

The alkaline rocks of the Bayankol complex cut Upper Proterozoic marbles and Lower Palaeozoic granitoids. The complex is oval in outline but a deep valley divides it into two separate outcrop areas. The northern area covers about 5 km² and has a complicated structure, the greater part being a complex of layered nepheline syenites. The layering comprises both alternating bands enriched or impoverished in mafic minerals and alternating layers with different mineral compositions; in

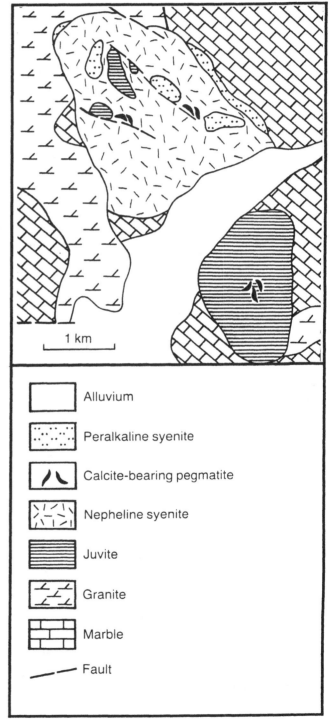

	Carbonatite
	Nepheline-cancrinite syenite
	Nepheline syenite
	Urtite
	Ijolite
	Plagiogranite
	Granodiorite
	Recrystallised marble
	Marble
	Fault

Fig. 172. Kharly (after Yashina, 1982, Fig. 8).

	Alluvium
	Peralkaline syenite
	Calcite-bearing pegmatite
	Nepheline syenite
	Juvite
	Granite
	Marble
	Fault

Fig. 173. Bayankol (after Kononova, 1962, Fig. 1).

both melanocratic and mesocratic layers there is a distinct trachytic texture. Two types of layering have been identified. In the lower part of the stratified complex an 'incompletely differentiated' series is developed which is represented by alternating melanocratic, mesocratic and leucocratic nepheline syenites. An increase of nepheline occurs in the upper part of leucocratic layers. The second type of stratified series is 'completely differentiated' and is common in the central and upper parts of the stratified complex. Within nepheline syenites, layers of juvite and urtite from 10–50 cm thick are present. The rocks richer in nepheline occur in the upper part of the leucocratic rocks near the boundary separating them from the overlying melanocratic nepheline syenites, which are occasionally malignites. In places nepheline–magnetite rocks occur instead of urtite. The nepheline syenites include nepheline (16–39%), microcline–perthite (49–59%), Na-hedenbergite (1–22%), lepidomelane (4–11%) and lesser amounts of sodalite, cancrinite, albite, titanomagnetite and apatite. In juvites the nepheline content increases with a commensurate decrease of the microcline–perthite (25–38%) and Na-hedenbergite. The northern part of the complex is rich in pegmatitic syenites mineralogically similar to the juvites. Nepheline syenitic pegmatites up to 15 × 4 m are encountered; they are located within an area 300 m long. The southern part of the complex covers 3 km². It has a more homogeneous composition, being composed essentially of urtite/juvite which consists redominantly of nepheline (80–90%). Post-magmatic albitization and calcitization has taken place. Calcite lenses and vein-like bodies 0.2–1 m thick occur. Apart from calcite, microcline–perthite, biotite, apatite and titanite are also present. Calcite from veins and lenses and from juvite gave $\delta^{18}O°/_{oo}$ SMOW values from 13.8 to 15.3 and $\delta^{13}C°/_{oo}$ PDB from −2.7 to −4.6. There appears to have been a mixed source for the calcite.

Economic The massif has a potential for the mining of nepheline as a raw material for alumina production.

Age K–Ar on nepheline gave 410 ± 16 Ma (Yashina and Borisevich, 1966) and Pb determinations on thorite 400 ± 80 Ma (Zykov *et al.*, 1961).

References Kononova, 1962 and 1976; Kononova and Yashina, 1984; Yashina and Borisevich, 1966; Zykov *et al.*, 1961.

---

20   VOSTOCHNO-HONCHULSKII

50°35′N; 96°46′E

Miaskitic nepheline syenite, mariupolite and augite-barkevikite syenite are present in this occurrence.

Reference Markin and Pavlenko, 1966.

---

21   UST'-HAIGASSKII   50°36′N; 96°48′E

This intrusion has a zonal structure. From the margin inwards the rocks are miaskitic nepheline syenites, amphibole diorite and aegirine essexite. The peralkaline rocks have a banded and migmatitic structure.

Reference Makhin and Pavlenko, 1966.

---

22   TEREKHOL   50°45′N; 97°35′E

Elongated approximately north–south, the 3 km² complex cuts Proterozoic marbles and granitized schists. It contains nepheline syenites in which marble xenoliths are preserved. The nepheline syenites vary from hastingsite-bearing varieties in the central part of the massif to

augite–hedenbergite ones at the margins, but almost everywhere they are albitized. Albitite bodies with biotite, arfvedsonite, astrophyllite and various accessory minerals including zircon, thorite, molybdenite and uraninite are common. In the vicinity of the contact marbles are changed to skarns and generally include diopside, biotite, and more rarely tremolite and cancrinite.

Reference Osokin *et al.*, 1974.

---

23   KHALTRIK   50°21′N; 97°04′E

Known to be alkaline but no details available.

Reference Makhin and Pavlenko, 1966.

---

24   CHARTIS   50°19′N; 96°00′E

Known to be alkaline but no details available.

Reference Makhin and Pavlenko, 1966.

---

25   PICHEKHOL   50°20′N; 96°29′E
Fig. 174

This complex cuts Upper Proterozoic marbles and schists and consists principally of pyroxene-nepheline syenites, that are albitized in several zones, and peralkaline syenites. It is thought that only the uppermost part of the massif is exposed, among which extensive areas of syenitized sandstones and marbles have been preserved. Both mesocratic and leucocratic nepheline syenites are present with nepheline contents of 15–25% in the central and southern parts of the massif but ecreasing to only 5–15% towards the northern and northeastern outer contacts; there is a concomitant increase in pyroxene northwards to 30% and even 40%. Pegmatites of 0.5–3 m in thickness are intensively developed in the massif and are composed of nepheline, microcline, pyroxene, lepidomelane and titanite with accessory zircon, magnetite, rinkolite, uraninite, britholite, thorite and thorianite.

Age Rb on thorianite gave 400 ± 20 Ma (Zykov *et al.*, 1961).

References Osokin *et al.*, 1974; Yashina, 1957; Zykov *et al.*, 1961.

---

26   SOLBELDER   50°20′N; 96°32′E

No details of this occurrence have been published.

Reference Makhin and Pavlenko, 1966.

---

27   DAKHUNUR   50°21′N; 96°45′E
Fig. 175

Ijolite and urtite form two bodies separated by Upper Proterozoic marbles. The western body of 1.5 × 0.35 km mainly comprises ijolites, which in the western and southern parts are enriched in calcite, with numerous very coarse-grained calcite veins. The calcite-bearing ijolites are cut by nepheline syenite veins. Calcite from calcite–silicate rocks and calcite veins has $\delta^{13}C°/_{oo}$ from +1.6 to −6.1. The eastern part of the complex is extremely variable in composition. There are extensive areas of pyroxenite among ijolite–urtites. In the southern part nepheline–zeolite pegmatites outcrop. The rocks of the ijolite–urtite series are composed of nepheline, with up to 85% in the urtites, and clinopyroxene, which is

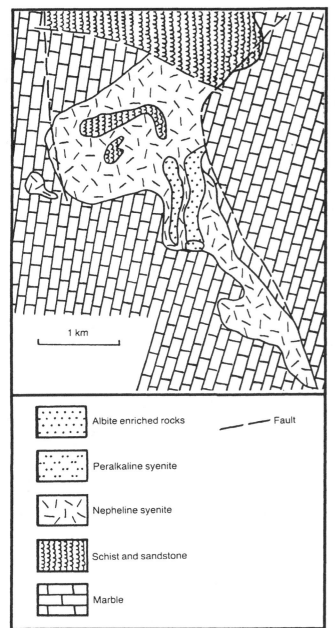

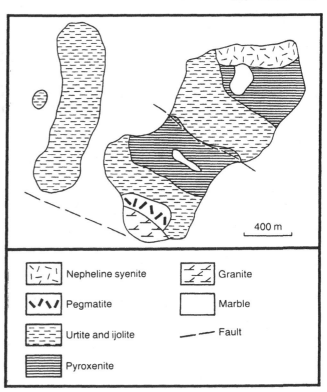

Fig. 175. Dakhunur (after Kononova, 1961, Fig. 5).

Fig. 174. Pichekhol (after Yashina, 1957, Fig. 5).

fassaitic in composition. Postmagmatic alteration of the urtite–ijolites resulted in the development of plagioclase, grossular and calcite. The most typical accessory minerals are apatite and titanite.

Economic The occurrence is a potential source of nepheline for alumina.

Age K–Ar on nepheline from urtite gave 402 ± 12 Ma (Kononova and Shanin, 1973).

References Kononova, 1961, 1962 and 1976; Kononova and Shanin, 1973.

---

28 TOSKUL   **50°18′N; 96°50′E**
                                            Fig. 176

The Toskul intrusion is funnel-shaped with steep outer contacts. Leucocratic, rather homogeneous, medium-

and coarse-grained, and occasionally pegmatitic, aegirine–hornblende– and trachytic hornblende–biotite-nepheline syenites are the dominant rock types. The pegmatites are concentrated in the central are and include varieties of nepheline syenite and nepheline-feldspar pegmatites. In the eastern part of the massif xenoliths of gabbro, that have been changed to diorite and syenite compositions, are preserved. The nepheline syenites generally contain 63–70% microcline perthite and 20% nepheline; hastingsite, biotite and clinopyroxene total not more than 8–13%.

Age K–Ar on biotite gave 330 ± 13 Ma and on nepheline 368 ± 14 Ma (Yashina and Borisevich, 1966).

References Yashina, 1957; Yashina and Borisevich, 1966.

---

29 UST'-KUNDUSS   **50°20′N; 97°21′E**

Known to be alkaline but no details available.

Reference Makhin and Pavlenko, 1966.

---

30 KHORYGTAG   **50°13′N; 97°11′E**

Known to be alkaline but no details available.

Reference Makhin and Pavlenko, 1966.

---

31 CHEKBEK   **50°16′N; 97°20′E**

Known to be alkaline but no details available.

Reference Makhin and Pavlenko, 1966.

---

32 AGASH   **50°13′N; 97°37′E**

This symmetrical stock of 4 × 3.5 km cuts Proterozoic marbles and schists. Lower Proterozoic gabbro and

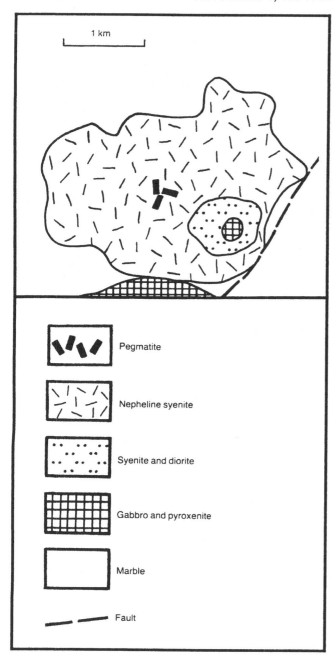

Fig. 176. Toskul (after Yashina, 1957, Fig. 4).

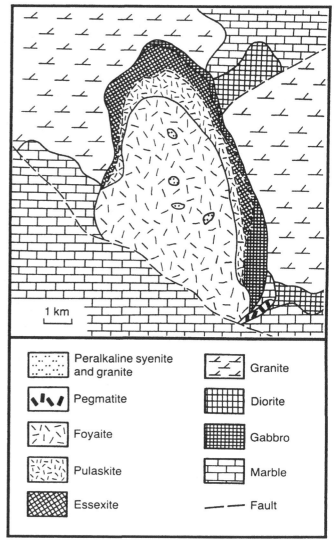

Fig. 177. Korgeredaba (after Yashina, 1982, Fig. 10).

diorite containing xenoliths of marble are also present. The principal rock type is biotite syenite with some quartz, but miaskitic nepheline syenite also occurs and contains hastingsite, diopside, augite, garnet and nepheline.

Reference Makhin and Pavlenko, 1966.

33 KORGEREDABA
50°05'N; 97°15'E
Fig. 177

Korgeredaba extends over 36 km² and truncates Upper Proterozoic marbles, Lower Palaeozoic gabbro–diorite and leucocratic granites of Devonian age. It was intruded in three phases. The first, and principal, phase is represented by hastingsite foyaites which occupy an area of 29 km² and have in the northern part a distinct zoned structure. The zoning has been produced by interaction of the foyaite with older gabbro–diorites. The latter have been intensively altered, which resulted in the formation of oligoclase essexite and pulaskite fenites. The zoned fenite aureole varies in width from 1.5 km in the northern part to several hundreds of metres in the western and eastern parts. Adjacent to the fenites at the periphery of the intrusive body there is a distinct outer contact zone of pulaskites which are variably fine- to coarse-grained and heterogeneously textured. The pulaskites have distinct boundaries with the fenites, which are of similar composition, but there is a gradual transition to the hastingsite foyaites. The second intrusive phase is represented by aegirine–arfvesonite foyaites and abundant peralkaline pegmatites containing endialyte, astrophyllite, rinkolite and other rare earth minerals. The third intrusive phase comprises dyke and stock-like bodies of peralkaline syenites and grano-syenites. At the contacts the hastingsite foyaites lose nepheline and become enriched in anorthoclase and quartz. Postmagmatic processes caused the development of albite-rich rocks with aegirine, which are referred to as mariupolites, and the crystallization of various rare-earth minerals including eudialyte, catapleite, astrophyllite and rinkolite.

Age K–Ar on biotite gave 304 ± 12 Ma (Yashina and Borisevich, 1966).

References Yashina, 1982; Yashina and Borisevich, 1966.

---

34 ULANERGE 50°07′N; 97°05′E
Fig. 178

The alkaline rocks of this complex cut Upper Proterozoic marbles and Palaeozoic granite dykes. It comprises predominantly hastingsite–nepheline syenites which are cut by pegmatite veins of nepheline syenite and dyke-like and stock-like bodies of quartz syenite and granosyenite. The nepheline syenites have a banded structure as indicated by the distribution of mafic minerals. The principal rock-forming minerals are nepheline (28–56%), microcline (25–55%) and hastingsite (6–25%) with lesser biotite, albite, calcite, possibly fluorite, titanite, apatite, titanomagnetite, pyrrhotite, andradite, graphite, pyrochlore and britholite. Nepheline syenite pegmatites are 0.5–3 m thick, up to 150 m long and are zoned. The centres are composed of giant crystal aggregates of microcline, nepheline, hastingsite, lepidomelane, cancrinite and cleavelandite (variety of albite), which forms nests and pockets. The nepheline syenites are sometimes replaced by albite and there is accessory zircon, biotite, pyrochlore and thorite.

Age K–Ar on biotite gave 322 ± 13 Ma (Yashina and Borisevich, 1966).

References Osokin *et al.*, 1974; Yashina, 1957; Yashina and Borisevich, 1966.

---

35 CHAKHYRTOI 50°10′N; 96°58′E
Fig. 179

Hedenbergite–nepheline syenites cut Upper Proterozoic marbles and form two main bodies. The southern one is represented by intensively albitized nepheline syenites and the northern one comprises numerous sub-vertical

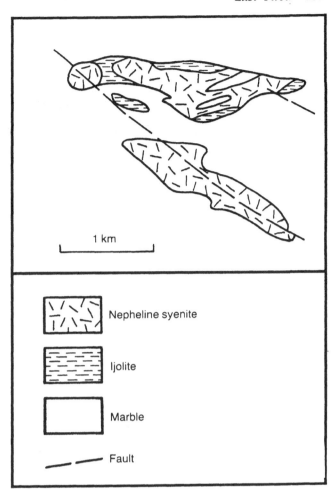

Fig. 179. *Chakhyrtoi (after Kononova, 1964, Fig. 1).*

injections of nepheline syenite intercalated with xenolithic marble blocks. The composition of the nepheline syenites is 35% nepheline, 35% microcline perthite and 15% Na-hedenbergite; there is some secondary alkali amphibole, rare plates of lepidomelane, calcite and typically accessory apatite and titanomagnetite. At the contact of nepheline syenites and marbles there are distinct reaction zones which vary in width from several centimetres to metres. The succession of zones are: marble–pyroxene-bearing recrystallized marble–pyroxenite–ijolite–ijolite with microcline–nepheline syenite. Chemical compositions of rocks, including rocks from the reaction zones, are available in Kononova (1964).

Reference Kononova, 1964.

---

36 CHIK 50°09′N; 96°43′E
Fig. 180

Ijolite–urtites make up several bodies cutting Upper Proterozoic marbles. The northeastern part of the main intrusion is a banded complex consisting of alternating bands of ijolite–urtite, ijolite and melteigite with amongst them intensively altered marble xenoliths. The thickness of the bands varies from several centimetres to several tens of metres and the bands differ in grain size with some variably grained and others medium- to coarse-grained and sometimes pegmatitic. The primary rock-forming minerals are nepheline, which in urtites comprise up to 90% of the rock, and fassaite. The

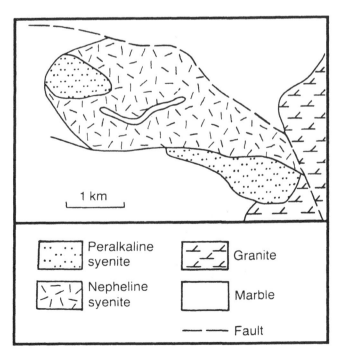

Fig. 178. *Ulanerge (after Yashina, 1957, Fig. 3).*

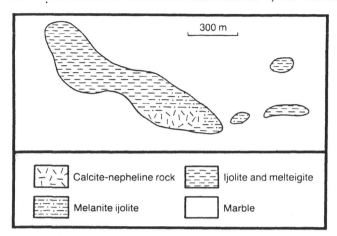

Calcite-nepheline rock

Melanite ijolite

Ijolite and melteigite

Marble

Fig. 180. Chik (after Kononova, 1961, Fig. 8).

southeastern part of the principal body contains calcite-bearing urtite, calcite–nepheline rocks and pure calcite rocks that look like carbonatites, but the calcite gives $\delta^{18}O°/_{oo}$ SMOW values from +14.7 to +16.4 and $\delta^{13}C°/_{oo}$ PDB from −0.9 to −2.0. Near the boundary of fassaite ijolites and urtites with calcite-bearing nepheline rocks melanite ijolites are found. They comprise nepheline, melanite and Na-hedenbergite and show evidence that fassaite has been replaced by melanite and Na-hedenbergite. In calcite-bearing urtites and calcite-nepheline rocks the role of secondary minerals is of greater importance with the replacement of nepheline by cancrinite (3–4%) and thomsonite. Apatite is the most common accessory mineral.

Economic The occurrence appears to be attractive for nepheline mining as a source of alumina.

Age K–Ar on nepheline from urtite gave 400 Ma (Kononova, 1976).

References Kononova, 1961 and 1976; Kononova and Yashina, 1984.

East Tuva references

DMITRE'EV, L.V. and KOTINA, R.P. 1966. Morphology and structural position of the Katun alkaline intrusion from East Sayan. *Sovetskaya Geologiya, Moskva*, 9: 106–23.
KONONOVA, V.A. 1961. *Urtite–ijolite intrusions from south- east Tuva and some problems of their genesis*. USSR Academy of Sciences, Moscow. 120 pp.
KONONOVA, V.A. 1962. The primary layered Ba'ankol intrusion of hedenbergite nepheline syenite. *In* O.A. Vorob'eva (ed.) *Alkaline rocks of Siberia*. 39–70. USSR Academy of Sciences, Moscow.
KONONOVA, V.A. 1964. On the interaction between nepheline syenites and marbles from the Chakhortoi injected intrusives field (south-east Tuva). In O.A. Vorob'eva (ed.) *Alkaline magmatism of the folded areas from the Siberian platform*. 182–206. Nauka, Moscow.
KONONOVA, V.A. 1976. *The Jacupirangite–urtite series of alkaline rocks*. Nauka, Moscow. 214 pp.

KONONOVA, V.A. and SHANIN, L.L. 1973. On the radiogenic argon in nepheline in connection with the question of how good it is for the dating of alkaline complexes. *In* G.D. Afanas'ev (ed.), *Geological-radiogenic interpretation of the divergence of age figures*. 40–52. Nauka, Moscow.
*KONONOVA, V.A. and YASHINA, R.M. 1984. Geochemical criteria for differentiating between rare-metallic carbonatites and barren carbonatite-like rocks. *Indian Mineralogist (Sukheswala Volume)*, 136–50.
KOTINA, R.P., KREMNEVA, M.A. and POPOVA, R.P. 1971. Some peculiarities of the metasomatic syenitization process and the formation of alkaline melts exemplified by the Katunsk massif (eastern Tuva). *Geokhimiya. Academiya Nauk SSSR, Moskva*, 980–91.
KOVALENKO, V.I. and POPOLITOV, E.I. 1970. *Petrology and geochemistry of the REE of the alkaline and granite rocks from north-east Tuva*. Nauka, Moscow, 158 pp.
KOVALENKO, V.I., OKLADNIKOVA, L.V., PAVLENKO, A.S., POPOLITOV, E.I. and FILIPOV, L.V. 1965. Petrology of the mid-Palaeozoic complex of granite and alkaline rocks from East Tuva. *In* B.M. Shmakin (ed.) *Geochemistry and petrology of the magmatic and metasomatic occurrences*. 55–149. Nauka, Moscow.
KUDRIN, V.S. 1962. Alkaline intrusions from north-east Tuva. *Sovetskaya Geologiya, Moskva*, 4: 40–52.
KUDRIN, V.S., KUDRINA, M.A. and SHURIGA, T.N. 1965. *Rare-metal metasomatic occurrences connected with midalkaline granites*. Nedra, Moscow. 146 pp.
MAKHIN, G.V. and PAVLENKO, A.S. 1966. The middle Palaeozoic (Sangilen) alkaline intrusive complex. *In* G.A. Kudr'avisev and V.A. Kuznetsov (eds) *Geology of the USSR* XXIX (part 1): 306–26. Nedra, Moscow.
OSOKIN, E.D., LAPIN, A.V., KAPUSTIN, Yu.L., POHVISNEVA, E.A. and ALTUHOV, E.N. 1974. Alkaline provinces of Asia. Siberian- Pacific group. *In* L.S. Borodin (ed.) *Principal provinces and formations of alkaline rocks*. 91–166. Nauka, Moscow.
*PAVLENKO, A.S. 1974. The Mongol-Tuva province of alkaline rocks. *In* H. Sorensen (ed.), *The alkaline rocks*. 271–93. John Wiley, London.
YASHINA, R.M. 1957. Alkaline rocks from southeast Tuva. *Izvestiya Akademii Nauk SSSR*, 5: 17–36.
YASHINA, R.M. 1963. On the contact reactions of the nepheline syenites and xenoliths of dolomite-bearing marbles (Arukty alkaline massive from southeast Tuva). *In* G.A. Sokolov (ed.) *Physico-chemical problems of the formation of rocks*, 2: 117–28. USSR Academy of Sciences, Moscow.
YASHINA, R.M. 1982. *The alkaline magmatism of folded-block areas*. Nauka, Moscow, 274 pp.
YASHINA, R.M. and BORISEVICH, I.V. 1966. The age of the alkaline rocks of East Tuva. *In* G.D. Afanas'ev (ed.) *Dating of the tectono-magmatic cycles and metallogenic stages*. 326–36. Nauka, Moscow.
ZYKOV, S.I., STUPNIKOVA, N.I., PAVLENKO, A.S., TUGARINOV, A.I. and ORLOVA M.P. 1961. The age of the intrusions from East Tuva and the Enisei Range. *Geokhimiya. Akademiya Nauk SSSR, Moskva*, 7: 547–60.

*In English

BAIKAL

In the mountain country of Baikal, of which the part lying east of Lake Baikal is known as Transbaikalia, only a few occurrences of alkaline rocks were known until the middle of this century. However, subsequently some 40 massifs with nepheline-bearing rocks and numerous peralkaline granites have been discovered. Within this huge province of some 1200 × 400 km the alkaline complexes form several distinct linear belts, the most northerly one extending for about 350 km and the Vitim belt for 400 km, although the similarity of the Komskii complex (Fig. 181, locality 52) to many of those of Vitim might indicate that this belt extends over 700 km. Most of the alkaline complexes are much older than the Baikal rift system, which extends for some 1500 km and is of Cenozoic age, although there is extensive basaltic volcanism related to the rifting which is, however, mostly concentrated outside the central rift system.

Taking into account the spatial distribution of the occurrences, their tectonic settings and the nature of the alkaline rocks, five sub-provinces have been defined. These are North and Central Baikal, Vitim and South and Southeast Baikal, and they are indicated on Fig. 181 by the letters A–E. These boundaries have been drawn not only according to the spatial distribution of the alkaline rocks, but also to reflect the different tectonic settings and nature of the alkaline rocks of each sub-province. A brief review, in English, of the relationship between the distribution of the alkaline rocks and structure is that of Altukhov et al. (1973).

North Baikal subprovince
1 Khorob
2 Gilindra
3 Bryzgunskii
4 Ovsak
5 Monyukan
6 Synnyr
7 Yaksha
8 Kudushkitskii
9 Burpalinskii
10 Akitskii
11–12 Goudzhekitskii
and Gorbylak
Central Baikal subprovince
13 Svyatonosskii
14 Tazheranskii
15 Slyudyanka
Vitim subprovince
16 Angidzhan
17 Bambuiskii
18 Tsipinskii
19 Pravo-Uliglinskii
20 Verkhne-Uliglinskii
21 Okunevskii
22 Chinskii
23 Snezhninskii
24 Saizhekonskii
25 Saizhenskii
26 Amalat
27 Gulkhenskii
28 Nizhne-Burul'zaiskii
29 Verkhne-Burul'zaiskii
30 Ipoloktinskii
31 Siriktinskii
32 Mukhalskii
33 Sherbakhtinskii
34 Ingurskii
35 Zimov'echinskii
36 Vitim volcanic field
37 Tuchinskii
38 Altanskii
South Baikal subprovince
39 Tunkinskoe volcanic field
40 Bartoiskoe volcanic field
41 Borgoiskoe volcanic field
42 Borgoiskii
43 Dabkhorskii
44 Nizhne-Ichetuiskii
45 Belogorskoe
46 Ortsekskii
47 Kharasunskii
48 Verkhne-Bulykskii
49 Botsi
50 Zormenikskii
51 Sukho-Khobol'skii
Southeast Baikal subprovince
52 Komskii
53 Khorinskii
54 Sredne-Oninskii
55 Bugutuiskii
56 Atkhinskii
57 Kukinskii
58 Ubukitskii
59 Kharitonovskii
60 Verkhne-Mangirtuiskii
61 Malo-Kunaleiskii

North Baikal Subprovince

The North Baikal subprovince of alkaline occurrences is 50 km in width and 350 km long and extends from the northern end of Lake Baikal to the central section of the Mama River. It is distributed along the axial part of the Baikal-Vitim dome. The country rocks are faulted Precambrian and Lower Cambrian. The alkaline intrusions are highly variable in size, the largest being the 564 km^2 Synnyr complex which, with the large Burpala intrusion, is located in the central part of the belt. The intrusions vary widely in shape, structure, composition and the depth to which they have been eroded. They may comprise single bodies or groups of bodies making up a single magmatic complex. According to petrographical, petrochemical and metallogenic features four associations can be distinguished within the North Baikal province (Zhidkov, 1990): (1) Yaksha-Synnyr (kalsilite and nepheline syenites, alkaline syenites) (2) Burpala (nepheline syenites, alkaline syenites, granosyenites) (3) Akit (alkaline syenites, granosyenites, alkaline granites and (4) Goudzhekit (nepheline syenites).

1 KHOROB   57°32'N 112°21'E

This occurrence of about 1.5 km^2 is composed of trachytic garnet–pyroxene peralkaline syenites.

Reference Zhidkov, 1968.

2 GILINDRA   57°30'N 112°30'E

The Gilindra complex covers about 35 km^2 and has a stock-like form cutting Upper Proterozoic granite gneisses. The complex evolved in two phases. During the first phase akerites, sviatonossites (andradite-rich peralkaline syenites) and granodiorites were emplaced, and during the second phase quartz akerites, syenites and adamellites. The akerites are hypidiomorphic textured rocks comprising oligoclase and oligoclase–andesine (An$_{24-30}$), microcline, aegirine–diopside (10–40% aegirine molecule); accessory minerals are titanite and Ti–Fe oxides. The sviatonossites are medium- and coarse-grained trachytic rocks with microcline phenocrysts and a groundmass composed of albite, aegirine–diopside, hornblende, andradite (up to 30%) and accessory titanite, zircon and apatite.

Reference Zhidkov, 1968.

3 BRYZGUNSKII   57°38'N; 113°07'E
(Uglinskii)

No details of this occurrence have been published.

Reference Kostyuk, 1990.

4 OVSAK   57°25'N 112°38'E

This occurrence consists of akerites, sviatonossites (andradite-rich peralkaline syenites) and other alkaline rocks with trachytic textures.

Reference Zhidkov, 1968.

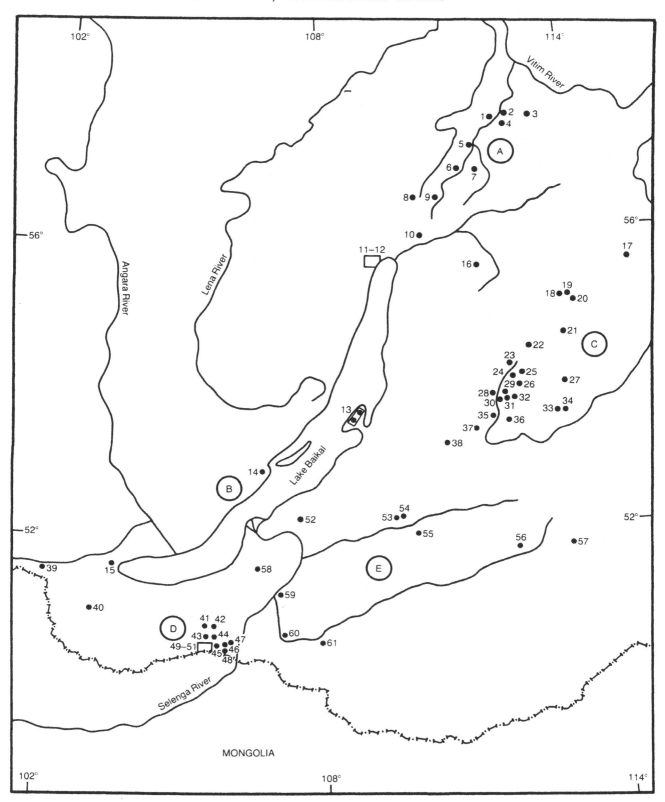

Fig. 181. Distribution of alkaline rocks and carbonatites in the area of Lake Baikal and the five sub-provinces into which the area has been divided. A North Baikal; B Central Baikal; C Vitim; D South Baikal; E Southeast Baikal.

5 MONYUKAN
57°14'N; 111°46'E

The small Monyukan massif of only some 2 km² comprises a number of isolated outcrops which are cut by younger alaskitic granites. It consists of garnet–pyroxene (occasionally with nepheline) trachytic peralkaline syenites. Veins have a granitic composition.

Reference Zhidkov, 1968.

---

6 SYNNYR
56°55'N; 111°20'E
Fig. 182

The Synnyr pluton is oval in shape, covers 564 km² and cuts Lower Cambrian limestones, conglomerates and sandstones and Palaeozoic granitoids. The massif has a zonal structure: the central part is made of trachytic alkaline syenites (pulaskites) around which are more extensive rings of pseudoleucitite and nepheline syenite. From the centre to the peripheral parts of the massif the following succession of rocks occurs: alkaline syenites – pseudoleucite syenites – pseudoleucite microcline syenites – pseudoleucite nepheline-microcline syenites – nepheline syenites. The central oval-shaped stock has an area of about 100 km² and is composed of trachytic pyroxene alkaline syenites which are mainly pulaskites. The alkaline syenites making up the stock have a relatively uniform composition, but nevertheless the inner parts are more leucocratic than the marginal ones, containing, besides microcline, only 3–7% of coloured minerals (aegirine–augite, hornblende). In the marginal parts of the stock up to 10% nepheline appears in the alkaline syenites and the proportion of coloured minerals is higher, usually with pyroxene up to 20%, and in places there is up to 15% of hornblende and up to 5% biotite. In parallel with these zonal changes the volume of accessories increases towards the margin, especially of titanite, magnetite and apatite. The inner ring of pseudoleucite rocks is adjacent to the pulaskite stock and forms a ring 0.5–4 km wide around it. These rocks are rather variable in composition and structure. One of the varieties of pseudoleucite rock is synnyrite which consists of dactylotypic intergrowths of orthoclase and microcline with kalsilite and nepheline, in which the ratio of K-feldspar to kalsilite ± nepheline is 61–67: 39–33. The K_2O in these rocks can reach 18–20%. In the pseudoleucite syenites microcline and nepheline are present as isolated crystals in pseudoleucite–micrographical intergrowths, so that pseudoleucite-microcline syenites and pseudoleucite-nepheline-microcline syenites can be distinguished. The proportion of pseudoleucite in these rocks is extremely variable and up to 50–60%. The marginal ring of nepheline syenites varies in width from 1–2 to 6–8 km. These nepheline syenites are miaskitic, medium- and coarse-grained rocks with allotriomorphic granular, occasionally indistinctly porphyritic, or poikilitic hypidiomorphic granular textures. Microcline (45–85%), infrequent orthoclase (sanidine), nepheline (5–45%), with very small mica inclusions, cancrinite, aegirine and mica (1–5%) are modally dominant. In foyaites the amount of biotite decreases to about 1%, but clinopyroxene increases up to 25–30%. Accessory minerals are apatite, magnetite, titanite, occasionally rutile, eudialyte, zircon, melanite, perovskite and baddeleyite. Initial Sr ratios of 0.7064 have been obtained from synnyrite and 0.7066 from nepheline syenite (Zhidkov, 1990). The order of emplacement of the various rock types is not clear and is being debated. According to Andre'ev (1981) the peralkaline syenites were formed before the feldspathoidal syenites. During the final stages a variable series of dykes were emplaced which include tinguaite, shonkinite, camptonite and monchiquite. At the outer contact of nepheline syenite with Cambrian marbles and dolomite, skarns having variable compositions are present. They can include forsterite–calcite and diopside–hedenbergite rocks with vesuvianite, phlogopite, spinel, wollastonite, epidote and magnetite. At contacts with terrigenous and volcanic rocks hornfelses with epidote, biotite, andalusite, corundum, staurolite and kyanite have been formed. For mineral and rock ompositions refer to Arkhangelskaya (1974), Andre'ev (1981) and Zhidkov (1990); σO^{18} values of 8.0–9.5 were obtained by Pokrovskii and Zhidkov (1993).

Economic Synnyrites are a potential ore for the production of alumina and potassium.

Age K–Ar on orthoclase from nepheline syenite gave 335 Ma; on biotite from nepheline syenites 204–349 Ma and on orthoclase from pulaskite 304 Ma. Pb–U–Th on zircon from nepheline syenite gave 350 Ma (Arkhangelskaya, 1974), and a Rb–Sr isochron on synnyrite gave 330±4 Ma and on nepheline syenite 311±1 Ma (Zhidkov, 1990). Andre'ev et al. (1991) obtained an age of 187±3 Ma from a five-rock Rb–Sr isochron, while Pokrovskii and Zhidkov (1993) obtained an age of 293±5 Ma from a whole-rock isochron.

References Andre'ev, 1981; *Andre'ev et al., 1991; Arkhangelskaya, 1974; Panina, 1972; *Pokrovskii and Zhidkov, 1993; Sveshnikova, 1984; Zhidkov, 1968 and 1990.

---

7 YAKSHA
56°55'N; 111°48'E
Fig. 183

From structural and morphological analysis and geophysical interpretation the Yaksha occurrence is apparently a plutonic dome built of alkaline rocks and granites and extending to a depth of not less than 10 km. The alkaline intrusion is in contact with Precambrian and Cambrian crystalline schists, gneisses and marbles and granitoid domes of late Palaeozoic age. The area of alkaline rock outcrops is about 100 km². The larger part of the massif is composed of alkaline syenites which consist predominantly of orthoclase and microperthite (60–80%), pyroxene (5–15%), biotite (3–12%), admixtures of albite and nepheline and accessories including titanite, magnetite and apatite and occasionally quartz. Because of wide variation in the quantity of nepheline (up to 20%), the rocks vary from alkaline syenites to foyaites. Nepheline, nepheline-kalsilite and kalsilite syenites are intimately mixed. Two structural-textural facies of the rocks are distinguished: massive and banded gneiss-like. Nepheline syenites (frequently pseudoleucite syenites) have a rather simple composition of orthoclase (50–70%), nepheline (15–30%), biotite (6–15%), pyroxene (up to 5%), albite (up to 3%) and accessory titanite, apatite, magnetite, garnet and zircon. Nepheline-kalsilite and kalsilite syenites (synnyrites) are rare. Usually leucocratic pseudoleucitic rocks with a distinctly orientated assemblage of dark-coloured minerals are predominant. They comprise orthoclase (65–75%), kalsilite (up to 2%), biotite (3–10%) with a small amount of garnet, pyroxene, nepheline and albite. Mineral and rock compositions are available in Orlova (1990) and Kashirin (1971) gives modal and chemical data, including REE, for part of the complex called Daoksha.

Age A Rb–Sr whole-rock isochron gave 313±11 Ma. K–Ar dating of amphibole in biotite–amphibole per-

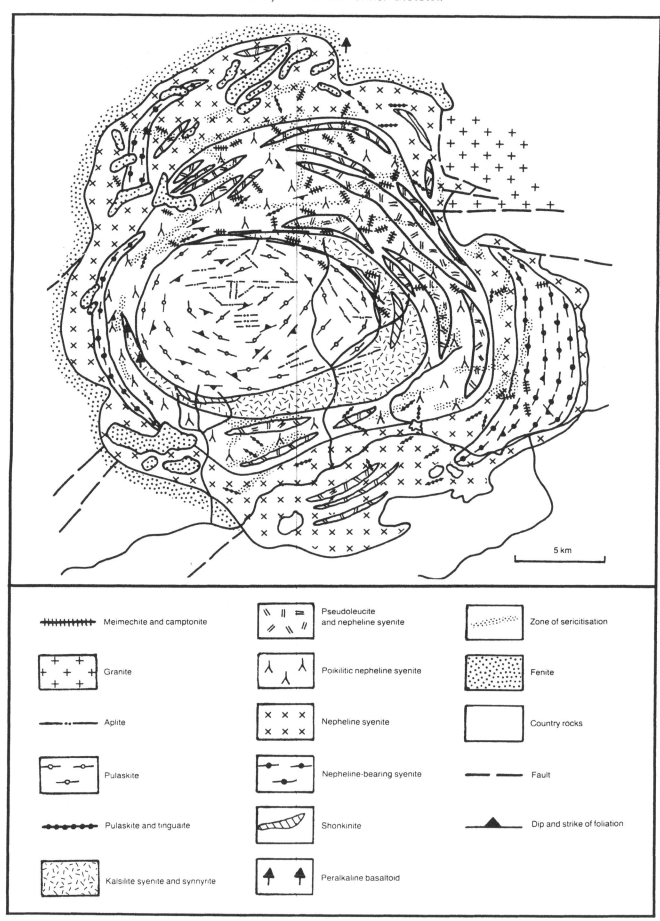

Meimechite and camptonite

Granite

Aplite

Pulaskite

Pulaskite and tinguaite

Kalsilite syenite and synnyrite

Pseudoleucite and nepheline syenite

Poikilitic nepheline syenite

Nepheline syenite

Nepheline-bearing syenite

Shonkinite

Peralkaline basaltoid

Zone of sericitisation

Fenite

Country rocks

Fault

Dip and strike of foliation

5 km

Fig. 182. Synnyr (after Zhidkov, 1990, Fig. 1b).

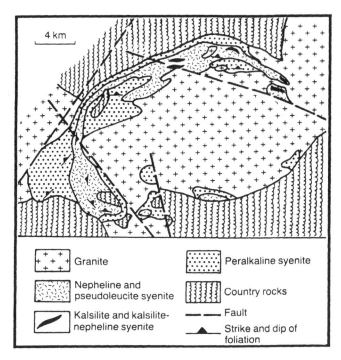

Granite		Peralkaline syenite	
Nepheline and pseudoleucite syenite		Country rocks	
Kalsilite and kalsilite-nepheline syenite		Fault	
		Strike and dip of foliation	

Fig. 183. Yaksha (after Zhidkov, 1990, Fig. 18).

alkaline syenite gave 310 ± 7 Ma, on biotite 283 ± 13 Ma, on feldspar 273 ± 12 Ma, on muscovite from muscovitized nepheline syenite 297 ± 15 Ma and on biotite 267 ± 12 Ma (Zhidkov, 1990).

References *Kashirin, 1971; Orlova, 1990; Zhidkov, 1990.

---

8 KUDUSHKITSKII   56°33′N; 110°20′E

No details of this occurrence have been published.

Reference Kostyuk, 1990.

---

9 BURPALINSKII   56°33′N; 110°45′E
(Burpala)   Fig. 184

The Burpalinskii alkaline complex occupies an area of about 220 km² and cuts sedimentary volcanogenic rocks of Proterozoic and Lower Cambrian age. The occurrence has a zoned concentric structure in which, starting from the outer contact, the rocks are granosyenites, quartz syenites, peralkaline syenites and pulaskites. The quartz-bearing rocks of Burpala, granosyenites and quartz syenites, are fine- and medium-grained, massive rocks which towards the centre of the complex are replaced by trachytic varieties. They comprise microcline (80–75%), oligoclase (An$_{28–29}$; 5–10%), quartz (5–10%), actinolite (5–7%), biotite (2–4%) and occasionally aegirine-diopside, arfvedsonite and accessory titanite and magnetite. In mesocratic varieties the dark-coloured minerals increase up to 25%. The peralkaline syenites and pulaskites are medium- and coarse-grained trachytic rocks of microcline-perthite (85–93%), aegirine-diopside (3–12%), biotite (0–3%), arfvedsonite, zeolitized nepheline, sodalite, albite, ferro-eckermannite and fluorite with minor titanite, ilmenite, zircon and apatite.

Among the pulaskites areas of foyaite up to 0.5×4 km and sodalite syenite are frequent. The foyaites are trachytic textured rocks consisting of microcline (65–70%), nepheline (12–20%), aegirine-augite (10%) and albite (5%). Pegmatites and dykes are also abundant in the occurrence. Syenite pegmatites are composed predominantly of microcline, aegirine-diopside and arfvedsonite with numerous accessory minerals including manganilmenite, titan-lavenite, caesium-bearing astrophyllite, eudialyte, ferrithorite and catapleite. Pegmatites of nepheline syenite are rare. They comprise microcline at the margins and nepheline–microcline aggregates with arfvedsonite, lepidomelane, ilmenite and eudialyte in central parts. The dyke series is represented by syenite-aplite, tinguaite, solvsbergite, grorudite and lamprophyres. Granites, granite-aplites and granite-pegmatites outcropping within the occurrence are of Upper Palaeozoic age and are not likely to be related to the alkaline complex. In hornfelses (fenites) at the outer contacts albitization-aegirinization zones with accessory melanite, cerite, chevkinite, caesium-bearing astrophyllite, sodium catapleite, lavenite and loparite are found, and Merlino *et al.* (1990) described the new mineral burpalite, a member of the cuspidine-wohlerite-lavenite family, from a fenitized sandstone in the western contact zone.

Age K–Ar on biotite from alkaline syenites gave 327 Ma and Pb–U–Th on zircon from syenite pegmatite 325 Ma (Arkhangelskaya, 1974).

References Andre'ev, 1981; Arkhangelskaya, 1974; *Merlino *et al.*, 1990; Panina, 1972; Zhidkov, 1965.

---

10 AKITSKII   56°01′N; 110°22′E
     Fig. 185

The Akitskii occurrence of 4.5 km² cuts Lower Palaeozoic granites and granosyenites. Arkhangelskaya (1974) determined the order of intrusion as peralkaline trachytic syenite, peralkaline aegirine granite and granosyenite and finally nepheline syenite. The peralkaline trachytic syenites are the most abundant rocks and are grey, leucocratic rocks of 75–90% microcline and aegirine-augite on which amphibole and biotite are developed; accessories include apatite, ilmenite and schorlomite. The peralkaline granites and granosyenites are fine-grained porphyritic rocks of a pinkish-grey. They comprise microcline perthite submerged in a groundmass of albite with quartz (10–20%) and aegirine; accessories are titanite, apatite, zircon and schorlomite. Nepheline syenites consist of microcline-perthite (35–70%) and nepheline (15–20%) in a fine-grained groundmass containing cancrinite, aegirine, arfvedsonite, biotite, titanite, apatite and ilmenite. The nepheline syenites contain partially resorbed remnants of the peralkaline syenites and granites. Five stages of postmagmatic mineralization are distinguished by Arkhangelskaya (1974): (1) microclinization, (2) albitization and aegirinization with the formation of albite-aegirine metasomatites, (3) a fluorine-carbonate stage involving development of ankerite, calcite, fluorite and rare-earth fluorine-carbonate minerals, (4) the formation of lithium-bearing minerals including zinnwaldite and taeniolite, and (5) a hypergene stage involving crystallization of cerussite, smithsonite, hematite and hydroxides of iron and manganese. Postmagmatic rocks form approximately north–south-trending zones which are up to 10 m wide and 400 m long. Two of them lie inside the complex and are enriched in rare minerals: others are located in the country rocks and these are poor in rare minerals. The rocks of these zones are composed of

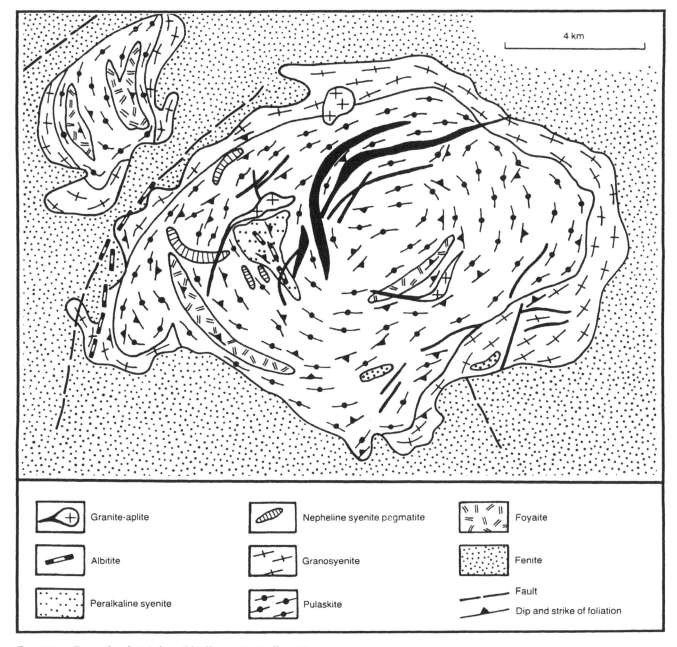

Fig. 184. Burpalinskii (after Zhidkov, 1960, Fig. 1).

microcline and albite with xenotime, taeniolite and pyrite. In the central parts of these zones small (up to 2 m diameter) nests of oligoclase with fluorite and ankerite with fluorite and a number of rare minerals occur, the most typical being parisite, xenotime, taeniolite, fluorite, ankerite and baryte.

Age K–Ar on biotite from peralkaline syenite gave 199 Ma (Arkhangelskaya, 1974).

References Andre'ev, 1981; Arkhangelskaya, 1974; Semenov *et al.*, 1974.

11–12 GOUDZHEKITSKII and GORBYLAK
                              55°43′N; 109°13′E
                                      Fig. 186

The occurrences of Goudzhekitskii (Fig. 186) and

Gorbylak, of 0.25 km², are located at the contact of Lower Proterozoic granites and granite–gneisses with crystalline schists and gneisses of the same age. The occurrences mainly comprise biotite–nepheline syenites amongst which bodies of biotite–riebeckite–nepheline syenite and melteigite are frequent. The rocks of these occurrencess generally have gneiss-like structures. The nepheline syenites are medium-grained grey rocks composed of microcline–perthite, nepheline, biotite, ilmenite, titanite, apatite and occasionally schorlomite. Secondary minerals are albite, cancrinite, muscovite, calcite, analcime and epidote. Microcline and nepheline are sometimes isolated as glomeroporphyritic aggregates which are surrounded by fine-grained microcline–nepheline–biotite rock. Metasomatic zones of biotite–albite rock are found.

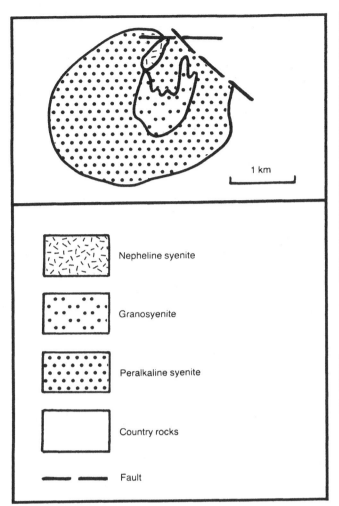

Fig. 185. Akitskii (after Arkhangelskaya, 1974, Fig. 13).

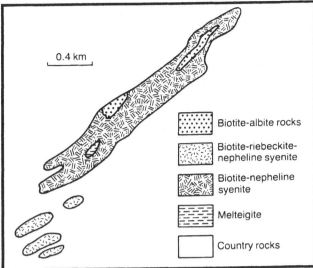

Fig. 186. Goudzhekitskii (after Arkhangelskaya, 1974, Fig. 12).

Age K–Ar on biotite from nepheline syenite gave 330 Ma (Arkhangelskaya, 1974).

References Andre'ev, 1981; Arkhangelskaya, 1974; Zhidkov, 1968.

Central Baikal subprovince

The few occurrences of alkaline rocks located close to the western and eastern shores of Lake Baikal (Fig. 181) lie within Precambrian formations and are of Precambrian and Palaeozoic age.

13 SVYATONOSSKII

53°39′N; 108°52′E
Fig. 187

Sited on the eastern shore of Lake Baikal, this intrusion is an irregularly shaped body which extends in a northeasterly direction for 30–35 km, conformably with the enclosing metamorphic rocks. Its width varies from 5 to 17 km and it has an area of about 230 km^2. The country rocks, which are Archaean or early Proterozoic, are represented by gneisses, schists, marbles and quartzites. The formation of the intrusion took place in two episodes, the first of which involved emplacement of aegirine–salite syenites and quartz syenites and the

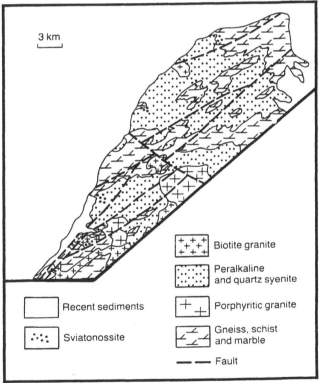

Fig. 187. Svyatonosskii (after Zanvilevich et al., 1985, Fig. 7.1).

second intrusion of biotite granites. Sviatonossites – garnet-bearing alkali syenites of variable composition – are genetically linked with the alkaline syenites and quartz syenites, although they form less than 1% of the area of exposed rocks. Alkali feldspar is predominant in the composition of alkaline syenites (45–69%), with plagioclase (15–32%), quartz (7–10%), aegirine–salite (1–4%), hastingsite and biotite. In addition to these minerals the sviatonossites contain about 30% of garnet in which andradite or almandine components are predominant. Compositions of rocks and minerals can be found in Zanvilevich et al. (1985).

Age Precambrian

Reference Zanvilevich et al., 1985.

---

14 TAZHERANSKII

52°52′N; 106°45′E
Fig. 188

The Tazheran occurrence comprises a number of intrusions lying within Precambrian schists, gneisses, calcite marbles and quartzites. Several stages have been established in the history of the intrusion, the first of which includes gabbro and gabbro–diorites, the second stage peralkaline syenite and later nepheline syenite, and the third stage veins, dykes and stocks of granite pegmatite. Pyroxenites and ijolites are present in the outer parts of the main intrusion, while at the outer contacts schists are altered to hornfelses, there is dedolomitization and recrystallization of marbles, and skarns are present. Desilication and alkalization have been recognized in granite pegmatites cutting alkaline rocks which, as a result, have been transformed into veins of aegirine syenite. The peralkaline syenites are the most widespread rocks in the massif and form an area of 1.5 × 1.5 km² within which are small, isolated bodies of gabbro–diorite. These syenites are medium- and coarse-grained rocks with trachytic textures and sometimes gneiss-like structures. They are leucocratic rocks comprising mainly alkali feldspar and albite–oligoclase with augite and biotite. Nepheline syenites form several bodies to the west of the principal intrusion where they are up to 1.4 km in length and 150 wide. In the south they are also well developed, the largest intrusion being

a stock of 0.5 × 0.5 km in which contacts against alkaline syenite are gradational. The nepheline syenites are generally trachytic rocks and consist of microcline perthite, albite–oligoclase, nepheline, aegirine–augite, hastingsite and biotite. Additional information can be found in Konev (1969).

Age K–Ar on biotite from nepheline syenite gave 350 Ma and from alkaline syenites 430 Ma (Konev, 1969).

Reference Konev, 1969.

---

15 SLYUDYANKA
(Malobystrinskii)

51°40′N; 103°28′E
Fig. 189

This complex, which is composed of gabbro–anorthosite, essexite, shonkinite and pyroxenite, lies within Precambrian gneisses, migmatites and marbles. It is cut by dykes of syenite and syenite–diorite. Pegmatitic granite and metasomatic diopside or phlogopite rocks are widely developed.

Economic Phlogopite has been mined since 1947 and lazurite is also extracted.

Reference Vasilyev et al., 1969.

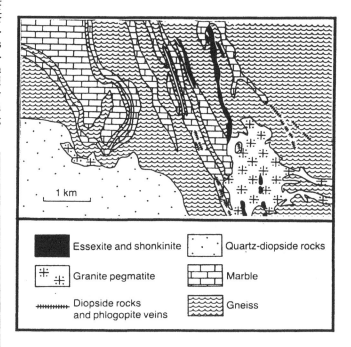

Fig. 189. Slyudyanka (after Vasilyev et al., 1969, Fig. 13).

Fig. 188. Tazheranskii (after Konev, 1969, Fig. 36).

Vitim subprovince

The Vitim alkaline subprovince is located (Fig. 181) within the Vitim plateau, lying to the east of Lake Baikal. The basement of the region was consolidated in Caledonian times. More than 20 alkaline intrusions were emplaced in the Middle Palaeozoic, the majority of them including nepheline-bearing rock types. Some of them are associated with Lower Palaeozoic (Tsipin and Chinsk) granitoids, whilst other occurrences (Saizhinskii and Gulkhenskii) are associated with basic rocks.

16 ANGIDZHAN 55°33′N; 111°40′E

Proterozoic sandstones and schists are partly covered by talus deposits in which large blocks of nepheline syenite occur. The nepheline syenites comprise nepheline, feldspar, hornblende, clinopyroxene (diopside, aegirine-augite) and biotite. The accessory minerals are ilmenite and apatite and secondary minerals include muscovite, analcime and chlorite. Chemical compositions of rocks are given by Andre'ev *et al.* (1969).

Reference Andre'ev *et al.*, 1969.

17 BAMBUISKII 55°36′N; 115°05′E
Fig. 190

The Bambuiskii massif consists of two bodies divided by the Bambu' river valley. They cut Lower Cambrian dolomites and dolomite-bearing limestones, which occur as numerous roof xenoliths. Their size varies from several metres to 700 × 250 m. The eastern massif covers 6 × 1.8 km and is predominantly composed of diorites and gabbro-diorites. In the upper part of the intrusion shonkinites and essexites are developed as an elongate body over 1 km in length. Hybrid rocks with garnet (sviatonossites), comprising aegirine–diopside, garnet, microcline and basic plagioclase, occur at the contact between shonkinites, essexites and carbonate sediments. The western massif, covering 4 × 1.5 km, is composed of alkaline biotite syenites, nepheline syenites and garnet syenites. Basic rocks are not extensively distributed in the western massif and occur only in contacts with carbonate sediments. In the contact zone the following

sequence of rocks is found, from the central part of the massif towards the contact, biotite syenite – biotite diorite – biotite–pyroxene gabbro – pyroxenite. Nepheline syenite comprises microcline–perthite (23–52%), albite (15–35%), nepheline (21–50%) and biotite (3–10)%. Hastingsite, aegirine–augite and calcite may be present and accessories include apatite, titanite, magnetite, zircon and fluorite.

Reference Andre'ev *et al.*, 1969.

18 TSIPINSKII 55°08′N; 113°30′E
Fig. 191

Tsipin is an oval-shaped body of about 5 km² which cuts lower Palaeozoic granitoids, schists and metamorphosed sandstones of Upper Proterozoic age. The massif is composed mainly of coarse-grained and pegmatitic nepheline syenite. At the northern and eastern contacts a band of fine-grained nepheline syenites 100–150 m wide has been mapped and these rocks also occur in the central part of the massif. Nepheline syenite comprises

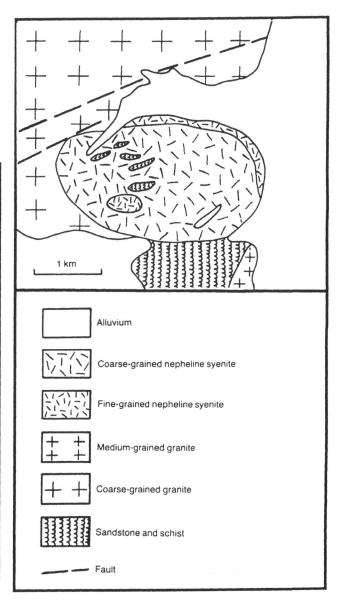

Sviatonossite
Shonkinite
Essexite
Pyroxenite
Gabbro-diorite
Diorite
Garnet syenite

Nepheline syenite
Peralkaline syenite
Biotite syenite
Granite
Dolomitic marble
Country rocks
— — Fault

Alluvium
Coarse-grained nepheline syenite
Fine-grained nepheline syenite
Medium-grained granite
Coarse-grained granite
Sandstone and schist
— — Fault

Fig. 190. Bambuiskii (after Andre'ev et al., 1969, Fig. 7).

Fig. 191. Tsipinskii (after Andre'ev et al., 1969, Fig. 17).

11–50% nepheline, 24–55% albite, 10–23% microcline and 20% lepidomelane. Along tectonic zones the alkaline rocks are extensively albitized and changed to ditroite, mariupolite, litchfieldite and miaskitic rocks. Among secondary and accessory minerals epidote, cancrinite, zircon, apatite, fluorite and Fe–Ti oxides are found. Data on the chemical compositions of rocks are available in Andre'ev *et al.* (1969).

Reference Andre'ev *et al.*, 1969.

19 PRAVO-ULIGLINSKII   55°08′N; 113°35′E

This massif is composed predominantly of peralkaline granites but lens-like isolated bodies of peralkaline syenite occur within them. Small bodies of nepheline syenite, generally 1–2 × 10–15 m but with one body 70 × 80 m, are to be found among the peralkaline syenites. The syenites and granites contain, besides K-feldspar, plagioclase, aegirine, hastingsite and riebeckite. Chemical analyses of the rocks are given by Sharackshinov (1984a).

Reference Sharackshinov, 1984a.

20 VERKHNE-ULIGLINSKII   55°03′N; 113°45′E

A series of stocks, up to 1.7 × 2.0 km, dykes and minor lens-like bodies of alkaline rocks are located over an area of 4.8 × 4.9 km and cut schists and marbles. Nepheline syenites comprise 11 small (0.1–0.8 km²) stocks and dykes, some cutting schists and others occupying the central parts of peralkaline syenite intrusions. Peralkaline syenites are subordinate and form small stocks, dykes, lenses of 550–60 × 200–250 m. The youngest rocks are dykes up to 1.5 m thick of fine-grained granite which cut all the sedimentary and magmatic rocks of the region. Chemical analyses will be found in Sharackshinov (1984a).

Reference Sharackshinov, 1984a.

21 OKUNEVSKII   54°39′N; 113°30′E
(Kapylyushy)   Fig. 192

The Okunevskii intrusion is composed of nepheline syenites and syenites which form interlayered injections

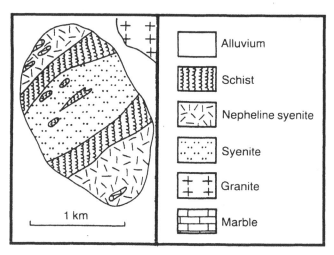

in biotite–feldspar schists. Nepheline syenites are grey, medium- to coarse-grained rocks which sometimes have a gneissose texture and consist of nepheline, microcline, plagioclase and biotite. Among accessory minerals zircon, apatite, magnetite, fluorite, pyrite and calcite are present. Secondary minerals are cancrinite, liebenerite and chlorite.

Reference Andre'ev *et al.*, 1969.

22 CHINSKII   54°31′N; 112°46′E

This alkaline massif, extending over 7.5 × 3.2 km, intrudes Lower Cambrian terrigenous carbonate sediments and lower Palaeozoic granitoids. The intrusion is composed mainly of peralkaline syenites but in the central part are some patches of nepheline syenite that occupy a total area of about 0.5 km². The composition of the peralkaline syenites is microcline–perthite, plagioclase, aegirine–augite, titanian garnet (up to 10%) and corundum with occasional quartz. Titanite, apatite, fluorite and Fe–Ti oxides are accessory. The nepheline syenites are medium- to coarse-grained massive rocks consisting of microcline (5–10%), microcline–perthite (55–60%), nepheline (20%), aegirine (10%), garnet (10%), biotite (3%), apatite, rutile, and Fe–Ti oxides. Analyses of the rocks will be found in Andre'ev *et al.* (1969).

Reference Andre'ev *et al.*, 1969.

23 SNEZHNINSKII   54°12′N; 112°18′E
(Inakit)   Fig. 193

The 3 × 6 km massif intrudes a schist–carbonate sequence of Upper Proterozoic age as well as Palaeozoic

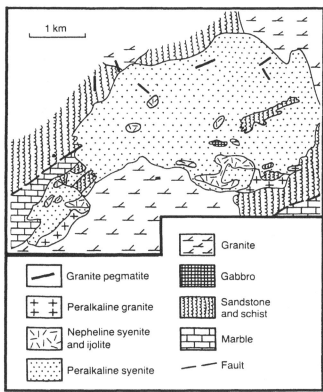

Fig. 192. *Okunevskii (after Andre'ev et al., 1969, Fig. 5).*

Fig. 193. *Snezhninskii (after Sharackshinov, 1984a, Fig. 16).*

granitoids. The intrusion comprises mainly peralkaline syenites with trachytic textures and small nepheline syenite bodies, with a maximum size of 250 × 900 m. Masses of ijolite–melteigite occur within the peralkaline syenites and are cut by nepheline syenite veins. Xenoliths of country rocks are common. Dykes, stocks and veins of fine-grained granite, aplite, granite porphyry, syenite porphyry and granite pegmatite are widespread in the massif. Chemical compositions are given by Sharackshinov (1984a).

References Konev, 1982; Sharackshinov, 1984a.

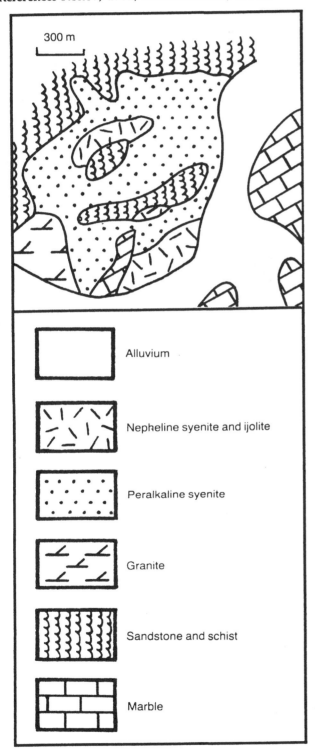

Fig. 194. *Saizhekonskii (after Konev, 1982, Fig. 4).*

24 SAIZHEKONSKII

54°04′N; 112°19′E
Fig. 194

The Saizhekonskii massif is about 1.5 × 2.5 km and intrudes Upper Proterozoic schists and limestones and granitoids of the Lower Proterozoic. The central part of the massif is composed principally of trachytic peralkaline syenites with lens-shaped bodies of juvite, urtite and feldspar-bearing urtite within them. In the central part of the massif large (up to 200 × 700 m) schist and marble xenoliths are preserved. Chemical compositions are available in Sharackshinov (1984a).

References Konev, 1982; Sharackshinov, 1984a.

25 SAIZHENSKII

54°07′N; 112°27′E
Fig. 195

This oval-shaped complex of 4 × 5 km cuts Proterozoic limestones and gabbros and Lower Palaeozoic syenites, xenoliths of which are preserved within the intrusion. The southern part of the massif has a complicated composition, being composed of pyroxenites, gabbro–pyroxenites and melanocratic gabbro amongst which coarse trachytic rocks are dominant. The total area of the gabbro–pyroxenite body is approximately 8 km². Close to the margins of the area of gabbro–pyroxenite

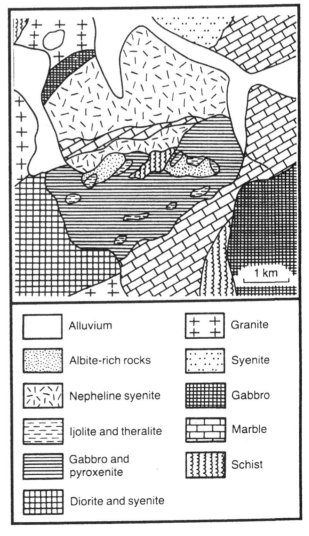

Fig. 195. *Saizhenskii (after Konev, 1982, Fig. 4).*

there is a series of vein-like and stock-like bodies of ijolite and theralite up to 200 × 400 m, of which the former are the more abundant. Several intrusions of alkaline gabbroids are found outside the massif. The northern part of the massif consists predominantly of nepheline syenites which generally contain hastingsite. Small areas of aegirine and liebenerite syenite are also common. At the gabbro-pyroxenite contact a zone of gneiss-like and pegmatitic hastingsite-nepheline syenite, biotite-nepheline syenite and congressite have been identified. The zone extends eastwards from the contact. Near this zone coarse xenoliths of marbled limestones are present, together with phyllitic schists. Three stages of intrusion have been distinguished in the history of the massif. They are: (1) titanaugite pyroxenites, gabbro-pyroxenites and melanocratic gabbro, (2) urtite-jacupirangite and leucocratic theralite, and (3) nepheline syenite, cancrinite and liebenerite syenites, congressite and aegirine syenite. Among postmagmatic rocks can be found in places albite-lepidomelane rocks as well as small veinlets, lenses and stocks (up to 200 × 300 m) of calcite rocks, which are regarded by some authors as carbonatites. The alkaline rocks of the massif are cut by diorite-syenites and granites that are likely to be Mesozoic in age.

Age K-Ar dating of kaersutite from pyroxenite gave 492 ± 40 Ma; nepheline from theralite 386 ± 12 Ma, nepheline from ijolite-urtite 294 ± 12 Ma and biotite from nepheline syenite pegmatite 306 ± 12 Ma (Andre'eva, 1982).

References Andre'ev, *et al.*, 1969; Andre'eva, 1982; Konev, 1962 and 1982.

---

26 AMALAT   53°57′N; 112°23′E
Fig. 196

This is a group of lens- and sheet-like bodies within an area of about 1.0 × 1.5 km which cut Upper Proterozoic

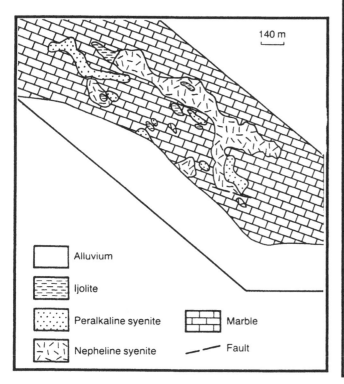

marbles. Geophysical data indicate that at depth these minor bodies coalesce into a monolithic intrusion. Nepheline syenites are the most extensively developed rock type of the occurrence among which foyaites and miaskitic varieties are present, the latter being particularly concentrated in the marginal zones where they are gneissose in appearance. Smaller areas are occupied by members of the urtite-jacupirangite series in which nepheline and pyroxene are irregularly distributed. Lamprophyres are also found.

References: Andre'ev *et al.*, 1969; Konev, 1982.

---

27 GULKHENSKII   53°55′N; 113°28′E
Fig. 197

Gulkhenskii comprises minor, steeply dipping sheet-like bodies stretching for 4 km and cutting Proterozoic marbles intercalated with schists and gabbro-diorites. Two phases in the formation of the massif have been identified, the first involving emplacement of pyroxenites and gabbro-pyroxenites with minor peridotite schlieren, and the second intrusion of nepheline-pyroxene rocks of the urtite-jacupirangite series with small amounts of theralite. Olivine gabbro is composed of olivine, labradorite, augite and hornblende with occasionally hypersthene and Ti-Fe oxide minerals. Ijolites

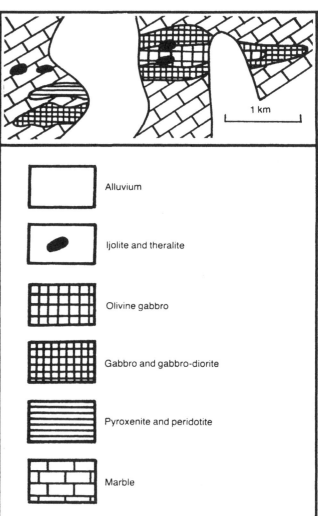

Fig. 196. *Amalat (after Sharackshinov, 1984a, Fig. 12).*   Fig. 197 *Gulkhenskii (after Konev, 1982, Fig. 4).*

consist of nepheline and a pyroxene which is reported to be fassaite (30%), but hornblende (10%) and garnet are also present, together with calcite and apatite. In leucotheralites nepheline and andesine dominate over the mafic minerals (fassaite, hornblende, biotite). The rocks have been described by Sharackshinov (1984a) and Konev (1982); mineral compositions are given by Andre'eva and Troneva (1986).

References Andre'eva and Troneva, 1986; Konev, 1964 and 1982; Sharackshinov, 1984a.

---

28 NIZHNE-BURUL'ZAISKII   53°45'N; 111°55'E
Fig. 198

Nizhne-Burul'zaiskii covers 2.3 km × 220–250 m and intrudes marbled limestones of Upper Proterozoic age. From drilling it appears that the intrusion forms a vertical sheet and extends to at least 450 m. It was emplaced in two phases: gabbro, gabbro–pyroxenite and pyroxenite; and an urtite–jacupirangite series. Nepheline syenite veins are rather rare and there are some younger leucocratic granite dykes. Within the massif there is a distinct zonality. The periphery is composed of jacupirangites, which are gradually replaced towards the centre first by melteigites and then by ijolites. Within the ijolites urtite forms isolated bodies 225–350 m across. The rocks of the urtite–jacupirangite series are medium-grained and display trachytic textures. They consist of varying proportions of nepheline and clinopyroxene together with amphibole (hastingsite, hornblende), garnet, biotite, oligoclase, calcite, apatite, magnetite and some secondary minerals including zeolites. The clinopyroxenes are variable in composition and may be salite, fassaite or hedenbergite. The gabbros and gabbro-pyroxenites are usually considerably altered to amphibolite. The nepheline syenites which form veins are medium- to coarse-grained rocks composed of oligoclase (15–30%), orthoclase (40–45%), nepheline (20–25%), rare grains of aegirine–augite partly replaced by amphibole and garnet. Data on the chemical compositions of rocks will be found in Andre'ev *et al.* (1969) and Sharackshinov (1975) and mineral compositions are given in Andre'eva and Troneva (1986).

Economic Urtite suitable for mining as a source of high quality alumina is present (Andre'eva and Troneva, 1982).

Age K–Ar on urtite gave 432 ± 16, on amphibolitized gabbro 464 ± 20 and on nepheline syenite 325 ± 10 Ma (Andre'eva, 1982).

References Andre'ev *et al.*, 1969; Andre'eva, 1982; Andre'eva and Troneva, 1986; Konev, 1982; Sharackshinov, 1975.

---

29 VERKHNE-BURUL'ZAISKII   53°51'N; 112°01'E
Fig. 199

This intrusion extends over some 2.7 × 7 km and includes primary liebenerite syenites, which cover about 70% of the area, biotite–muscovite, cancrinite and nepheline syenites, juvites, urtites, ijolites and melteigites. The massif mostly intrudes Upper Proterozoic limestones and Cambrian sandstones and quartzites and probably Lower Palaeozoic to Upper Proterozoic gabbroids and granites are also involved. The rocks of the urtite–melteigite series comprise nepheline, aegirine-augite and a small amount of amphibole (hastingsite, arfvedsonite, riebeckite), garnet, titanite, calcite, biotite, apatite, magnetite and secondary minerals including cancrinite and zeolites. Unaltered nepheline syenites are

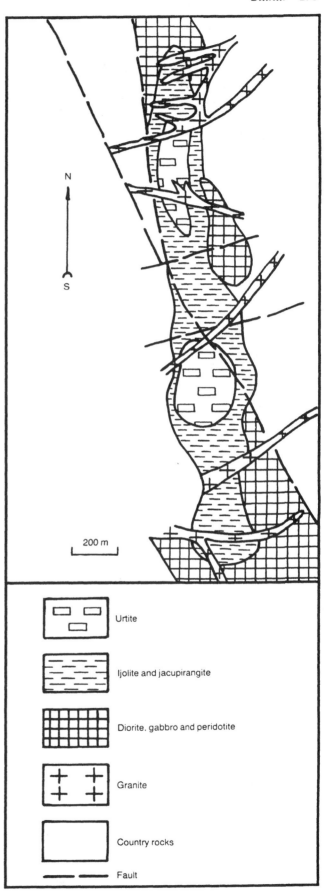

Urtite

Ijolite and jacupirangite

Diorite. gabbro and peridotite

Granite

Country rocks

——— Fault

Fig. 198. Nizhne-Burul'zaiskii (after Sharackshinov, 1984a, Fig. 3).

preserved in small areas and composed of alkali feldspar (45–50%), nepheline (3–40%), aegirine–augite (6–15%), hastingsite and alkali amphiboles (0–20%), biotite (0–15%); calcite, garnet, titanite, magnetite, apatite and secondary cancrinite and zeolite are also present. Chemical compositions will be found in Sharackshinov (1984a) and Andre'eva and Troneva (1986).

References Andre'ev *et al.*, 1969; Andre'eva and Troneva, 1986; Sharackshinov, 1984a.

---

30   IPOLOKTINSKII   53°50'N; 111°50'E

At this locality alkaline rocks cut Upper Proterozoic marbles and comprise two small bodies of 50 × 100 and 100 × 200 m. They contain urtites and ijolites; small areas of melteigite and theralite and nepheline syenite veins are found. The rocks comprise nepheline (3–60%), clinopyroxene (40–80%), hastingsite, garnet, calcite (up to 3%) and cancrinite. Microcline, plagioclase and scapolite are occasionally found.

References Konev, 1982; Sharackshinov, 1975.

---

31   SIRIKTINSKII   53°49'N; 112°06'E
                                                      Fig. 200

This stock-like massif intrudes crystalline limestones and amphibole schists of upper Proterozoic age. The massif predominantly comprises peralkaline syenites and medium-grained nepheline syenites. At the southern outer contact small areas of ijolite and urtite are present among the nepheline syenites. For the most part peralkaline granites are concentrated near the marginal parts of the complex. In the central area there is a 600 × 100 m body composed of peralkaline granite and peralkaline syenite. The paragenesis of the alkaline minerals of all the alkaline rocks are very similar. Mafic

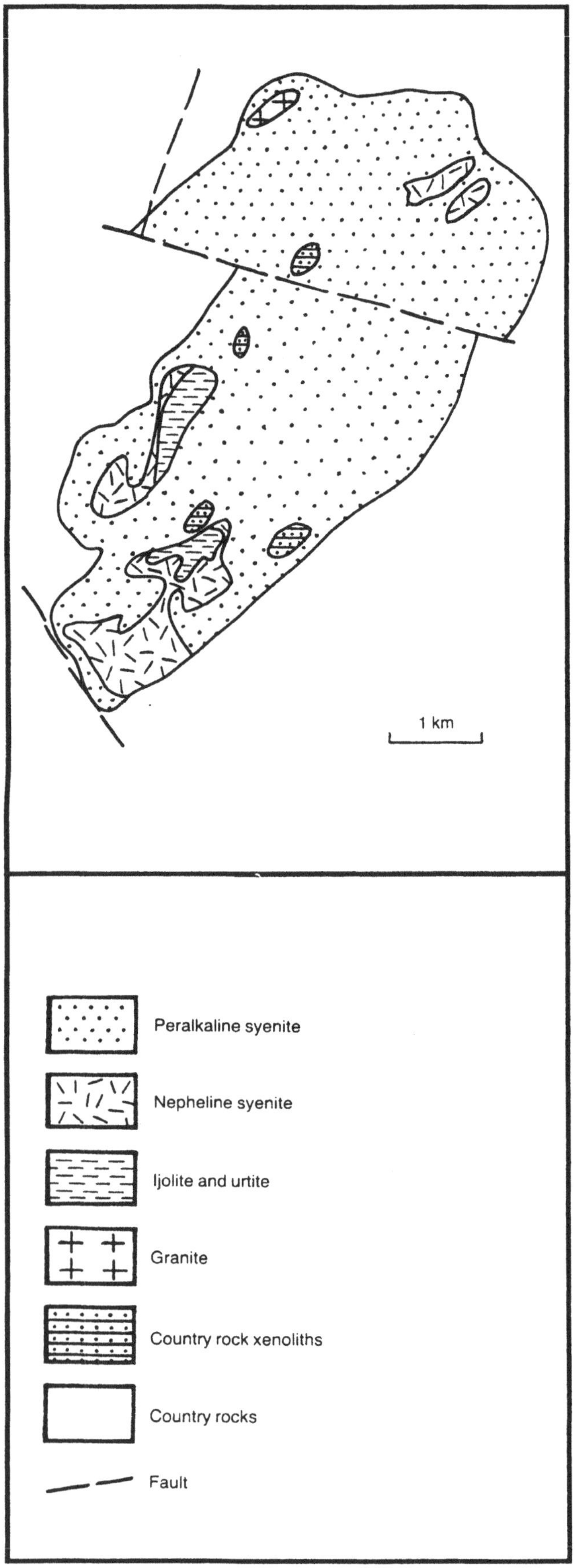

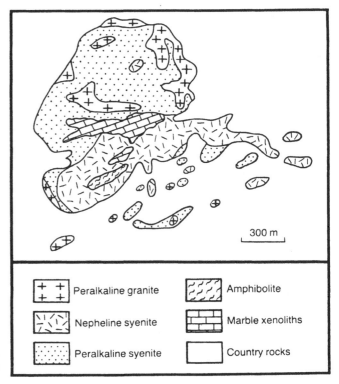

Fig. 200. Siriktinskii (after Sharackshinov, 1984a, Fig. 8).

Fig. 199. Verkhne-Burul'zaiskii (after Sharackshinov, 1984a, Fig. 7).

minerals are biotite, aegirine and riebeckite, the syenites are dominantly alkali feldspar rocks, and quartz and nepheline are diagnostic of the granites and nepheline syenites. A vein facies is represented by granite–aplites and peralkaline pegmatites. Postmagmatic alteration caused albitization and microclinization which change the rock compositions throughout the massif, but mainly along faults.

References Andre'ev *et al.*, 1969; Sharackshinov, 1984a.

---

32 MUKHALSKII   53°49′N; 112°16′E
Fig. 201

At the present erosion level only a small part (200–400 × 1500 m) of the intrusion can be seen, the rest being hidden beneath a cover of Neogene basalts. Drilling indicates that the intrusion extends over an area of at least 11.5 × 3 km but the margins have not been detected. It is one of the most extensive alkaline intrusions of the region and cuts predominantly marbles and dolomites of Upper Proterozoic age. It comprises various nepheline–pyroxene rocks. In the centre of the massif leucocratic rocks, including urtites, monmouthites and congressites, are dominant, but towards the margins ijolites and melteigites are predominant. Occasionally juvite and nepheline syenite are found. The textures of the urtite and ijolite are hypidiomorphic granular, sometimes with idiomorphic nepheline. These rocks comprise nepheline (10–85%), clinopyroxene (3–65%), which varies widely in composition, amphibole (10%), calcite (1–20%), garnet and titanite (2%). The rocks are partly replaced in places by albite and microcline. Xenoliths of country rock limestones up to about 50 × 100 m have been preserved within the intrusion. Information on the chemical composition of the rocks is available in Sharackshinov (1975) and Andre'eva and Troneva (1986).

Economic The urtites of the massif are regarded as a potential source for high-quality alumina (Andre'eva and Troneva, 1986).

References Andre'eva and Troneva, 1986; Konev, 1982; Sharackshinov, 1975 and 1984b.

---

33 SHERBAKHTINSKII   53°36′N; 113°12′E
Fig. 202

This intrusion, with an area of about 210 km², breaks through early Proterozoic metamorphic formations as well as granitoids of early to mid-Palaeozoic age. A considerable part of the area of the massif is overlapped by Neogene subalkaline basalts and basanites. The intrusion consists mainly of syenites, both peralkaline and subalkaline, as well as peralkaline granites. The syenites cover some 180 km² and veins and dykes are widespread, the latter being granite pegmatites with riebeckite, granite porphyries and grorudites.

Age Permian–Triassic according to geological evidence.

Reference Zanvilevich *et al.*, 1985.

---

34 INGURSKII   53°36′N; 113°10′E
Fig. 202

This 140 km² massif, like Sherbakhtinskii (No. 33), cuts early Proterozoic metamorphic rocks and Palaeozoic granites and is partly covered by Neogene basalts. The intrusion is characterized by the concentric disposition of the rock units: biotite granite in the centre and peralkaline granites around the margin. The peralkaline

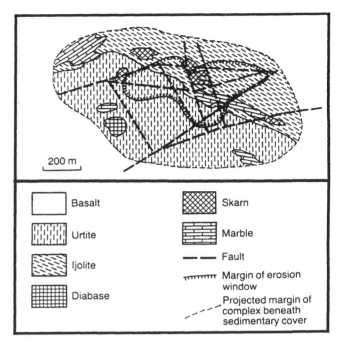

Fig. 201. Mukhalskii (after Sharackshinov, 1984a, Fig. 11).

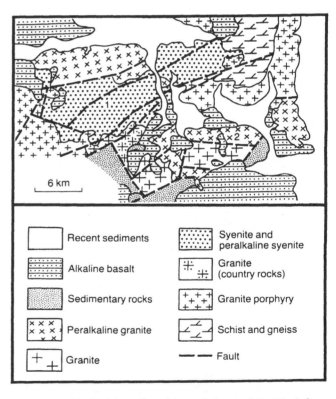

Fig. 202. Sherbakhtinskii (1) and Ingurskii (2) (after Zanvilevich et al., 1985, Fig. 2.10).

granites occupy about 100 km² and veins and dykes are abundant and include peralkaline granite pegmatites, granite porphyries and syenite porphyries. The peralkaline granites are coarse-grained, sometimes porphyritic and composed of alkali feldspar, quartz, aegirine and riebeckite.

Age Geological evidence indicates a Permian–Triassic age.

Reference Zanvilevich *et al.*, 1985.

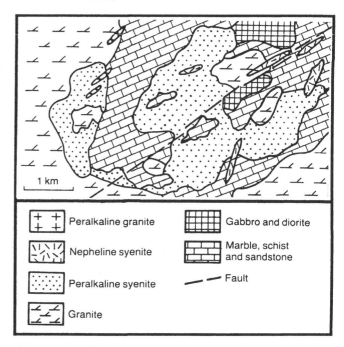

35 ZIMOV'ECHINSKII   53°35′N; 111°45′E
Fig. 203

About 80% of the massif is composed of medium- and coarse-grained hastingsite syenites, which sometimes contain quartz. It cuts limestones intercalated with silt-stones and sandstones, which are altered to hornstones in the vicinity of the contact, and early Palaeozoic gabbro and gabbro–diorites. Porphyritic granites and syenites are extensively developed in the region. Nepheline syenite bodies, which can reach 100 m across, occur at the boundary between hastingsite syenites and lime-stones. Within most areas of the peralkaline syenites a large number of country rock xenoliths occur which vary in size up to 500 × 200 m. Nepheline syenites are grey and dark-grey rocks, which are often trachytic and banded. They consist of microcline-perthite (23–70%), albite (7–39%), nepheline (16–30%), hastingsite (6–30%), aegirine–augite (0–10%) and biotite (0–5%). Accessory minerals include apatite, zircon, titanite, calcite and Ti–Fe oxide minerals. Postmagmatic altera-tion caused the development of garnet (grossularite-andradite), cancrinite and zeolite. Porphyritic urtites and monmouthites, with phenocrysts up to 2 cm across, and congressites are also found in the massif, and are distinguished by their dark minerals with monmouthites characterized by the presence of hastingsite, urtites by clinopyroxene and congressites by biotite.

Reference Sharackshinov, 1984a.

Fig. 203. Zimov'echinskii (after Sharackshinov, 1984a, Fig. 3).

36 VITIM VOLCANIC FIELD   53°40′N; 112°01′E
Fig. 204

On the Vitim plateau basaltic flows extend over approx-imately 7000 km² (Fig. 204); this is the most extensive area of Cainozoic volcanic activity in the Baikal region. Miocene alkali olivine basalts with subordinate basa-nites and pyroclastic rocks are the most abundant and fill tectonic depressions and river valleys with lava sequences 200–400 m thick (Ionov *et al.*, 1993a). The youngest flows are Pliocene basanites and hawaiites. There are more than 20 volcanic centres of Miocene age and these consist of scoria cones with small lava flows. A single occurrence of alkali picritic tuff has been found. In the Dzhilinda Depression two volcanogenic–sedimentary series are present. The lower series is 200 m thick, of which the volcanic rocks com-prise some 36%. The upper volcanogenic–sedimentary series is 150–160 m thick and volcanic rocks comprise more than 90%. At the bottom of the volcanic sequence mildly alkaline olivine basalts are about 3 m thick and are covered by 1 m of siltstones and a 4 m thick boulder bed. There follow two basanite layers with a total thickness of 38 m, which were succeeded by sandstones (23 m) and siltstones. There then follow three series con-sisting of basanites (90–100 m), basalts (50–60 m) and hawaiites (40 m). The final stage of the volcanic activity was the production of scoria cones which are about 80–90 m high. Small peridotite xenoliths and pyroxene megacrysts occur in both Miocene and younger basalts throughout the field. A very detailed account of these xenoliths, including analyses of the host rocks, is given by Ionov *et al.* (1993a).

Age Miocene and Pliocene. K–Ar determinations on alkaline basalts from various stratigraphic levels gave

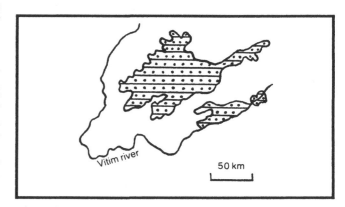

Fig. 204. Cenozoic basalts of the Vitim Plateau (after Abromovich et al., 1988.

ages from 14 ± 1.2 Ma to 4 ± 0.6 Ma (Rasskazov and Batyrmurza'ev, 1985).

References *Ionov *et al.*, 1993a; Rasskazov and Batyrmurza'ev, 1985.

37 TUCHINSKII   53°30′N; 111°26′E
Fig. 205

The Tuchinskii massif, of 2.3 × 0.7 km, lies between porphyritic and medium-grained leucocratic granites to the northwest and marbles with hornfelsed sandstones of Lower Cambrian age to the southeast. About 70% of the massif is composed of nepheline syenite and hastingsite and pyroxene alkaline syenites. The alkaline syenites are developed principally along the periphery of the massif, forming an external zone up to 300 m wide. Urtites (monmouthites) are found among the nepheline syenites in the form of small lens-like bodies up to 160 × 80 m. A characteristic feature of the intrusion is the inter-leaving of alkaline and nepheline syenites. Hastingsite

Legend for Fig. 203:
- Peralkaline granite
- Nepheline syenite
- Peralkaline syenite
- Granite
- Gabbro and diorite
- Marble, schist and sandstone
- Fault

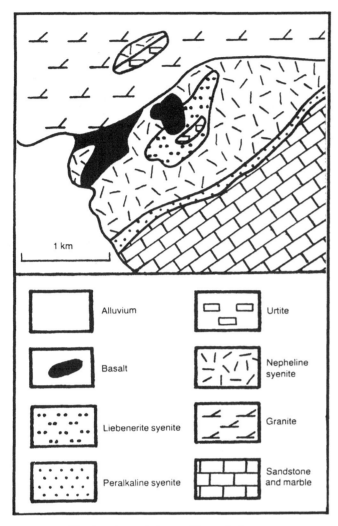

Fig. 205. Tuchinskii (after Sharackshinov, 1984a, Fig. 2).

syenites are present in the zone of nepheline syenites, where they form sheet-like lenses 60–70 m thick. Nepheline syenites, also in the form of lenses, occur in the peralkaline syenite of the peripheral zones of the massif, while small bodies of juvite, monmouthite and urtite up to 100 × 200 m are also found. The massif abounds in hornstone xenoliths. Moreover, from the preserved layering in the abundant xenoliths, it is possible to determine their original stratigraphic positions. In the vicinity of the main Tuchin massif there are two smaller intrusions, the northern one of which is composed of massive nepheline and alkaline syenites, amongst which numerous limestone xenoliths have been preserved.

Reference Sharackshinov, 1984a.

---

38 ALTANSKII   53°18′N; 110°45′E

Nepheline syenites and aegirine–augite syenites are present in the form of talus blocks over an area of 1 km².

Reference Sharackshinov, 1984a.

South Baikal subprovince

South of Lake Baikal (Fig. 181) vigorous volcanic activity took place during Mesozoic and Cainozoic times. Numerous volcanic centres and fields are composed predominantly of basalts, basanites and hawaiites, with the greatest volumes being erupted in Cainozoic times. They are known not only from south of the lake, mainly in the basin of the Dzhida River, e.g. the Bartoiskoe volcanoes, but also to the west, the volcanoes of the Tunkin Depression, and to the east. To the south and east of the Borgoi Depression are found numerous intrusions of nepheline syenite which are situated in the large (about 300 km²) Yenkhor massif, which consists of biotite syenites, granosyenites, diorites and syenite–diorites and these form elevated ridges that frame the volcanic Borgoi Depression, which is of Cretaceous age.

---

39 TUNKINSKOE VOLCANIC FIELD
51°36′N 102°00′E
Fig. 206

Small cinder cones and lava flows make up the Tunkinskoe volcanic field. The lavas have vesicular and massive structures and in some places are covered by typical agglutinates. Analyses indicate that these rocks are alkali olivine basalts with normative nepheline up to 8% and are somewhat differentiated (Mg 54–58). The possibility of the presence of tholeiitic basalts in the Tunkinskoe field is at present a matter of debate. In Kononova *et al.* (1993) it is stressed that amongst the rocks studied so far tholeiites have not been recognized. The volcanic cones are not large: the height generally varies between 40 and 50 m, rarely reaching 125 m. Most of the cones are dome-shaped with flat or rounded summits. According to geophysical data the base of the basalt pile of the Tunkinskoe field is at 1,500–2,000 m below sea level. Assuming that the initial height of the Khamar-Daban ridge over the Tunkin Depression reached 500–700 m at the beginning of volcanic activity, the total amplitude of vertical movements due to deformation associated with the igneous activity attained 3,500 m.

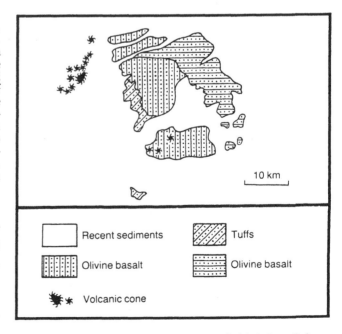

Fig. 206. The Tunkinskoe volcanic field (after Belov, 1963, Fig. 90).

Age Quaternary–Neogene.

References Belov, 1963; Kiselev, 1981; Kononova *et al.*, 1993.

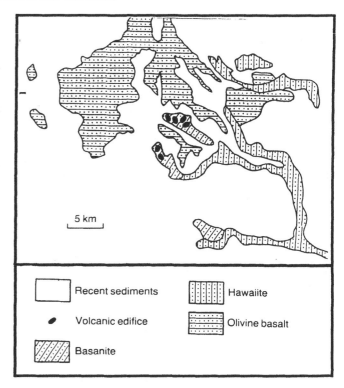

40 BARTOISKOE VOLCANIC FIELD

51°00′N; 103°03′E
Fig. 207

The Bartoiskoe volcanic field is one of the numerous volcanic fields of the Baikal rift system. The maximum volcanic activity took place in the Miocene and during the Pliocene and Holocene the intensity of volcanic activity decreased markedly. The magmatic and tectonic activity were not synchronous in so far as the rate of sinking of the bottom of the rifts and the growth of a dome were slight in the Oligocene to early Pliocene, that is during the period of maximum volcanic activity. However, from the Pliocene up to the present the growth of rift depressions and domes has sharply increased, although volcanic activity has stopped. The lavas extend beyond the rifts, and lava fields of different ages occupy three geomorphological levels in the present relief. The following groups, based on geomorphology, can be distinguished: (1) 'summit' moderately alkaline olivine basalts. The overall thickness of the lava flows is about 400–500 m and they have an age of 27 Ma; (2) 'valley' lava flows of hawaiite are widespread: they have ages of about 3 Ma; (3) the youngest 'bed' lava flows and small cinder-pumice volcanic cones have an age of 1.4 Ma. The volcanic rocks of the last stage are basanites or olivine tephrites, and sometimes contain leucite. In the cinder-pumice cones xenoliths of spinel lherzolite and augite and anorthoclase megacrysts are found, as are more rarely Ti-biotite megacrysts. The volcanic rocks change in composition between the first and third stages, notably in the increase in the content of alkalis and the elements Zr, Nb, RE, Sr, Pb, etc., while the degree of silica undersaturation increases. In terms of mineral composition the proportion of modal alkali feldspar and normative nepheline in hawaiites (second stage) increases (5–8%), modal leucite appears and the content of normative nepheline in the basanites of the concluding stage of volcanic activity increases to between 10 and 20%. The composition of the lavas leads to the conclusion that they are differentiated products of mantle melts, and that there was an increasing degree of the metasomatic transformation of the mantle source concomitant with a decrease of the extent of melting with time. Geophysical data indicate that the lithosphere in the region of Bartoiskoe volcanoes had a thickness of only 60–75 km and amongst the xenoliths both hydrous and 'dry' peridotites are present in approximately equal numbers. The isotopic composition of Sr and Nd in the mantle xenoliths seems to indicate that the mantle in this region was heterogeneous. Detailed data on the compositions of the rocks can be found in papers by Kononova *et al.* (1986, 1987, 1988 and 1993).

Age Neogene–Holocene, K–Ar determinations record activity at 27 Ma, 3 Ma and 1.4 Ma (Kononova *et al.*, 1988).

References Ionov *et al.*, 1993b; Kononova *et al.*, 1986, *1987, 1988 and 1993.

Legend (Fig. 207): Recent sediments; Volcanic edifice; Basanite; Hawaiite; Olivine basalt

Fig. 207. The Bartoiskoe volcanic field (after Belov, 1963, Fig. 97; with additions by Kononova et al., 1986).

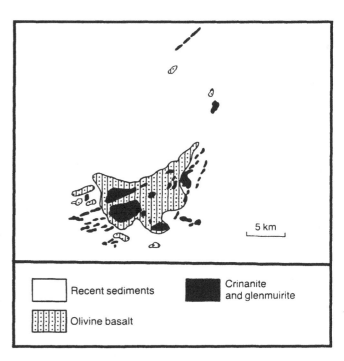

Legend (Fig. 208): Recent sediments; Olivine basalt; Crinanite and glenmuirite

Fig. 208. The Borgoiskoe volcanic field (after Belov, 1963, Fig. 60).

41 BORGOISKOE VOLCANIC FIELD

50°44′N; 105°20′E
Fig. 208

The Borgoiskoe volcanic field consists of thick, more than 5 km, accumulations of effusive and sedimentary rocks into which laccoliths and sills have been emplaced. It is partly covered by Recent sediments. Within the sedimentary rocks, which consist of conglomerates, siltstones, sandstones and limestone, occur basic tuffs and moderately alkaline basalt lavas which are of hawaiite type. The latter crop out extensively over the basin, where they form a cover 120 m thick. The lavas comprise dozens of flows with thicknesses up to about 10 m. The flows in the central part of the depression are horizontal. Approximately simultaneously with the eruption of the lavas laccoliths and sills of crinanite and glenmuirite were intruded. The laccoliths are clearly reflected in the topography, forming hills up to 100–150 m high. The crinanites are analcime dolerites containing olivine (4–11%), plagioclase (25–41% and An_{53-62}), titanaugite (10–20%), analcime (up to 20%), alkali feldspar (up to 20%) and biotite (4–5%). In glenmuirites the biotite content rises sharply, up to 19%, and the clinopyroxene content falls to a maximum of 2%. The usual accessories in the crinanites and glenmuirites are apatite and Fe–Ti oxides. The compositions of rocks and minerals are given by Belov (1963).

Age K–Ar on whole rocks gave 113 ± 10 Ma (Kononova *et al.*, 1993).

References Belov, 1963; Kononova *et al.*, 1993.

---

42 BORGOISKII

50°44′N; 105°40′E
Fig. 209

The Borgoi intrusion, with an area of about 2.6 km², is the largest manifestation of nepheline syenites in the region of the Dzhida River. In the north the intrusion is in contact with extrusive rocks and Recent sediments and in the south with metamorphic diorite. The nepheline syenites grade into syenites, but usually the transition zone does not exceed 10 cm in width. Nepheline syenites, with only moderate nepheline contents of about 4–5%, are the most widespread rocks of the intrusion. Albitization is widely developed and mariupolite zones are found accompanying pegmatites of nepheline syenite. Zones of syenite replaced by muscovite are encountered

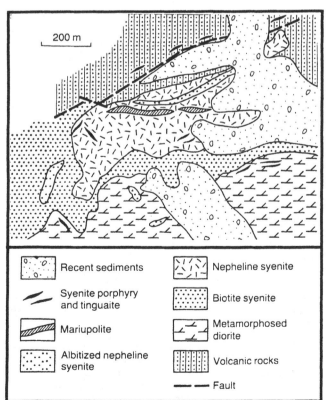

Recent sediments	Nepheline syenite
Syenite porphyry and tinguaite	Biotite syenite
Mariupolite	Metamorphosed diorite
Albitized nepheline syenite	Volcanic rocks
	Fault

Fig. 209. Borgoiskii (after Kuznetsova, 1975, Fig. 2).

in some places. Within the nepheline syenites xenoliths of biotite schist, and relics of metamorphosed extrusive rocks, as well as chains of lens-like bodies of syenite porphyry, tinguaite and solvsbergite have been identified. The solvsbergites consist of alkali feldspar (30–75%), albite (20–49%), aegirine (0–13%), hornblende (0–16%) and biotite (0–7%); sodalite, muscovite, fluorite, lavenite, apatite and titanite are also present. The tinguaite is usually present as dykes which have a porphyritic texture with large alkali feldspar phenocrysts. The groundmass of these rocks contains nepheline (7%), albite (42%), aegirine (4%), biotite (6%) and hornblende. Small blocks, up to 30 cm across, of ijolite have been found and consist of aegirine 49%, nepheline 43%, albite 4%, alkali feldspar 2%, biotite, titanite, fluorite and apatite. Thermometry on gas–fluid inclusions in nepheline in nepheline syenite indicates temperatures varying from 80 to 780°C (Panina, 1972).

Age Whole rock K–Ar determinations on nepheline syenite gave 188 Ma (Panina, 1972).

References Andre'ev *et al.*, 1969; Kuznetsova, 1975; Panina, 1972.

---

43 DABKHORSKII

50°40′N; 104°33′E

Dykes of tinguaite and solvsbergite cut albitized nepheline syenites. The massif lies amongst biotite and hornblende syenites.

Age Mesozoic.

Reference Andre'ev *et al.*, 1969.

---

44 NIZHNE-ICHETUISKII

50°40′N; 105°40′E
Fig. 210

A number of bodies of nepheline syenite comprise this occurrence and extend over a total area of about 3.7 km² within coarse-grained biotite syenites. They have sinuous contacts and vary widely in size, the largest body being about 1,000 × 250 m. The nepheline syenites have massive, foliated or heterogeneous structures and the nepheline tends to form phenocrysts 3–5 cm in diameter within a background of fine-, medium- and sometimes coarse-grained rock. The modal composition of these rocks is alkali feldspar (55–70%), nepheline (20–30%), biotite (2–3%), riebeckite (2–5%), andesine (0–5%), garnet (1–2%), titanite, fluorite, apatite and some secondary minerals, including liebenerite, cancrinite, epidote and zoisite. Albitization of the nepheline syenites is widespread and accompanied by the formation of mariupolites. In the mariupolites the nepheline content varies between 15 and 30% and aegirine or biotite and hastingsite are also present. The bodies of mariupolite contain nests and vein-like areas of nepheline syenite pegmatite up to 1 × 0.2 m.

Age Whole rock K–Ar determinations gave 108–137 Ma (Kuznetsova, 1975).

References Andre'ev *et al.*, 1969; Kuznetsova, 1975.

---

45 BELOGORSKOE

50°32′N; 105°48′E

Nepheline syenites form several small bodies which are concentrated along tectonic zones within a body of syenite porphyries. Gradual transitions from nepheline-free syenite porphyries to rocks of tinguaite type are observed. Sometimes nests occur, rarely exceeding 1 × 2 m, of coarse-grained nepheline syenite and albite-nepheline rock with a nepheline content of 2–3%. There

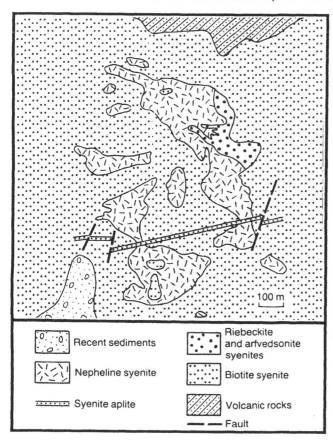

Fig. 210. Nizhne-Ichetuiskii (after Andre'ev et al., 1969, Fig. 43).

are also albite–nepheline–sodalite rocks with sodalite contents of up to 20–25%.

Age Mesozoic – by analogy with other nepheline syenites of the region.

Reference Kuznetsova, 1975.

46 ORTSEKSKII   50°30'N; 105°56'E

Several outcrops of nepheline syenite occur within biotite syenites.

Age Mesozoic – by analogy with other alkaline massifs of the region.

Reference Kuznetsova, 1975.

47 KHARASUNSKII   50°37'N; 106°00'E

Nepheline syenites form several small bodies of 10 × 15, 4 × 30 and 10 × 120 m amongst biotite or aegirine-hornblende syenites. The nepheline content is variable and generally between 5 and 25%. Small areas of coarse-grained nepheline syenite are found. Sodalite forms thin veins in the nepheline syenites.

Age Mesozoic – by analogy with other nepheline syenites of the region.

Reference Kuznetsova, 1975.

48 VERKHNE-BULYKSKII   50°29'N; 105°56'E

This is a small area of nepheline syenite occurring within biotite syenites.

Age Mesozoic – by analogy with other alkaline massifs of the region.

Reference Kuznetsova, 1975.

49 BOTSI   50°30'N; 105°37'E
(Enkhorskii)

This occurrence consists of two bodies of nepheline and liebenerite syenite that cut through coarse- and medium-grained muscovitized syenites which grade into biotite syenites at depth – as indicated by drilling. The nepheline syenite bodies do not exceed 10 × 5 m and are composed mainly of alkali feldspar (40–50%), albite (5–20%), nepheline (2–50%), biotite (5–15%) and aegirine (5–20%). Accessory minerals are represented by titanite, zircon and apatite; overall they do not exceed 1–2%. The nepheline syenites vary from fine- to coarse-grained and pegmatitic varieties. The texture of the rocks may be gneiss-like or massive.

Age Whole rock K–Ar determinations on nepheline syenite gave 172 Ma (Panina, 1972).

References Kuznetsova, 1975; Panina, 1972.

50 ZORMENIKSKII   50°27'N; 105°37'E

A small group of lens-like bodies of albite–microcline–nepheline rocks which cut biotite syenites is situated in the same massif of biotite syenites as the Botsi occurrence.

Age Mesozoic – by analogy with other massifs of the region.

Reference Kuznetsova, 1975.

51 SUKHO-KHOBOL'SKII   50°27'N; 105°39'E

This is a small body of nepheline and cancrinite syenites which occurs in biotite–hornblende syenites.

Age Mesozoic – by analogy with other massifs of the region.

Reference Kuznetsova, 1975.

Southeast Baikal subprovince

Southeastern Transbaikalia is an area in which are developed many intrusions of alkaline and subalkaline granites and syenites (Zanvilevich et al., 1985; Sheinmann et al., 1961) with only the Komskii intrusion (locality 52) consisting of nepheline syenite. A few occurrences of alkali granite of Precambrian age are known, but it was not until the Permian and Triassic periods that very large complexes were emplaced. Over 350 such intrusions with areas of 1–2 km² to 600–700 km² are known, but not all are alkaline and few have been investigated in detail. They form a belt stretching for up to 1,500 km with an average width of about 200 km. It is only partly located within southeastern Transbaikalia, continuing towards the northeast as the Vitim Province and south-westwards into northern Mongolia. There are numerous Palaeozoic intrusions of diorite, gabbro–diorite, grano-diorite and plagiogranite and extrusives including trachyandesite, dacite and rhyolite and this igneous activity was completed during the Permian and Triassic by the emplacement of intrusions of enhanced alkalinity. Only the most well studied and typical massifs of alkaline granite and syenite are described here (Fig. 181), there being many occurrences for which there is little or no information available. The fullest account is to be found in the monograph of Zanvilevich *et al.* (1985), which gives data on the compositions of rocks and minerals of alkaline granites and syenites as well as of other igneous rocks of Transbaikalia.

52 KOMSKII   52°17′N; 107°26′E
Fig. 211

Komskii consists of several small intrusions spread over an area of 3.5 × 1.5 km. It is located between areas of Upper Palaeozoic granite and syenite and consists of two types of nepheline syenite. One is medium-grained and displays a gneissic structure and the other is a coarse-grained, leucocratic biotite–hastingsite and aegirine–augite–nepheline syenite with a massive structure. The latter is situated in the central area and covers 1.5 × 0.4 km. Xenoliths of fenitized schist and marble occur in the nepheline syenites but are unknown in outcrop. The youngest rocks of the complex are fine-grained syenites and granosyenites, in some of which aegirine–augite occurs.

Reference Konev, 1982.

53 KHORINSKII   52°17′N; 109°36′E
Fig. 212

This is one of the largest igneous complexes in the region, its area exceeding 650 km². It is a heterogeneous pluton with a poorly defined zonal structure. The approximately circular outer zone is several hundreds of metres to 5 km wide. It is composed of moderately alkaline rocks, including syenites and quartz syenites (of the first stage), with subordinate amounts of granite (the second stage). The internal zone is more or less symmetrical and here peralkaline granites (of the third stage) are predominant. Two to three phases of intrusion have been identified in each stage, each of which is completed by the intrusion of dykes. The country rocks are granodiorites, granites and Palaeozoic syenites, as well as trachytes and trachyrhyolites which are provisionally referred to the late Palaeozoic to early Mesozoic. The peralkaline granites of the first phase are coarse- and medium-grained and are characterized by an abundance of

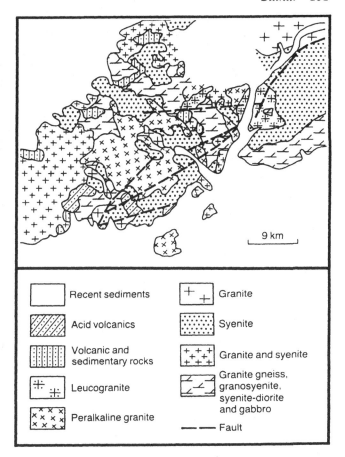

Fig. 212. Khorinskii and the southwestern part of Sredne-Oninskii (after Zanvilevich et al., 1985, Fig. 2.4).

miarolitic cavities and by some heterogeneity of structure which manifests itself in the form of frequent schlieren of relatively finer grained or pegmatitic granites. Peralkaline granites of the second phase of intrusion are fine- and medium-grained and porphyritic.

Age Rb–Sr isochron determinations on granites gave 250 ± 11 Ma (Zanvilevich *et al.*, 1985). Geological evidence indicates a Triassic to Jurassic age.

Reference Zanvilevich *et al.*, 1985.

54 SREDNE-ONINSKII   52°17′N; 109°41′E
Fig. 212

This occurrence is located immediately northeast of the Khorinskii complex and repeats all the features of its structure and composition.

Age Triassic–Jurassic

Reference Zanvilevich *et al.*, 1985.

55 BUGUTUISKII   52°00′N; 110°00′E
Fig. 213

Bugutuiskii is situated in the central part of an area of alkaline granites. All told, about 30 occurrences have been identified in this area and rare-metal mineralization is associated with some of them. The Bugutuiskii intrusion is elliptical in shape and extends in a northeasterly direction for some 20 km. It is intruded into Proterozoic gneisses, early Palaeozoic granites and Permian–Triassic

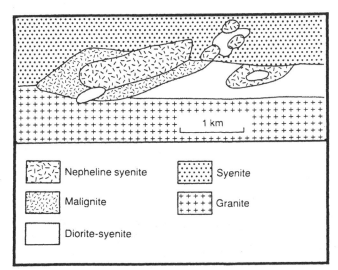

Fig. 211. Komskii (after Konev, 1982, Fig. 5).

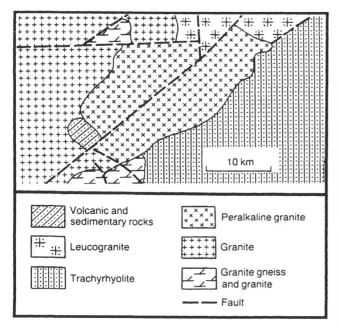

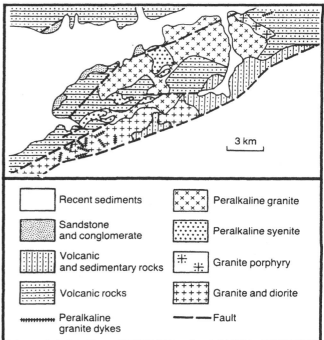

Fig. 213. Bugutuiskii (after Zanvilevich et al., 1985, Fig. 2.11).

Fig. 214. Atkhinskii (after Zanvilevich et al., 1985, Fig. 2.8).

trachyrhyolites. The massif is composed of peralkaline granites and quartz syenites of two stages of intrusion which differ chiefly in grain size, the earlier rocks being characteristically coarser. Peralkaline granites of the first stage, which are coarse- and medium-grained, are composed predominantly of alkali feldspar, quartz (about 25%), aegirine and alkali amphibole. The rocks of the second stage – quartz syenites and granites – are fine- and medium-grained, with mafic minerals represented by aegirine and riebeckite (1 to 5%). There are dykes of granite porphyry, quartz syenite porphyry and syenite which extend for up to 1 km from the main complex.

Age Triassic–Jurassic – from geological evidence.

Reference Zanvilevich *et al.*, 1985.

---

56 ATKHINSKII   51°44′N; 112°06′E
Fig. 214

This occurrence is associated with a system of faults and extends over about 22 km in a zone which is roughly 4–6 km wide. Generally the main intrusion truncates volcanic rocks, including trachyrhyolites, comendites, trachybasalts and trachyandesites, as well as normal granites. The margins of the intrusion are very irregular, which is partly due to the approximately horizontal contact with the roof, which is close to the present erosion level. At the contacts the country rocks are hornfelsed and in some places the metamorphism has led to the development of K-feldspar in the enclosing rocks which are also intersected by dyke-like apophyses of fine-grained peralkaline granite. The intrusion is composed mainly of aegirine–riebeckite granites which are of a lilac or light grey colour. They were preceded by coarse-grained peralkaline syenites which have remained intact within granites in the form of bodies up to 2.5 × 1 km². Two textural varieties are distinguished amongst the peralkaline granites: coarse- to medium-grained and medium- to fine-grained. The latter, as a rule, forms veins and dykes from 5–10 cm up to 10 m wide, which cut the coarser-grained granites. The peralkaline granites

contain alkali feldspar, aegirine, riebeckite and quartz. Prismatic crystals of plagioclase (An_{33-36}) are poikilitically included in K-feldspar in the peralkaline syenites and colourless clinopyroxene is surrounded by an alkali amphibole fringe.

Age Geological evidence indicates a Permian–Triassic or Triassic age.

Reference Zanvilevich *et al.*, 1985.

---

57 KUKINSKII   51°44′N; 113°14′E
Fig. 215

The Kukinskii intrusion extends in a northeasterly direction for 22–24 km and is about 10 km wide. It is emplaced in granodiorites, porphyritic granites and volcanics of late Palaeozoic and early Mesozoic age. Coarse- and medium-grained aegirine–riebeckite granites are predominant. A rectangular body of later fine-grained peralkaline granite has been mapped in the centre of the pluton. Dykes, up to 1.5 m wide, of peralkaline aplite, pegmatitic granite, granite porphyry and quartz syenite porphyry are known and are concentrated in the vicinity of the outer contacts.

Age Permian–Triassic or Triassic.

Reference Zanvilevich *et al.*, 1985.

---

58 UBUKITSKII   51°37′N; 106°32′E
Fig. 216

This complex has very irregular, sinuous outlines and an area of 150 km². The country rocks are granites, syenites and quartz syenites of middle and late Palaeozoic age, as well as Eocambrian limestones and schists and Permian trachytes and trachyrhyolites The principal rock type of Ubukinskii is a coarse-grained, moderately alkaline syenite in which, in places, there are xenoliths of microdiorite. The complex is cut by

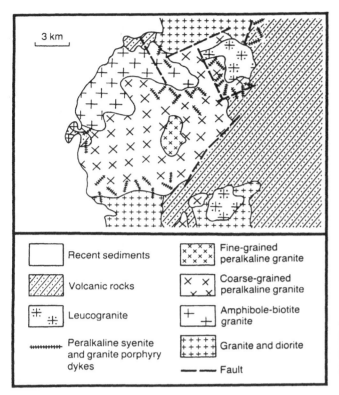

Fig. 215. Kukinskii (after Zanvilevich et al., 1985, Fig. 2.9).

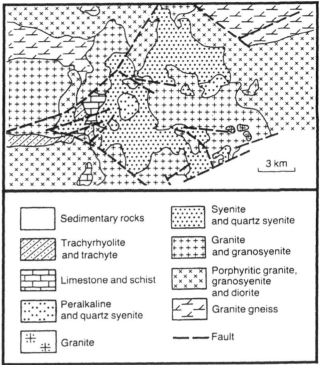

Fig. 216. Ubukitskii (after Zanvilevich et al., 1985, Fig. 2.7).

numerous veins of leucogranite, aplite and granosyenite. Later rocks are represented by relatively fine-grained peralkaline and subalkaline syenites, quartz syenites and granites. The peralkaline syenites are leucocratic rocks in which the dark-coloured minerals are riebeckite and rarely biotite.

Age Triassic–Jurassic.

Reference Zanvilevich et al., 1985.

---

59 KHARITONOVSKII   51°12′N; 107°03′E

Fig. 217

This occurrence is wedge-shaped and has an area of about 230 km². It cuts through a variety of plutonic, volcanic and volcanic–sedimentary formations of Proterozoic and Palaeozoic age. Three stages in the formation of the massif are identified. The first stage involved the intrusion of coarse-grained, moderately alkaline syenites, fine- and medium-grained porphyritic syenites and two feldspar leucogranites. The second stage encompassed moderately alkaline granites, quartz microsyenites and microgranites and the third stage coarse- and medium-grained peralkaline syenites and quartz syenites, medium- and fine-grained syenites, quartz syenites and dykes of alkaline syenite, quartz syenite and granite. All these rocks are compositionally similar with alkali feldspar predominant and dark minerals represented by aegirine and riebeckite with biotite in subordinate amounts.

Age K–Ar determinations on alkali amphibole from peralkaline syenites of the third stage gave 224 ± 8 and 231 ± 9 Ma, while amphibole from moderately alkaline syenites of the first stage gave dates varying from 195 ± 9 to 294 ± 11 Ma (Zanvilevich et al., 1985).

Reference Zanvilevich et al., 1985.

60 VERKHNE-MANGIRTUISKII   50°38′N; 107°10′E

Fig. 218

This is a small intrusion situated within early Proterozoic gneisses and crystalline schists. It is formed by peralkaline granites that have a fine-grained texture and clearly expressed orientation of the rock-forming minerals. Widely developed pegmatites lie parallel to the general fabric and, less frequently, cross-cut it; they range in thickness from several centimetres to 0.5 m. The granites are allotriomorphic, homogeneous rocks composed of coarse (2–5 mm) prismatic, and less frequently isometric, perthite or antiperthite (Or_{44-61}), quartz (20–24%), plagioclase (5–10%; An_{13-17}) and up to 3% aegirine–salite (40–68% of the aegirine molecule) and katophorite. Accessory minerals are magnetite, titanite, apatite, zircon and orthite.

Age Geological evidence indicates a Precambrian age.

Reference Zanvilevich et al., 1985.

---

61 MALO-KUNALEISKII   50°32′N; 107°58′E

Fig. 219

Peralkaline syenites and quartz syenites form an approximately circular intrusion with an area of 120 km². The enclosing rocks include rhyolitic lavas and tuffs, normal syenites and granosyenites and late Palaeozoic granites. The peralkaline syenites and quartz syenites are medium- and coarse-grained leucocratic rocks formed principally of alkali feldspar which sometimes encloses crystals of plagioclase. The mafic minerals are represented mainly by katophorite with aegirine or biotite in subordinate amounts. Remnants of diopside are present in the nuclei of katophorite crystals. Numerous dykes of peralkaline quartz syenite, granite and microgranite are present.

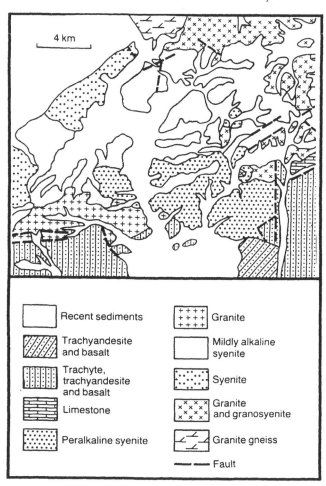

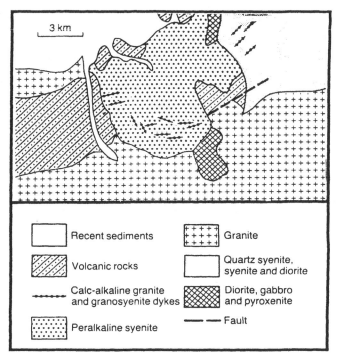

Fig. 219. *Malo-Kunaleiskii (after Zanvilevich et al., 1985, Fig. 2.5).*

Age K–Ar determinations on amphibole from peralkaline syenite gave 233 Ma (Zanvilevich *et al.*, 1985). A Rb–Sr isochron on rocks and minerals gave an age of 233 ± 5 Ma (Shergina *et al.*, 1979).

References Shergina *et al.*, 1979; Zanvilevich *et al.*, 1985.

Fig. 217. Kharitonovskii (after Zanvilevich et al., 1985, Fig. 2.3).

Baikal references

ABRAMOVICH, G.Y., MITROFAMOV, G.L., POL'AKOV, G.V. and MRENOV, P.M. (eds) 1988. *Map of the magmatic associations of south-east Siberia and northern Mongolia. 1:1,500,000.* Irkutsk.
*ALTUKHOV, Ye.N., KOZLOVA, T.K. and MOROZOV, L.N. 1973. Distribution of alkalic rocks in structures of the Baikal Highlands. *Doklady Earth Science Sections, American Geological Institute,* 210: 65–7.
ANDRE'EV, G.V. 1981. *Petrology of the association of the potassium, nepheline and peralkaline syenites.* Nauka, Novosibirsk, 85 pp.
*ANDRE'EV, G.V., POSOKHOV, V.F. and SHALA-GIN, V.L. 1991. The age of the Synnyr intrusion. *Geochemistry International,* 28(12): 99–102.
ANDRE'EV, G.V., SHARAKSHINOV, A.I. and LITVI-NOVSKY, B.A. 1969. *Intrusions of nepheline syenite of Western Transbaikalia.* Nauka, Moscow. 188 pp.
ANDRE'EVA, E.D. 1982. General features and age of the formation of some alkaline-gabbro massives from the Vitim Plateau. *Petrology and metallogeny of the magmatic associations.* 253–62. Nauka, Moscow.
ANDRE'EVA, E.D. and TRONEVA, N.V. 1986. Rock-forming minerals of the alkaline rocks from the Vitim Plateau. *The special features of rock-forming minerals from magmatic rocks.* 148–65. Nauka, Moscow.
ARKHANGELSKAYA, V.V. 1974. *Rare-metal alkaline complexes of the southern margin of the Siberian platform.* Nedra, Moscow, 128 pp.
BELOV, I.V. 1963. *The trachybasalt formation of the Baikal region.* USSR Academy of Sciences, Moscow. 371 pp.

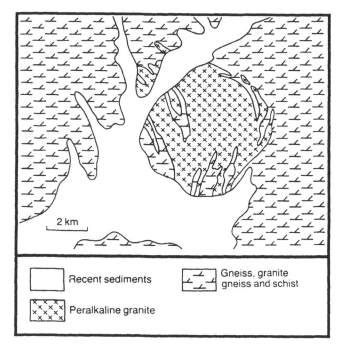

Fig. 218. Verkhne-Mangirtuiskii (after Zanvilevich et al., 1985, Fig. 7).

*IONOV, D.A., ASHCHEPKOV, I.V., STOSCH, H.-G., WITT-EICKSCHEN, G. and SECK, H.A. 1993a. Garnet peridotite xenoliths from the Vitim volcanic field, Baikal region: the nature of the garnet-spinel peridotite transition zone in the continental mantle. *Journal of Petrology*, 34: 1141–75.

IONOV, D.A., KRAMM, U., STOSCH, H.-G. and KOVALENKO, V.I. 1993b. Evolution of the upper mantle beneath the southern Baikal rift zone: a Sr–Nd isotope study of xenoliths from the Bartoy volcanoes. *Magmatism of rifts and sedimentary belts*. 211–33. Nauka, Moscow.

*KASHIRIN, K.F. 1971. Petrochemical and geochemical characteristics of nepheline syenites of northern Baykal highland. *International Geology Review*, 13: 1330–8.

KISELEV, A.I. 1981. Cainozoic volcanism of central and eastern Asia. *In* N.A. Logachev and S.I. Sherman (eds) *Problems of fissure tectonics*. 71–86. Nauka, Siberian Branch of the Academy of Sciences, Novosibirsk.

KONEV, A.A. 1962. *Petrography of the alkaline ultrabasic and alkaline basic rocks from the Sa'zen and Gulkhen plutons*. East-Siberian Publishing House, Irkutsk. 140 pp.

KONEV, A.A. 1964. The complex of ultrabasic and alkaline basic rocks from the Vitim Plateau. *Geology of the USSR*, 35(1): 434–40. Moscow.

KONEV, A.A. 1969. The Tazheran massif of gabbro, alkali and nepheline syenites. *In* E.P. Pavlovsky (ed.) *Geology of the Baikal Region*. 121–7. Institute of the Earth's Crust, Irkutsk.

KONEV, A.A. 1982. *Nepheline-bearing rocks from the Sayan-Baikal area*. Nauka, Siberian Branch of the Academy of Sciences, Novosibirsk. 201 pp.

KONONOVA, V.A., KELLER, J. and PERVOV, V.A. 1993. Continental basaltic volcanism and the geodynamic evolution of the Baikal-Mongolian region. *Magmatism of rifts and sedimentary belts*. 234–64. Nauka, Moscow.

KONONOVA, V.A., PERVOV, V.A. and KELLER, J. 1986. Continental Cainozoic volcanism in Dzhidin (USSR) and Khangai (Mongolia) volcanic fields. *Izvestiya Akademii Nauk SSSR, Seriya Geologiya*, 22: 53–68.

*KONONOVA, V.A., PERVOV, V.A., DRYNKIN, V.I., KERSIN, A.L. and ANDREYEVA, Ye.D. 1987. Rare-earth and rare elements in the Cenozoic basic volcanics of Transbaikalia and Mongolia. *Geochemistry International*, 12: 32–46.

KONONOVA, V.A., IVANENKO, V.V., KARPENKO, M.I., ARAKELYANTS, M.M., ANDREEVA, Ye.D. and PERVOV, V.A. 1988. New data on the K–Ar age of Cainozoic continental basalts of the Baikal Rift system. 1988. *Doklady Akademii Nauk SSSR*, 303: 454–8.

KOSTYUK, V.P. 1990. Geologic-structural position of potassium alkaline formations in the Baikal-Stanovoy rifting system. *In* G.V. Pol'akov and V.V. Kepezinskas (eds) *Potassium alkaline magmatism of the Baikal-Stanovoy rifting system*. 10–24. Nauka, Siberian Branch of the Academy of Sciences, Novosibirsk.

KUZNETSOVA, F.V. 1975. *Nepheline syenites of the margin of the Borgoi depression*. Nauka, Siberian Branch of the USSR Academy of Sciences, Novosibirsk. 92 pp.

*MERLINO, S., PERCHIAZZI, N., KHOMYAKHOV, A.P., PUSHCHAROVSKII, D.Y., KULIKOVA, I.M. and KUZMIN, V.I. 1990. Burpalite, a new mineral from Burpalinski massif, north Transbajkal, USSR: its crystal structure and OD character. *European Journal of Mineralogy*, 2: 177–85.

ORLOVA, M.P. 1990. Mesozoic stage of magmatism. *In*

G.V. Pol'akov and V.V. Kepezinskas (eds) *Potassium alkaline magmatism of the Baikal-Stanovoy rifting system*. 65–123. Nauka, Siberian Branch of the Academy of Sciences, Novosibirsk.

PANINA, L.I. 1972. *Mineral-genetic characteristic features of some alkaline massifs of the Baikal area*. Nauka, Siberian Branch of the USSR Academy of Sciences, Novosibirsk, 127 pp.

*POKROVSKII, B.G. and ZHIDKOV, A.Ya. 1993. Origin of the ultrapotassic rocks of the Synnyr and southern Sakun massifs (Transbaikal area) from isotopic data evidence. *Petrology*, 1: 195–204.

RASSKAZOV, S.V. and BATYRMURZA'EV, A.S. 1985. Cenozoic basalts from the Vitim Plateau and determination of their ages. *Geologiya i Geofizika*. Novosibirsk, 5: 20–7.

SEMENOV, E.I., ESKOVA, E.M., KAPUSTIN, Yu.L. and HOMYAKOV, A.P. 1974. *The mineralogy of the alkaline massifs and their deposits*. Nauka, Moscow. 248 pp.

SHARACKSHINOV, A.O. 1975. *Petrology of nepheline syenite from the Vitim Plateau*. Nauka, Siberian Branch of the Academy of Sciences, Novosibirsk. 154 pp.

SHARACKSHINOV, A.O. 1984a. *Alkaline magmatism of the Vitim Plateau*. Nauka, Siberian Branch of the Academy of Sciences, Novosibirsk. 183 pp.

SHARACKSHINOV, A.O. 1984b. The Mukhal deposit – a new genetic type of nepheline ore. *Geologiya Rudnykh Mestorozhdenii*, 1: 89–92.

SHEINMANN, Yu.M., APELTSIN, F.R. and NECHAYEVA, Ye.A. 1961. Alkaline intrusions, their location and mineralization associated with them. *In* A.I. Ginzburg (ed.) *Geology of the deposits of rare elements*. 12–13. Gosgeoltekhizdat, Moscow.

SHERGINA, Yu.P., MURINA, G.A., KOZUBOVA, L.A. and LEBEDEV, P.B. 1979. The age and some genetic features of the rocks of the Kunalei Complex in Western Transbaikalia according to the data of the Rb–Sr method. *Doklady Akademii Nauk SSSR*, 246: 1199–202.

SVESHNIKOVA, E.V. 1984. Potassium alkaline series of magmatic rocks. *In* V.A. Kononova (ed.) *Magmatic rocks*. 185–213. Nauka, Moscow.

VASILYEV, Ye.P., VISHNYAKOV, V.N. and REZNITSKY, L.Z. 1969. Geology of the area of the Slyudyanka phlogopite deposits. In E.P. Pavlovsky (ed.) *Geology of the Baikal Region*. 63–8. Institute of the Earth's Crust, Irkutsk.

ZANVILEVICH, A.N., LITVINOVSKY, B.A. and ANDRE'EV, G.V. 1985. *The Mongolian-Transbaikalian alkali-granitoid province (geology and petrology)*. Nauka, Moscow. 232 pp.

ZHIDKOV, A.Ya. 1960. Differentiated pluton of alkaline rocks in the North Baikal region. *Proceedings on the geology and ore deposits of eastern Siberia, VSEGEI, Leningrad*, 32: 119–25.

ZHIDKOV, A.Ya. 1961. New North Baikal alkaline province and some features of nepheline enrichment of its rocks. *Doklady Akademii Nauk SSSR*, 140: 181–4.

ZHIDKOV, A.Ya. 1965. *Synnyr and Burpala alkaline intrusives from North Baikal*. Ph.D. thesis, University of Leningrad. 210 pp.

ZHIDKOV, A.Ya. 1968. Apatite-bearing alkaline intrusions of the North Baikal area. In V.I. Smirnov *et al.* (eds) *Apatite*, 126–32. Nauka, Moscow.

ZHIDKOV, A.Ya. 1990. Palaeozoic stage of magmatism. In G.V. Pol'akov and V.V. Kepezinskas (eds) *Potassium alkaline magmatism of the Baikal-Stanovoy rifting system*. 32–64. Nauka, Siberian Branch of the Academy of Sciences, Novosibirsk.

*In English

ALDAN

The Aldan alkaline province, which includes more than 30 alkaline massifs of a wide range of compositions and ages, extends over 1000 km in an east–west direction in the southeastern part of the Siberian platform, and occupies an area of over 200,000 km^2 (Fig. 220). The alkaline intrusions are concentrated predominantly within the confines of the Aldan shield which, in the south and east, is bordered by a system of deep faults associated with Phanerozoic fold belts. During the Mesozoic these fault systems were apparently Benioff zones separating the palaeocean from the continental shield. At present the Aldan shield forms a small plateau.

The geological structure of the various blocks of the Aldan shield vary from east to west. They were formed in the earliest stages of the development of the Earth's crust during lower Archaean times. In the lower Archaean rocks of the western and eastern blocks of the Aldan shield basic crystalline schists predominate, while in the central block the proportion of such schists decreases sharply.

Changes in the composition of the alkaline igneous rocks can also be traced in an east–west direction: from east to west an increase in the alkalinity of the igneous rocks takes place, while a rise of the K:Na ratio is also observed. Within the western and central parts of the Aldan province potassium-rich rocks of enhanced alkalinity predominate, including leucitites, lamproites,

pseudoleucite syenites, peralkaline granites, shoshonites and latites. In the eastern part of the Aldan province, however, the rocks are predominantly members of the K–Na series. This difference of the potassium to sodium ratio in the magmatic rocks from different parts of the province is found in the magmatic products of all epochs, beginning with the lower Proterozoic. The lateral variation of the alkaline rocks in the Stanovoy ridge orogenic area is considered to be the result of involvement during late Jurassic and early Cretaceous times in a Benioff zone. The Aldan rift is apparently part of this zone and was formed along the ancient Stanovoy suture (Maksimov, 1982).

Taking into account the east–west variations in the compositions of the alkaline rocks, three subprovinces of the western (Nos 1–7), central (Nos 8–25) and eastern Aldan (Nos 26–32) can be distinguished (Fig. 220). The greatest number of alkaline complexes is in the central Aldan subprovince (Fig. 223), where the numerous occurrences are predominantly of Mesozoic age and are concentrated within the Aldan shield along its northern margin. The Archaean crystalline basement and the overlying platform rocks, within the area of the central Aldan subprovince, are cut by numerous faults, which were repeatedly activated in the Proterozoic and Mesozoic, so that the whole region is a mosaic of radial blocks. Down-faulted blocks have preserved the plat-

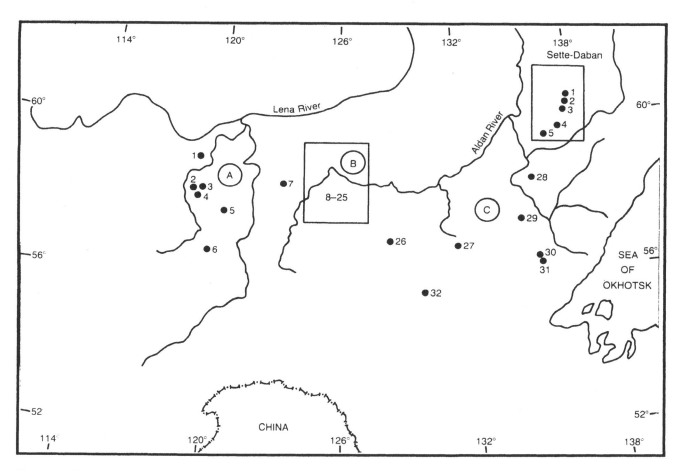

Fig. 220. *Distribution of occurrences of alkaline rocks and carbonatites of the Aldan Province. Western (A), central (B – see also Fig. 223) and eastern (C) sub-provinces are distinguished. The location and occurrence numbers of the Sette-Daban Province are indicated (after Kost'uk, 1990, Fig. 5, with author's additions).*

form cover of Lower Cambrian shales, limestones and sandstones. The magmatic rocks are represented by multiple complexes of central type (e.g. Inagli, Yakokut) and sills and breccia pipes. The extent of the alkaline occurrences varies from 4 to 440 km². Mixed volcanic-plutonic complexes, with relicts of extrusive volcanic rocks preserved in subsidence calderas (e.g. Tommot, Dzhekonda, Yllymakh, Yakokut) are widely developed. The main manifestations of alkaline magmatism took place during the Mesozoic and three stages of activity with ages of 142–147, 133–137 and 120–124 Ma have been established (Bogatikov *et al.*, 1991). An early general review, in English, is that of Maksimov (1973), and there is a brief introduction, also in English, in Butakova (1974).

1	Molbo	17	Inagli
2	Bolshemurunskii	18	Tommot
3	Malomurunskii	19	Seligdar
4	Dogaldynskii	20	Yakokut
5	Khani	21	Mrachnaya pipe
6	Yuzhnosakunskii	22	Rododendron
7	Dzhangylakh	23	Zarya
8	Yagodka pipe	24	Dzhekonda
9	Anomaly 4B pipe	25	Yllymakh
10	Kaila pipe	26	Lomam
11	Onye	27	Arbarastakh
12	Lyzhnaya pipe	28	Ingili
13	Ryabinovyi	29	Konder
14	Ryabinovaya pipe	30	Sybakh
15	Ykhukhtinskii	31	Chad
16	Strelka	32	Tokko

1 MOLBO   59°05′N; 118°49′E

This dyke, 1.3–1.5 m thick, is formed from basic leucite-phlogopite lamproite. It cuts Lower Cambrian marls, which overly the Precambrian basement, along sharp contacts and the enclosing rocks are metamorphosed to hornfelses and chloritized. The dyke has a zoned structure with the central part porphyritic and the outer part aphyric. The porphyritic lamproite of the centre is formed from rare altered phenocrysts of olivine, leucite and laths of phlogopite. The fine-grained outer part comprises mainly phlogopite and polygonal grains of pseudoleucite with microlites of altered olivine, clinopyroxene, magnetite and apatite set into a finer grained mass of phlogopite, chlorite and possible alkaline amphibole. The mica is both ferruginous phlogopite and biotite with a very low aluminium and titanium content (about 3% TiO_2).

Age Ages of 157–158 Ma (Orlova, 1990) and 92–122 Ma (Makhotkin *et al.*, 1989) have been determined.

References Makhotkin *et al.*, 1989; Orlova, 1990.

Murun

The Murun complex comprises three major intrusions (Fig. 221), namely Bolshemurunskii (locality 2), Malomurunskii (locality 3) and Dogaldynskii (locality 3). Malomurunskii, which alone is commonly referred to as Murun, is the largest of the three and by far the most important because of the wide range of rock types, including ultra-potassic varieties containing kalsilite, many of which are layered.

2 BOLSHEMURUNSKII   58°23′N 118°55′E
Fig. 221

This massif has an area of about 6 km² and is formed predominantly of alkaline syenites (pulaskites) but with biotite–aegirine and aegirine–nepheline syenites also present. Xenoliths of the enclosing country rocks are abundant within the alkaline rocks.

Reference Orlova, 1990.

3 MALOMURUNSKII   58°24′N; 119°04′E
(Murun)   Fig. 221

Malomurunskii is elliptical in shape, narrowing towards the northeast. The complex was formed as a result of several major magmatic phases: (1) extrusion of alkaline trachytes and leucite phonolites, (2) emergence of a lopolith-like, stratified intrusion which is cut by feldspathoidal and alkaline syenite pegmatites, (3) formation of necks and extrusion of alkaline basaltoids and trachytes and (4) emplacement of concentric and radial dykes of pseudoleucite syenite porphyry (tinguaite porphyries), alkaline lamprophyres and aplitic syenites. In different parts of the complex numerous relicts of fenitized country rocks are found.

The Malomurunskii lopolith consists of rocks which form five distinct stratified series. They are characterized by the regular change of the composition of rocks from bottom to top, with the lower parts of the lopolith consisting of more melanocratic and less alkaline rock varieties. The upper and middle parts of the units are formed by more leucocratic, highly alkaline, more iron- and titanium-rich varieties of feldspathoidal syenites and alkaline pyroxenites. It is presumed that the pluton was formed by processes of crystal differentiation from an alkaline-basaltoid melt enriched in K, Ti and volatile components. The five rock series which comprise the pluton are named and identified according to the predominant rock types within them. From the bottom to the top these are: (1) bottom pyroxenitic series including biotite clinopyroxenite, shonkinite and fergusite; (2) lower alkaline gabbroic series with biotite clinopyroxenite, yakutite, kalsilite–feldspathic clinopyroxenite, shonkinite, fergusite and melanosynnyrite; (3) middle alkaline syenitic series comprising three sub-series, (a) alkaline-gabbroic of malignite, shonkinite and fergusite (b) synnyritic of melano- and meso-synnyrites and (c) alkaline-syenitic of alkaline poikilitic syenite and kalsilite syenite; (4) upper nepheline syenitic series of nepheline syenite and aegirinite or alkaline pyroxenite; (5) marginal stratified malignitic series including malignite, shonkinite and kalsilitic and pseudoleucitic syenites. Rocks of the fifth series are located in the southwestern part of the complex which is topographically the highest area. The stratification, lamination and thicknesses of the layers within the different series in the various parts of the complex are different. They are displayed best of all in rocks of the marginal and lower stratified series, but they are often obscured by later mylonitization, development of a schistosity and aegirinization. The layering in the rocks of the bottom pyroxenite series is less well developed, but here too along with the pyroxenites are thin, intercalated layers of olivine clinopyroxenite, shonkinite and fergusite. Within each laminated series several units are observed. Melanocratic rocks, notably clinopyroxenite, alkaline gabbroids or fergusite, are taken as representing the beginning of each unit. No fewer than eight micro-units can be identified within the middle series. Usually the rhythmic units have thicknesses ranging from 20–30 m to 60–80 m and include (from bottom to top): alkaline and biotite pyroxenite or aegirinite, shonkinite and malignite, fergusite and malasynnyrite, nepheline and kalsilite syenites and alkaline syenite. The thickness of the layers of alkaline pyroxenites in the rhythmic units decreases from bottom to top with a concomitant increase in thickness of the leucocratic layers.

A wide aureole of contact metamorphic rocks, both skarns and fenites, and including charoites and carbonate–silicate rocks, occurs around the Malomurunskii lopolith. The skarns are highly variable in composition and include diopside–garnet (andraditic), diopside–phlogopite, melilite–forsterite and forsterite–diopside assemblages. The enclosing Proterozoic limestones have been metamorphosed and are replaced by K-richterite and tetraferriphlogopite resulting in the formation of highly alkaline skarns. The aureole of fenites and fenitized rocks is widest in the southern part of the complex where it reaches three kilometres. In this area also are developed carbonate–silicate rocks – calcitic and benstonitic carbonatites with aegirine and K-feldspar and quartz-carbonate veins. The deposits of charoite are spatially and genetically related to areas of intense faulting, dyke emplacement and fenitization. The intrusion of calcite carbonatites are considered to coincide with the formation of the southwestern part of the fenite zone at the southern end of the complex and they are located amongst dolomites and fenitized quartz sandstones. They take the form of veins and sheets 0.3–0.5 m thick which are conformable with the stratification of the intensively fenitized enclosing rocks. The latter are enriched with aegirine, richterite and tinaksite. Benstonite–silicate rocks are present as layers and lensoid bodies having both coarse and fine fragmental structures. In some areas they form chambers and tube-like zoned bodies 0.2–0.8 to 1 m in diameter and also form typical cross-cutting veins. The carbonate–silicate rocks display zonal structures which are manifested by the gradual change from the enclosing aegirine-microcline fenites through feldspar rock with benstonite to essentially benstonitic carbonatites, which also contain microcline and aegirine. These rocks are also enriched in barium carbonate, strontium carbonate and sulphides of Fe, Cu, Pb and Zn together with other accessory minerals. The rocks of Malomurunskii described as carbonatites are variously ascribed to a magmatic or metasomatic origin. An outcrop of calcite-benstonite–aegirine rock looked at by one of the authors (ARW), was considered to be undoubtedly of metasomatic origin. The occurrences of charoite are found in the southern part of the fenite aureole of the complex, where they form lens-shaped bodies up to 1 km across, 22 such bodies being outlined on a map accompanying Biryukov and Berdnikov (1993, Fig. 1). The rock comprises charoite (20–90%), aegirine, tinaksite, quartz, feldspar, calcite, K-arfvedsonite, pectolite, canacite and many other minerals (Biryukov and Berdnikov, 1993, Table 2). The majority of investigators consider the charoite mineralization to be metasomatic but there are also proponents of a magmatic origin, the field relationships not being necessarily diagnostic. Prokof'yev and Vorob'yev (1992) have investigated the P-T conditions of formation of Sr–Ba carbonatites and charoite using fluid inclusions in quartz. A general account, in English, giving much petrological and mineralogical data is that of Biryukov and Berdnikov (1993).

Olivine-bearing basic lamproites in the form of dykes have been encountered in boreholes in the central parts of the Malomurunskii complex and in its contact zone amongst the Precambrian basement rocks. The dykes cut through alkaline syenites and produce a clearly defined hardened zone in the contact metamorphosed rocks. They are olivine-bearing diopside-phlogopite basic lamproites which are coarse- and medium-grained porphyritic rocks with large crystals of olivine and laths of phlogopite in a fine-grained matrix of microlites of diopside, small laths of dark-brown phlogopite and poikilitic grains of feldspar. Practically all of the dykes are altered to some extent with the production of potassic richterite which develops at the expense of olivine and sometimes forms, with laths of tetraferriphlogopite, radial aggregates around relicts of fresh olivine. A small amount of alkaline amphibole may also be present in the main body of the rock. Partial rims of bright green aegirine are developed around diopside grains while in some cases the lamproite is totally replaced by a coarse-grained aggregate of tetraferriphlogopite plus richterite and clinopyroxene. The olivine of these rocks has the composition Fo_{90} with a high content of Ca but low Cr. The mica is zoned with cores of phlogopite and borders of ferruginous phlogopite. The diopsidic pyroxene has an Mg:Fe ratio of 90–86 and a low titanium content. Diopside–phlogopite-richterite basic lamproites are also found in the charoitic rocks as numerous xenoliths and as dykes. They are composed of ferruginous phlogopite, potassic richterite set in a medium-grained matrix of feldspar, mica, richterite and pyroxene. Shchadenkov et al. (1989) consider that the pyroxenite and shonkinite of Malomurunskii is the intrusive equivalent of the hypabyssal lamproites of the area. Detailed information on the composition of rocks (78 analyses) and minerals for the whole complex can be found in Orlova (1988b and 1990) and Orlova et al. (1988). Pokrovskiy and Vinogradov (1991) present isotope data for Sr, O, C and H of a wide range of intrusive silicate rocks and carbonatites as well as for country rocks. Samsonova and Donakov (1968) describe in some detail the kalsilite-bearing rocks and give analyses of this mineral.

Economic The lilac-coloured charoite-bearing rock is exploited for jewellery and polishing (Vladykin, 1986), the deposit being investigated in detail at the present time. Biryukov and Berdnikov (1993, Fig. 1) give a general map of the charoite locality showing 22 named deposits. Kalsilite-feldspar rocks are a potential source for the production of K and Al.

Age K–Ar on biotite from kalsilite syenite gave 130 ± 6 Ma and on biotite from alkaline syenite 130 ± 6 to 134 ± 5 Ma (Orlova, 1990; Orlova et al., 1986a). Bogatikov et al. (1991) obtained K–Ar ages on phlogopite from lamproite of 132 ± 6, 133 ± 6 and 138 ± 6 Ma.

References *Biryukov and Berdnikov, 1993; Bogatikov et al., 1991; Orlova, *1988a, *1988b and 1990; Orlova and Shchadenkov, 1988; Orlova et al., 1986a and 1992; Pokrovskiy and Vinogradov, 1991; *Prokof'yev and Yorob'yev, 1992; Samsonova and Donakov, 1968; Shchadenkov et al., 1989; Vladykin, 1986.

4 DOGALDYNSKII

58°00′N; 119°00′E
Fig. 221

This intrusion, with an area of 8 km², has a zonal structure and grades, from the periphery to the centre, from alkaline gabbroids (malignite and shonkinite) to feldspathoidal syenites and peralkaline syenites.

Age K–Ar on biotite from alkaline syenite gave 174 ± 12 Ma and alkali feldpar from peralkaline syenite 181 ± 5 Ma (Orlova, 1990; Orlova et al., 1986a).

References Orlova, 1990; Orlova et al., 1986a.

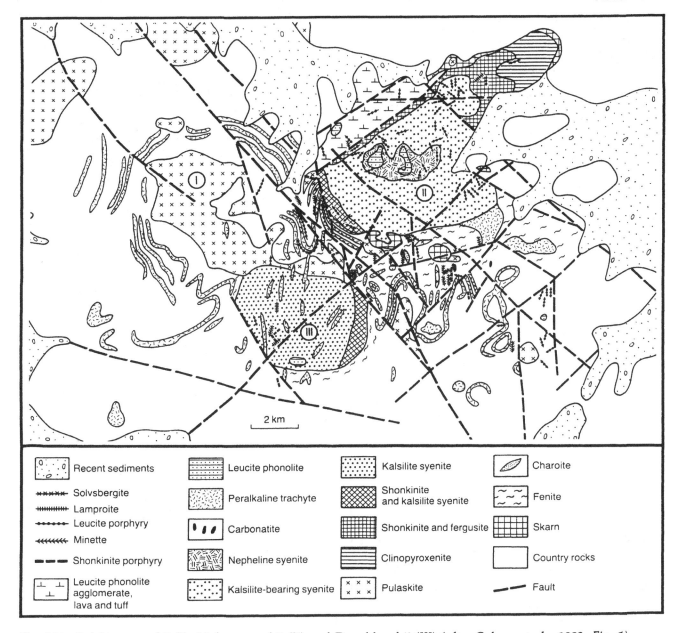

Legend:

- Recent sediments
- Solvsbergite
- Lamproite
- Leucite porphyry
- Minette
- Shonkinite porphyry
- Leucite phonolite agglomerate, lava and tuff
- Leucite phonolite
- Peralkaline trachyte
- Carbonatite
- Nepheline syenite
- Kalsilite-bearing syenite
- Kalsilite syenite
- Shonkinite and kalsilite syenite
- Shonkinite and fergusite
- Clinopyroxenite
- Pulaskite
- Charoite
- Fenite
- Skarn
- Country rocks
- Fault

Fig. 221. Bolshemurunskii (I), Malomurunskii (II) and Dogaldynskii (III) (after Orlova et al., 1992, Fig. 1).

5 KHANI   57°54′N; 120°22′E

Olivine–diopside ultrabasic lamproites have been discovered in a massif of phlogopite pyroxenites. The lamproites are porphyritic rocks with large (1–2 cm) fresh olivines set in a medium-grained mesostasis of phlogopite admixed with a small amount of diopside. A gradual transition from the Khani lamproite to the intrusive phlogopite pyroxenites can be observed. Such transitions are accompanied by a reduction in the overall amount of olivine concomitant with an increase in the content of diopside.

Age K–Ar determinations on phlogopite from lamproite gave 1818±25 and 1870±25 Ma (Bogatikov *et al.*, 1991).

Reference Bogatikov *et al.*, 1991.

6 YUZHNOSAKUNSKII   56°47′N; 119°39′E
(Sakun)   Fig. 222

Many of the intrusions of this 10 km² complex take a somewhat arcuate form and cut through Precambrian metamorphosed sedimentary rocks and granite-gneisses and partly through upper Palaeozoic granitoids. In the north, potassic alkaline rocks are in contact with Palaeozoic quartz syenites. Within the complex layering is well developed and two series have been distinguished. The lower series is represented by potassic mesocratic rocks and the upper series by highly potassic leucocratic rocks. The mesocratic series rocks, which extend over 6 km², are rather more abundant than the leucocratic series (3 km²). There is a rhythmic repetition of rock types. The mesocratic series consists of pulaskite (70%), lusitanite (10%), leucocratic alkaline syenite, pseudoleucite and nepheline syenites (10%) and micaceous pyroxenite (5–7%). Shonkinite, fergusite, malignite and synnyrite are subordinate. Within the repeated units a distinct rock type generally prevails. For instance, at the bottom of the mesocratic series lusitanites are usually found which are succeeded by micaceous pyroxenites and then by pulaskites. Another feature which is observed in the structure of the mesocratic layers is that the mafic minerals, both within individual layers and through the sequence as a whole, decrease upwards. In addition to microlayering the upper salic series has a stratification, which is not clearly seen macroscopically, due to the non-uniform distribution of feldspathoids which are finely intergrown with K-feldspar. The upper series is formed predominantly by four rock types: (1) nepheline-bearing syenites, often with a pseudo-leucitic texture, (2) alkaline syenites, which are essentially trachytic textured pulaskites, and massive tonsbergites, (3) kalsilite-bearing syenites (synnyrites), and (4) feldspathoidal syenites or micaceous microcline metasomatic rocks. The complex alternation of these rocks can be traced in a section having a thickness of about one kilometre. This diverse group of rocks comprises essentially six minerals: K-feldspar (orthoclase and microcline), nepheline, kalsilite, clinopyroxene (aegirine-bearing diopside), lepidomelane and garnet (andradite–melanite with a small admixture of grossular component). Many rocks contain porphyrocrysts and ovoids of pseudoleucite up to 5–10 cm in diameter. They are of a variable composition and comprise kalsilite, nepheline and K-feldspar. Details of the chemistry of these rocks can be found in Zhidkov (1990). Pokrovskii and Zhidkov (1993) obtained an initial $^{87}Sr/^{86}Sr$ ratio of 0.70428, $\delta^{18}O$ values of 3.8–5.9 for the dominant igneous rock types and δD of -75 to -87 for pyroxenite and pseudoleucite syenite, but notably lower values of -100 to -122 for synnyrite and feldspathoidal syenite.

Age K–Ar determinations on biotite from synnyrite gave 330 ± 18 Ma and from biotite pyroxenite 321 ± 8 Ma (Zhidkov, 1990). Pokrovskii and Zhidkov (1993) obtained 288 ± 5 Ma from a Rb–Sr whole rock isochron.

References *Pokrovskii and Zhidkov, 1993; Zhidkov, 1990.

---

7 DZHANGYLAKH   58°52′N; 122°35′E

This is a 4 km diameter intrusion which has a zonal structure. Three intrusive phases have been distinguished: (1) malignites, which occupy the outer part of the complex, (2) larvikites, which form an inner ring and (3) a central stock consisting of amphibole syenite porphyry. Small dykes and lenses of nepheline syenite cut the rocks of phases (1) and (2). Minor areas of pseudoleucite porphyry, that could be remnants of extrusive rocks, are found. The nepheline syenites comprise prismatic grains of anorthoclase, anhedral grains of nepheline, biotite, aegirine or aegirine–augite and alkali amphiboles, either arfvedsonite or riebeckite. Accessory minerals include titanite, apatite, titanomagnetite and brown garnet.

Reference Bilibin, 1958.

---

8 YAGODKA PIPE   58°49′N; 125°04′E

This is a breccia pipe, with the uppermost explosive products still preserved, which cuts Lower Cambrian

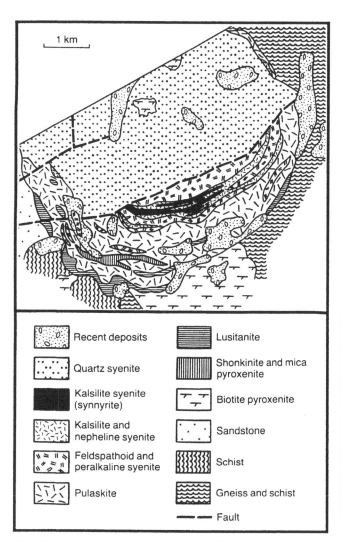

1 km

Recent deposits		Lusitanite	
Quartz syenite		Shonkinite and mica pyroxenite	
Kalsilite syenite (synnyrite)		Biotite pyroxenite	
Kalsilite and nepheline syenite		Sandstone	
Feldspathoid and peralkaline syenite		Schist	
Pulaskite		Gneiss and schist	
		Fault	

Fig. 222. Yuzhnosakunskii (after Zidkov, 1990, Fig. 20).

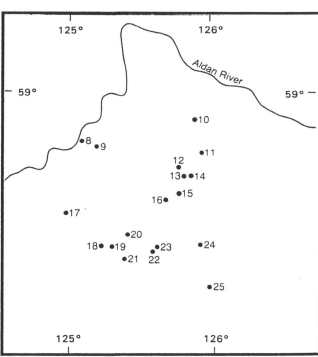

Fig. 223. Distribution of occurrences of alkaline rocks and carbonatites in the central Aldan sub-province.

limestones. The igneous products of the pipe are close to ultrabasic olivine–phlogopite lamproites.

Reference Bogatikov *et al.*, 1991.

9 ANOMALY 4B PIPE   58°48′N; 125°12′E

This pipe contains highly magnesian olivine–phlogopite leucitites which throughout the entire pipe are present as an autolith breccia. The cores of the autoliths have zoned structures in which the central parts are formed of fine-grained olivine leucitite while the margins are of dark, very fine-grained rocks corresponding in composition to the same olivine leucitite. The autoliths are set in a matrix of either olivine leucitite of the same composition or of a lapilli or crystal tuff cement. There are abundant round and angular xenoliths of Upper Jurassic leucite phonolite, which are not known from surface outcrop, as well as Lower and Upper Mesozoic alkali feldspar syenites.

Reference Bogatikov *et al.*, 1991.

10 KAILA PIPE   58°53′N; 125°52′E

With a diameter of 600–700 m Kaila is one of the largest breccia pipes of the central Aldan. It intrudes Precambrian crystalline rocks of the Aldan shield basement and dolomites of lower Cambrian age, and cuts two small dykes of olivine lamproite. The larger part of the pipe is formed by crystal tuff breccias and xenolith–tuff breccias comprising a heterogeneous mixture of small fragments of intrusive, or autolithic, fine-grained lamproite, and individual large crystals of altered olivine and feldspar aggregates which are cemented by very fine-grained clayish material, which usually comprises up to 30–40% of the volume of the breccia. Large blocks of intrusive lamproite, which reach several metres and sometimes tens of metres in diameter, are sometimes found within the breccia. Lamproites from these large blocks have a coarser texture than lamproites from fragments and autoliths as well as lamproites of the early dykes intruded before emplacement of the pipes. The fragmental lamproites are fine-grained porphyritic rocks with phenocrysts of olivine (10–12%), which are pseudomorphed by talc and saponite, chrome–diopside (7–15%) and phlogopite (5–12%). The fine-grained groundmass consists of microlites of diopside, tiny laths of phlogopite, rare altered olivine grains, aggregates of poikilitic K-feldspar and glass, which is partly or fully replaced by vermiculite and saponite. Pseudoleucite is present in the groundmass of eruptive breccia fragments. Blocks of quartz sandstone and Precambrian crystalline rocks also occur in the breccias and it is noteworthy that also found amongst the eruptive breccias are xenoliths which resemble middle Jurassic carbonaceous shales. Such shales are not found in the area of the pipe and must have been derived from a level higher than the present erosion surface. Similar relationships of entrained blocks have been described from kimberlite pipes in northern Siberia and testify to the convective character of the emplacement process during the formation of the eruptive breccias.

Age K–Ar on phlogopite from lamproite gave 133 ± 6 to 170 ± 5 Ma (Makhotkin *et al.*, 1989).

References Bogatikov *et al.*, 1991; Makhotkin *et al.*, 1989.

11 ONYE   58°46′N; 125°56′E

Four intrusive bodies cover an area of about 20 km². They consist of aegirine and aegirine–augite syenites

which are accompanied by numerous dykes of solvsbergite and tinguaite.

Reference Bilibin, 1958.

12 LYZHNAYA PIPE   58°42′N; 125°46′E
                                         Fig. 224

This breccia pipe of about 200 m diameter is located on the northern margin of the Ryabinovyi complex (locality 13). It cuts Archaean basement rocks and alkaline syenites of the Ryabinovyi intrusion. The pipe consists of olivine–diopside–phlogopite basic lamproites (mg 77) with phenocrysts of olivine, pseudomorphed by talc and saponite (10–20%), chrome–diopside (7–15%) and phlogopite (5–12%). The fine-grained groundmass comprises microlites of diopside, phlogopite, K-feldspar and altered glass.

Reference Bogatikov *et al.*, 1991.

13 RYABINOVYI   58°40′N; 125°49′E
                                         Fig. 224

This rather irregularly shaped complex has an area of about 50 km² and involves both volcanic and plutonic rocks. The enclosing rocks are granite gneisses and granites which are fenitized adjacent to the margins of the complex. In the contact zone, as well as in the roof, are blocks and fragments of limestones of the Cambrian

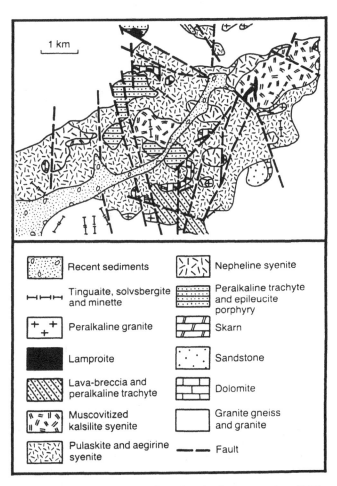

Fig. 224. Ryabinovyi (after Kochetkov et al., 1989, Fig.1). The Lyzhnaya lamproite pipe lies at the northern edge of the map.

basement cover, which have been altered to marble and skarns, and Jurassic sandstones that have been hornfelsed. The complex includes a considerable diversity of volcanic, plutonic and hypabyssal rocks which alternated temporally but the major stages of magmatism, according to geological evidence, are as follows. The first stage is represented by quartz and alkaline peralkaline trachytes. They are encountered only in the form of xenoliths in other rocks. The second stage comprises volcanic rocks which include peralkaline trachyte, pseudoleucite phonolite and their brecciated equivalents. They have been preserved at the summits of mountain peaks and form a partial cover to the laccoliths. The third stage encompasses aegirine syenite and pulaskite, nepheline syenite and alkaline syenite porphyries. They form about 80% of the area of the complex and are extensively metamorphosed and replaced by K-feldspar, which may comprise up to 80–90% of the rock, albite and aegirine. The ratio of primary rock to metasomatic products varies from 70:30 to 50:50. The metasomatized rocks are characterized by the abundance of accessory minerals including apatite, magnetite, titanite, melanite, leucoxene and rarer xenotime. The fourth and fifth stages include alkaline trachytes and their tuff equivalents, lava breccias, lamproite, solvsbergite porphyry, grorudite, nordmarkite and peralkaline granites; many of these rock types are present in the form of dykes. The nepheline syenites contain alkali feldspar (70–80%), nepheline (4–5%), aegirine (11–20%) and biotite and albite in small quantities. In pulaskites the content of feldspar increases up to 87% and, in addition, albite (7%) and clinopyroxene (9%) are present; accessories include magnetite and apatite. Aegirine syenites are formed in the main of alkali feldspar (up to 75%) and clinopyroxene (17–20%) with small amounts of albite, biotite and accessory magnetite, apatite, titanite, fluorite and garnet. In addition to K-feldspar (80%) and clinopyroxene (5%) the nordmarkites contain up to 15% quartz. More detailed data on the composition of the rocks can be found in papers by Orlova (1990), Kravchenko (1972), Kochetkov *et al.* (1989) and Eremeyev *et al.* (1993).

Economic There are manifestations of molybdenum-copper and gold mineralization (Kochetkov *et al.*, 1986 and 1989).

Age K–Ar determinations on K-feldspar for rocks of the second stage gave 212 ± 4, 179 ± 11 and 205 Ma; for the rocks of the third stage 150 ± 14, 143 ± 1, 140 ± 3 Ma and for rocks of the fourth and fifth stages 120 and 108 Ma (Kochetkov *et al.*, 1989). Whole rock Rb–Sr isochrons gave 137.8 ± 4.8 Ma for pyroxenite, 138.0 ± 0.3 Ma for shonkinite, 149.4 ± 0.7 for nepheline syenite and 146 ± 29 Ma for aegirine syenite (Eremeyev *et al.*, 1993).

References *Eremeyev *et al.*, 1993; Kochetkov *et al.*, 1986 and 1989; Kravchenko, 1972; Orlova, 1990.

---

14 RYABINOVAYA PIPE   58°40′N; 125°51′E

The Ryabinovaya pipe cuts through a dense swarm of dykes, the overall length of which is about 1 km. Individual dykes have a complex zonal structure with the marginal parts composed of a magmatic breccia which consists of a mixture of angular fragments of coarse-grained alkaline syenite in a fine-grained cement of olivine–diopside lamproite. The relative proportion of cement and syenite debris is highly variable. The width of the marginal contamination zone is usually 2–3 cm but varies depending on the thickness of the dyke. A sharp change in the appearance and composition of the rocks takes place in the central section of

dykes which are more than 10 m thick. Xenoliths of syenite and lamproite olivine phenocrysts are now absent and the texture becomes medium-grained and slightly porphyritic. Compositionally the rocks of this central zone correspond to phlogopite shonkinite porphyries and highly magnesian phlogopite minettes. The Ryabinovaya pipe proper is composed of autolith breccias with a magmatic cement and corresponding in composition with olivine–diopside lamproites (Mg 73–75%). The pipe rocks are fine-grained and porphyritic with large, first-generation phenocrysts of olivine and smaller second-generation olivine phenocrysts in the matrix. The olivine has a composition of Fo_{92-94}, with a NiO content of 0.5%, and comprises 15–25% of the rock. Chromite inclusions with Cr_2O_3 values of 61% are present in the olivine. The fine-grained matrix consists of microlites of diopside, patches of feldspar, in which it is sometimes possible to recognize relict aggregates of pseudoleucite, and tiny crystals of dark-brown mica (1–3%).

Age K–Ar on biotite from lamproite gave 137 ± 6 Ma and 142 ± 6 Ma (Makhotkin *et al.*, 1989).

References Bogatikov *et al.*, 1991; Makhotkin *et al.*, 1989.

---

15 YUKHUKHTINSKII   58°38′N; 125°45′E

This laccolith is composed of aegirine syenite and pulaskite with some augite syenite. The youngest rocks are aegirine granites which form small stocks.

Reference Bilibin, 1958.

---

16 STRELKA   58°33′N; 125°40′E

Strelka is an intrusion of nepheline syenite of about 1 km diameter which cuts Archaean gneisses and Cambrian dolomites. The nepheline syenites have a trachytic texture and are formed of alkali feldspar (27–60%) and albite (17–36%) with some aegirine/aegirine–augite and nepheline and accessory magnetite, titanite, apatite, anatase, garnet and sulphides. The nepheline syenites are often albitized and in such rocks accessory eudialyte, apatite, zircon, ramsayite, baddeleyite, lamprophyllite, loparite, lomonosovite, Ca-rinkolite and labuntsovite are to be found.

References Kravchenko *et al.*, 1982; Orlova, 1990; Osokin *et al.*, 1974.

---

17 INAGLI   58°32′N; 124°59′E
                                                Fig. 225

The Inagli complex is situated in the western part of the central Aldan (Fig. 223) in the periphery of the Inagli magmatic dome. The complex has a ring form and an area of about 30 km². The enclosing rocks include sandstones of Upper Proterozoic age, Lower Cambrian dolomites and dolomitized limestones and Precambrian crystalline rocks. The Lower Cambrian sediments, which are generally subhorizontal, are slightly arched near the Inagli intrusion to form a dome-like structure 10–12 km in diameter. The Archaean basement is raised by about 200–300 m. Sills and laccoliths of trachydacite of Lower Jurassic age are present in sedimentary rocks overlying platform rocks east of the massif. Small stocks of potassic alkaline granite and sills of syenite porphyry of Lower Cretaceous age are also known in the area. Inagli is a complex, multi-phase pluton in which five structural units have been identified: (1) The core of the complex consists of dunites with rare intercalated layers of wehrlite. Drilling has indicated that the dunites persist to a depth of 900 m. The olivine of the dunites has the composition Fo_{92-93} with the unusual CaO content of up to 0.5–0.66% and a low nickel content. Pyroxenes from the ultramafic rocks are mostly diopside. (2) The periphery of the ultrabasic core has a complex structure and consists of metamorphosed phlogopitized dunite, phlogopite wehrlite and wehrlite. Olivine from this series of rocks has a lower CaO content. Blocks of dolomite, as well as a large number of sills of lamproite, similar in composition to the Upper Jurassic lamproites of the Yakokut complex (locality 20), are found in the peripheral zone. (3) The inner intrusive ring zone consists of differentiated sub-vertical ring-dykes of olivine pyroxenite, missourite and shonkinite. These ring-dykes cut the dunites of the core. (4) The outer ring zone, which is 10–500 m wide, consists of shonkinite, melanocratic syenite, syenite and pulaskite. The shonkinite contains xenoliths of dunite, peridotite and limestone. (5) Ring zones of metasomatic rocks and pegmatites. The metasomatic rocks have several mineral associations, including phlogopite with chrome–diopside and phlogopite plus chrome–diopside and K-feldspar; the phlogopite has >14% Al_2O_3. The pegmatites consist mainly of feldspar and magnesioarfvedsonite in association with agpaitic accessory minerals and are concentrated mainly in zones of albitization and microclinization and include lamprophyllite, leucosphenite, batisite and innelite. Chemical compositions of rocks and minerals can be found in Orlova (1990) and Bogatikov *et al*. (1991).

Economic A deposit of bright-green chrome-diopside is worked for jewellery (Orlova *et al*., 1986b) and a vermiculite deposit has been discovered and investigated.

Age K–Ar determinations on biotite from malignite gave 129 ± 5 Ma and on biotite from pyroxenite 136 ± 5 Ma; alkali feldspar from malignite gave 137 ± 5 Ma and from pulaskite 112 ± 5 Ma (Orlova, 1990).

References Bogatikov *et al*., 1991; Glagolev and Korchagin, 1974; Korchagin, 1966; Kravchenko *et al*., 1982; Orlova, 1990; Orlova *et al*., 1986b; Sveshnikova and Eremeyev, 1982.

---

18 TOMMOT   58°23′N; 125°13′E
Fig. 226

This is an asymmetric complex of both extrusive and intrusive rocks which are emplaced in Archaean granites and granite gneisses. The extrusive rocks occupy a depression, up to 950 m deep, in the Archaean basement along the western part of the massif. The volcanic series is subdivided into three parts. The lowest one consists of feldspar leucitites which are overlapped with discordance by peralkaline trachytic lavas and breccias, which constitute the central part of the series. The uppermost part consists of peralkaline trachytic ignimbrites. Leucitite and leucite tephrite dykes cut the ignimbrites. In the eastern part of the complex, peralkaline syenites are dominant. There is a body of larvikites situated parallel to the margin of the complex, which is characterized by strongly orientated tablets of alkali feldspar. The larvikite is porphyritic, the main minerals being K-feldspar (50–71%) and clinopyroxene (10–25%); accessory minerals are amphibole, biotite, titanite, quartz, apatite, magnetite, zircon, rutile, baddeleyite and sulphides. These alkaline rocks are cut by small bodies of micromonzonite consisting of K-feldspar, plagioclase, pyroxene and amphibole. As well as larvikite there is an intrusion of pulaskite, which is a highly

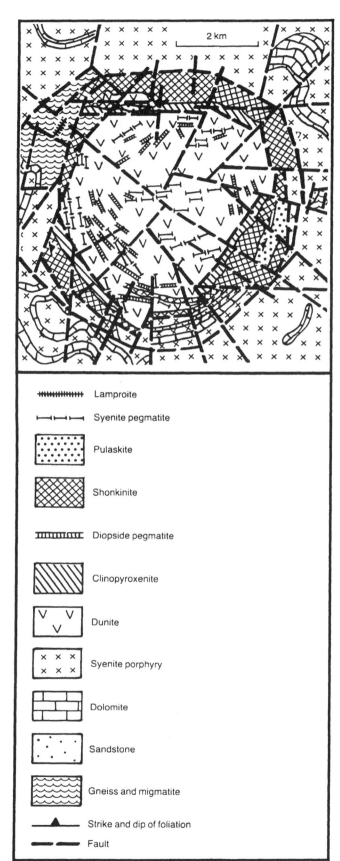

Lamproite

Syenite pegmatite

Pulaskite

Shonkinite

Diopside pegmatite

Clinopyroxenite

Dunite

Syenite porphyry

Dolomite

Sandstone

Gneiss and migmatite

Strike and dip of foliation

Fault

Fig. 225. Inagli (after Orlova, 1990, Fig. 24).

Legend:

- Tinguaite, bostonite and epileucitite
- Augite syenite
- Nepheline syenite
- Larvikite
- Muscovite syenite
- Pulaskite
- Hornblende syenite
- Monzonite and diorite-syenite
- Epileucitite
- Syenite porphyry
- Peralkaline trachyte
- Jurassic sandstone
- Cambrian dolomite
- Granite, schist and gneiss
- Fault

1 km

Fig. 226. Tommot (after Maksimov, 1973, Fig. 3).

leucocratic peralkaline syenite. The centre of the complex is formed of weakly alkaline, two feldspar syenites which form an irregularly-shaped stock. The composition of the syenites varies from the centre, where the rocks are leucocratic amphibole syenites, to the margin which consists of mesocratic amphibole-pyroxene syenite and biotite–pyroxene monzonite. The pyroxene in these rocks is principally salite. The youngest rocks of the massif are tinguaite, epileucitite, lamprophyres and melanite-bearing peralkaline syenites, and these form small dykes and irregular bodies within the extrusive rocks. According to K–Ar age determinations, syenite porphyries are of the same age. Sodium-rich metasomatic rocks that contain much albite and aegirine form linear bodies among the nepheline syenites and their pegmatites contain agpaitic accessory minerals including loparite, innelite, lomonosovite, Ca-rinkolite and labuntsovite.

Age K–Ar determinations on whole rocks vary from 167 ± 5 Ma to 125 ± 3 Ma (Orlova, 1990). New K–Ar determinations on minerals (amphibole, mica and alkali feldspar) and whole rocks give ages from 146 to 153 Ma, but mica from monzonite (141 ± 5 Ma) and from larvikite (135 Ma) gave younger ages. For the volcanic rocks ages range from 148 to 153 Ma (Pervov *et al.*, 1991).

References Kravchenko *et al.*, 1982; Orlova, 1990; Pervov *et al.*, 1991.

---

19 SELIGDAR

58°24′N; 125°18′E
Fig. 227

The Seligdar apatite ore field was formed during the Proterozoic and developed in a zone 6–7 km wide and 20–25 km long; Fig. 227 only depicts part of this field. The country rocks are Archaean quartzites, schists, granites and gneisses, with remnants of the platform cover, represented by Cambrian dolomites and sandstones, also preserved in the region. The ancient rocks are cut by a series of faults, associated with which are serpentine–chlorite and quartz–chlorite–sericite metasomatites, as well as dykes of Mesozoic syenite porphyry, pseudoleucite porphyry, kersantite and dykes and small stock-like bodies of shonkinite, which are located both in the main complex and around it. Vaseilenko *et al.* (1982) have reconstructed the original composition of the complex and have suggested that the following magmatic-metasomatic events were involved in its genesis. In the late Proterozoic era a basic potassic alkaline intrusive complex of central type was emplaced at Seligdar. Earlier peridotite–missourite phases of this complex contained on average 4–6% of P_2O_5 and in schlieren up to 20% P_2O_5. In the final stages of intrusive activity shonkinite dykes were emplaced. It is surmised that differentiation of the melt took place with the generation of an aqueous calcium carbonate pneumatolytic phase which initially, through the reaction of a carbonate–aqueous composition, led to the alteration of the rocks to greenstones. The mineralogy of these rocks is:

orthoclase + plagioclase + biotite + apatite + magnetite + chlorite + phengite + hematite + carbonate or

orthoclase + plagioclase + apatite + magnetite + albite + phengite + chlorite + hematite + carbonate + leucoxene or

plagioclase + phengite + chlorite + epidote + quartz + carbonate + hematite.

At lower temperatures (400–500°C) carbonate auto-

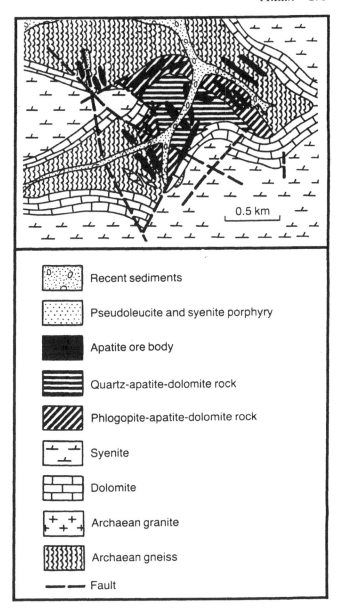

Recent sediments	
Pseudoleucite and syenite porphyry	
Apatite ore body	
Quartz-apatite-dolomite rock	
Phlogopite-apatite-dolomite rock	
Syenite	
Dolomite	
Archaean granite	
Archaean gneiss	
Fault	

Fig. 227. Seligdar apatite ore deposit (after Vaseilenko et al., 1982, Fig. 21).

metasomatism brought about further dolomitization and subsequent calcitization of the rocks. During these metasomatic processes phosphorus was inert but carbonate, alkalis, silicon and magnesium were lost into the country rocks, resulting in the formation of K-feldspar, quartz and wollastonite metasomatites. In the course of this process the apatite–carbonate rocks of Seligdar were generated. These rocks consist principally of apatite, dolomite and calcite with less hematite (martite), quartz and phlogopite and accessory monazite, zircon, rutile, tourmaline, orthite, topaz, pyrite, chalcopyrite, galena and magnesite. The proportions of the rock-forming minerals vary within very wide limits. There are a number of ideas as to the generation of the Seligdar apatite deposits. According to Entin (1966) the principal role belongs to high-temperature apatite-bearing carbonatites, and the igneous association is pursued further by Entin *et al.* (1985). In contrast, Bulakh *et al.* (1985) conclude that the phosphate has been mobilized from the surrounding country rocks. A notable feature of the

Seligdar apatite is the abundance of micro-inclusions of monazite, serpentine, carbonate, chlorite, titanite and mica. The apatite of the apatite–carbonate rocks is typically a red colour, which is due to hematite inclusions. The compositions of ores, rocks and some minerals can be found in the monograph by Vaseilenko *et al.* (1982) and chemical data on apatite and C and O isotope measurements on carbonate rocks in Smirnov (1978).

Economic Because of the favourable location close to the Baikal–Amur Railway there were plans to bring the Seligdar apatite deposits into commercial production with a beneficiation plant capable of handling 4 million tonnes per annum (Krasilnikova and Ilyin, 1989). Reserves are estimated at 300 million tonnes running 6–8% P_2O_5. The prospective ore is an apatite–dolomite rock with 10–15% apatite and up to 60–65% dolomite, but there are leaner ores of 0.6–2 and 8–10% apatite (Smirnov, 1976).

Age K–Ar determinations on phlogopite from the contact zone with the ore body gave dates of 2038, 1980, 1885 and 1844 Ma while the Pb–Pb method on apatite from apatite–carbonate ores gave about 2000 million years (Smirnov *et al.*, 1975). A Pb–Th isochron on apatite gave an age of 930 Ma (Vaseilenko *et al.*, 1982).

References *Bulakh *et al.*, 1985; Entin, 1986; Entin *et al.*, 1977 and *1985; *Krasilnikova and Ilyin, 1989; *Smirnov, 1976 and 1978; Smirnov *et al.*, 1975; Vaseilenko *et al.*, 1982.

---

20 YAKOKUT
58°27′N; 125°29′E
Fig. 228

The elliptical, 28 km² Yakokut complex cuts through Lower Cambrian dolomites and Jurassic terriginous deposits. The complex comprises both volcanic and plutonic rocks including dykes of alkaline and moderately alkaline comagmatic rocks. The volcanic rocks have been preserved as relics within a deeply eroded caldera which has been heavily faulted. At the present level of erosion the volcanic rocks occupy the ground of lowest relief and have a maximum thickness of about 500 m. Three volcanic units have been distinguished. The lowermost horizon, some 200 m in thickness, is formed by lavas of feldspar epileucitite and leucitite. The middle unit, 200–250 m thick, is composed of lavas, lava breccias and tuffs of various leucitites and biotite leucite melaphonolites with thin intercalations of argillite and tuffaceous sandstone. The uppermost unit consists of lavas and lava breccias of leucite phonolite and trachyphonolite with rare layers of sandy limestone. The thickness of the upper unit is 150–200 m. In the middle volcanic unit, near the southern contact, numerous bodies, apparently sills, of black, porphyritic lamproites occur. Lamproites of a similar composition have also been found in the outer contact zone of the complex. Here they are represented by small dykes. Some isolated lamproite dykes have been located in the northern part of the complex (Bogatikov *et al.*, 1986 and 1991). Within the Yakokut volcanics the moderately alkaline series, that is the trachytes, form the summits of bold peaks in the centre of the complex, but their age relationships with the strongly alkaline volcanics have not so far been established. The leucitites consist of microporphyritic pseudoleucite aggregates of orthoclase and albite (up to 30%) and 30% salite, in a groundmass consisting of microlites of orthoclase, albite and mica. Olivine-bearing leucitite consists of altered olivine (about 25%), which is replaced by mica, salite (30%) and microlites of orthoclase in the matrix. Phlogopite leucitite contains

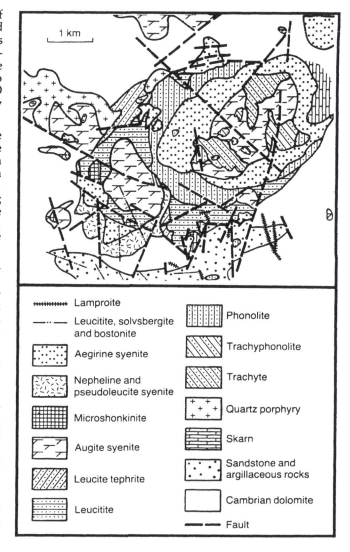

Lamproite

Leucitite, solvsbergite and bostonite

Aegirine syenite

Nepheline and pseudoleucite syenite

Microshonkinite

Augite syenite

Leucite tephrite

Leucitite

Phonolite

Trachyphonolite

Trachyte

Quartz porphyry

Skarn

Sandstone and argillaceous rocks

Cambrian dolomite

Fault

Fig. 228. Yakokut (after Maksimov, 1973, Fig. 4 with additional data from Bogatikov et al., 1991).

phenocrysts of phlogopite (20%) and diopside (10%) with orthoclase and albite in the groundmass. Biotite-leucite melaphonolite has phenocrysts of pseudoleucite and diopside (10%), while leucite phonolite has phenocrysts of orthoclase and mica and pseudoleucite aggregates. Three main series can be distinguished amongst the intrusive rocks of the complex. These are microshonkinites, which are the earliest intrusive series and apparently emplaced before extrusion of the volcanic units, arcuate bodies of nepheline and pseudoleucite syenite and peralkaline aegirine–augite syenites in the centre of the massif, which cut through the volcanic units. All these intrusive bodies are the upper parts of a large ring complex of alkaline rocks, which is not yet exposed. Moderately alkaline augite syenite forms a number of stocks which cut the peralkaline syenites. The intrusions of augite syenite are located both within and around the complex. Within the limits of the central complex and its contact zone a large number of approximately north–south-trending, 1–3 m thick dykes and several small, some 100 m diameter, subvolcanic stocks are known. They comprise olivine leucitite, leucitite, olivine–leucite–phlogopite phonolite, leucite–aegirine phonolite, bostonite, solvsbergite, syenite porphyry and lamproite.

Age K–Ar determinations on phlogopite from lamproite gave 142 ± 5 and 147 ± 6 Ma (Makhotkin *et al.*, 1989); K–Ar on volcanics, phlogopite and whole rocks, gave 137–141 Ma (Kanukov *et al.*, 1991).

References Bogatikov *et al.*, *1986 and 1991; Kanukov *et al.*, 1991; Makhotkin *et al.*, 1989; *Maksimov, 1973; Maksimov and Ugr'umov, 1971.

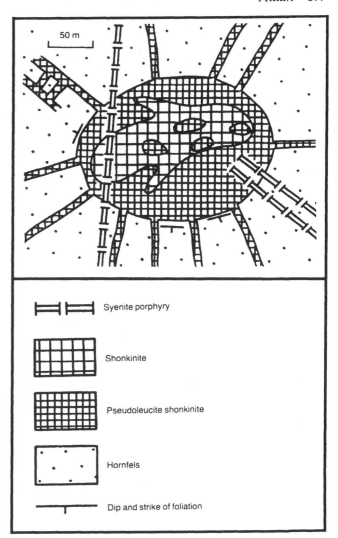

Fig. 229. *Rododendron (after Orlova, 1990, Fig. 30).*

References Eremeyev, 1987; Orlova, 1990.

21 MRACHNAYA PIPE   58°21'N; 125°22'E

This 1 × 0.1 km pipe is formed of tuff breccias and lamproite. The rock is composed essentially of devitrified glass in which are set phenocrysts of olivine and phlogopite. Small crystals of titanomagnetite (up to 10% TiO_2), and isolated grains of armalcolite (70% TiO_2), rutile and ilmenite have been identified. Generally the olivine phenocrysts, which comprise up to 40% of the rock, are pseudomorphed by chlorite, mica and other minerals. Details of the mineralogy and rock chemistry can be found in Bogatikov *et al.*, 1991.

Reference Bogatikov *et al.*, 1991.

22 RODODENDRON   58°22'N; 125°36'E
                                    Fig. 229

Rododendron is a circular stock, 200–300 m in diameter, which cuts sandstones of Jurassic age. The name derives from the abundance of rhododendrons growing in the area. In the central part it is formed by shonkinite and syenite around which are shonkinite porphyries and syenite porphyries. In the marginal parts of the intrusion volcanic formations, predominantly tuffaceous sandstones and tuff breccias, have been preserved. Compositionally the volcanic rocks are mainly trachytes, which are hornfelsed in the marginal parts of the stock, but leucite-bearing trachytes are also present. The trachytes are porphyritic rocks with phenocrysts of biotite, orthoclase, rarer sodic plagioclase (An_{20-30}) and sometimes apatite in a fine-grained trachytic base (50–60%). In the leucite-bearing trachytes phenocrysts are represented by salite, which is sometimes replaced by arfvedsonite. In the trachytic matrix, in which phenocrysts of leucite are replaced by sericite, phlogopite occurs and melanite, fluorite and titanite are patchily developed. The shonkinite porphyries are dark rocks with augite phenocrysts (30–40%) and a small amount of biotite (up to 10%) with large pseudoleucites, olivine and altered hornblende (6%). The groundmass is formed of K-feldspar, biotite, augite, plagioclase (An_{40-50}) and tiny grains of leucite, apatite and magnetite. The syenite porphyries are leucocratic coarse-grained rocks with phenocrysts of orthoclase, plagioclase, augite, hornblende, biotite and sometimes isolated olivine grains. The groundmass is orthoclase and plagioclase (An_{20-30}) with some augite, biotite and olivine. The central shonkinites are of two types: one with nepheline, the other with pseudoleucite. Apart from the pseudoleucite and nepheline, orthoclase, augite, phlogopite, isolated olivine grains, apatite, titanite, magnetite and pyrite also occur. The central syenites are formed mainly of orthoclase perthite and aegirine with accessory apatite, zircon, fluorite and magnetite and secondary biotite, albite, titanite and pyrite.

Age K–Ar dating of orthoclase from shonkinite porphyry gave 166 ± 5 Ma (Eremeyev, 1987).

23 ZARYA   58°23'N; 125°38'E

This is a small, circular stock formed mainly of shonkinites and syenites.

Reference Eremeyev, 1987.

24 DZHEKONDA   58°24'N; 125°55'E
   (Dzhekondinskii)   Fig. 230

A nearly circular complex about 4.5 km in diameter, Dzhekonda cuts through Lower Cambrian dolomites. Caldera subsidence has preserved a sequence, up to 650 m thick, of alkaline extrusive rocks in which are rare intercalations of silicic argillites. The extrusive rocks are situated in the central area of the complex with intrusives occurring as incomplete marginal zones and a central intrusion. The extrusive rocks occupy approximately 80% of the area of the complex and are represented predominantly by peralkaline trachytes and analcime porphyries, with tuffs and tuff breccias of the same rock types. Melanite syenites, represented by an arc-like body up to 200 m wide on the eastern and northern edges of

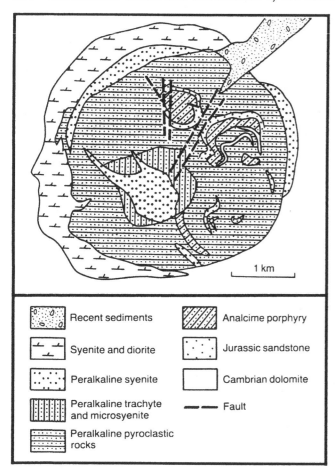

Fig. 230. Dzekonda (after Maksimov, 1973, Fig. 2).

the complex, are the earliest intrusive rocks. They have a trachytic and sometimes porphyritic texture and may be medium- or coarse-grained. They are composed of K-feldspar (75–88%), melanite (up to 12%) and alkali amphibole with smaller quantities of nepheline and albite; relicts of non-alkaline pyroxene and basic plagioclase are present. In the central part of the complex the extrusives are cut by a stock-like body of augite syenites which have a markedly trachytic character. Microcline is the dominant phase (60–80%) together with prismatic aegirine–augite, albite, biotite and amphiboles, as well as accessory titanite, apatite, magnetite, zircon, baddeleyite, rutile and fluorite. The youngest intrusive rocks are pyroxene syenites/diorites which form a broad, up to nearly one kilometre wide, horseshoe-shaped body extending around the complex on the western, southern and northern sides as well as small dykes which cut Cambrian dolomites to the north. They are composed of plagioclase (An_{30-32}) and K-feldspar, the feldspars making 50–60% of the rocks. The plagioclase is often zoned from andesine cores to oligoclase rims (An_{20-25}) with more basic phenocrysts in some porphyritic rocks (up to bytownite). The mafic minerals are pyroxene (0–6%) and hornblende (0–6%); accessories include zircon, apatite, titanite, melanite, lamprophyllite and corundum. Numerous dykes of solvsbergite, aplite and leucocratic granosyenites are developed amongst the extrusive rocks. Xenoliths of dolomite included in the alkaline rocks are totally altered to skarns.

References *Maksimov, 1973; Osokin *et al.*, 1974.

25 YLLYMAKH   58°15′N; 125°59′E
Fig. 231

This mixed volcanic and intrusive complex of central type has an area of about 40 km². It is circular and has a zonal structure which is complicated by a complex pattern of faults. Extrusive formations play a prominent role and are represented by pseudoleucite phonolites and peralkaline trachyte and their tuffs, which occupy zones in the western and eastern parts of the complex. The volcanics often contain small xenoliths and nodules of rocks of ultrabasic composition. The intrusive rocks cut through the volcanics and the earliest group includes malignites, melanocratic syenites and monzonites. The shonkinites and malignites form arcuate bodies and small stocks in the marginal parts and near to the complex. The malignites contain xenoliths of wehrlite and pyroxenite in which the olivine is more magnesian (Fo_{74}) than in the malignite (Fo_{48}). A later intrusive phase is represented by nepheline and pseudoleucite

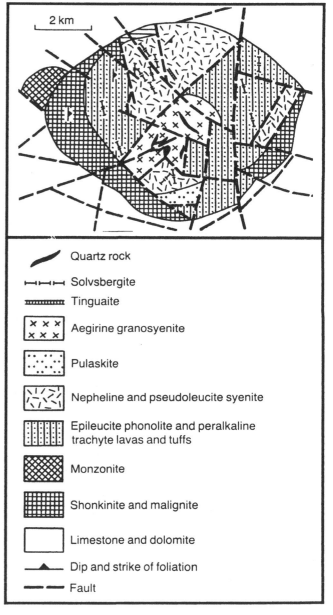

Fig. 231. Yllymakh (after Orlova, 1990, Fig. 27).

syenites which form large fault-bounded blocks in the interior of the complex as well as lens-like bodies of nepheline-bearing pegmatites. Both small xenoliths and large blocks (1.5–2 m) of feldspathized volcanic rocks are found in them. Throughout the complex the syenites are intensively sericitized, and near late granites they are albitized. Dykes of pseudoleucite tinguaite occur within the volcanics and alkaline gabbroids. Alkaline leuco-cratic syenites (pulaskites) are younger than the nepheline syenites and contain xenoliths of melanocratic syenite, malignite, monzonite and phonolite. A central stock consists of aegirine granite and granosyenite, which cuts across the alkaline leucocratic syenites. Dykes of granite porphyry and granosyenite aplite are spatially associated with the granitoids. Intense meta-somatic alteration of many rock types is manifest, but particularly in the volcanics, including replacement of dark minerals by aegirine, zeolitization and sericitization of feldspathoids and microclinization and albitization. In different syenites and in malignites postmagmatic changes have led to extensive recrystallization and the transformation of orthoclase into microcline and its replacement by albite. Albite-rich metasomatic rocks that were formed in linear zones among the nepheline syenites contain agpaitic accessory minerals including lamprophyllite, eudialyte, rosenbushite, loparite, lomo-nosovite, rinkolite and labuntsovite. Chemical analyses of rocks will be found in Orlova (1990) and Bilibin (1974); Chepurov (1973) describes melt inclusions in pseudoleucite.

Economic Gold values of 3.64 p.p.m. in aegirine granite and 1.48 p.p.m. in pseudoleucite syenite have been found (Shnai and Orlova, 1977).

Age K–Ar determinations on malignite and shonkinite gave 165 ± 5 Ma, on melanocratic syenite 163 ± 5 Ma, on monzonite 152 ± 5 Ma, on trachyte 50 ± 4 Ma, on pseudoleucite phonolite 142 ± 5 Ma, on nepheline and pseudoleucite syenite 158 ± 5 Ma, on alkaline syenite (pulsaskite) 133 ± 4 Ma, on granosyenite 131 ± 4 Ma and on aegirine granite 133 ± 4 Ma (Orlova *et al.*, 1986a; Orlova, 1990).

References Bilibin, 1974; *Chepurov, 1973; El'yanov and Moralev, 1974; Kravchenko *et al.*, 1982; Orlova, 1990; Orlova *et al.*, 1986a; Shnai and Orlova, 1977.

---

26 LOMAM
57°07′N; 128°05′E
Fig. 232

This occurrence is represented by several stocks the largest of which has an area of 1.5 km². The outcrops of alkaline rocks are isolated and are known as Lomam's stocks. According to new observations (N.V. Vladykin, pers. comm.) the alkaline rocks form a circular body around a central massif of calc-alkaline granite and syenite that is called 'Bilibin'. The alkaline rocks cut the Cambrian and Lower Jurassic country rocks and include micaceous peridotite, missourite, shonkinite and fergu-site. The main rock-forming minerals are olivine of variable composition, diopside, biotite and phlogopite, leucite and pseudoleucite and K-feldspar; amphibole (richterite–tremolite) is found in shonkinites and fergu-sites; kalsilite, and much more rarely nepheline, are found in altered leucite; apatite, titanite, chromite and magnetite are present as accessory minerals. The amount of apatite reaches a maximum in some melanocratic rocks in which P_2O_5 values of more than 2% occur. The micaceous peridotites consist of olivine, clinopyroxene and phlogopite in equal amounts; inclusions of leucite about 0.05 mm in diameter occur in the pyroxene. Olivine-rich xenoliths or lenses are present in the mica-

ceous peridotites, the olivine having a higher Mg content than that of the peridotite olivine. The proportion of mafic minerals decreases in the missourites and, apart from leucite inclusions in pyroxene, up to 30% of leucite occupies the area between the mafic minerals. Primary melt inclusions have been found in olivine. Shonkinites and fergusites, that grade into melanocratic pseudo-leucite syenites, have a similar suite of minerals but in different proportions. The shonkinites are rather variable in composition, consisting of pseudoleucite with feldspar, kalsilite and muscovite in the proportions 6:2:1. The pseudoleucite and unaltered leucite crystals have well defined crystal shapes and contain circular inclusions of K-feldspar and nepheline. The pyroxene is diopside, olivine varies from Fo_{92} in micaceous perido-tite to Fo_{65-63} in fergusite; phlogopite has 22–24% MgO in the micaceous peridotites and biotite 17–18% in the shonkinites and fergusites. From their whole rock and mineral compositions, phlogopite verite and phlogopite missourite could be the plutonic facies of lamproites.

Age K–Ar determinations on phlogopite from lamproite gave from 119 ± 5 to 125 ± 5 Ma (Bogatikov *et al.*, 1991).

References Bogatikov *et al.*, 1991; Khitrunov, 1990; Panina, 1990; Vavilov *et al.*, 1986.

---

27 ARBARASTAKH
56°45′N; 131°00′E
Fig. 233

Lying within Precambrian gneisses, granites and schists, the Arbarastakh complex has a striking concentric struc-ture. It is a 7×5 km stock which is composed domi-nantly of micaceous magnetite pyroxenite. All younger rocks lie within the pyroxenite stock or within the enve-loping fenite aureole. A sickle-shaped body of apatite-magnetite–forsterite rocks (phoscorite) is situated in the centre of the complex. In the outer parts of the stock and in the fenite aureole small arcuate bodies of syenite, ijolite and nepheline–pyroxene rocks are concentrated. A large number of arcuate sheets of carbonatite, which are concentrically arranged about the centre of the com-plex and younger than the syenites and ijolites, are pre-sent throughout. Vertical dykes of picrite are radially disposed. The pyroxenites are only exposed in a few places. They are generally much replaced by mica and amphibole and are permeated by a network of veins of differing ages. The contacts of the pyroxenites with the fenites are sharp. Fenite fragments, up to 3 m³, are found in the outermost pyroxenite zone and these blocks are surrounded by mica–pyroxene reaction borders. The pyroxenites are formed by diopside, titanomagnetite and accessory apatite. The fenite aureole is up to 0.8 km wide. Nepheline–pyroxene and albite–pyroxene rocks do not form substantial independent bodies but occur as large numbers of small veins, lenses and bands in a broad ring zone. Ijolite–melteigite, ijolite porphyry and kalsi-lite ijolite porphyry form dykes in pyroxenites and in fenites. Alkaline syenites form numerous veins, amongst which two age groups can be distinguished: the earlier group are cancrinite alkaline syenites (cancrinite + alkali feldspar + pyroxene ± biotite) and the later group consists of biotite and pyroxene alkaline syenites which contain calcite impregnations (K–Na feldspar + pyroxene ± biotite ± calcite). Syenites containing what is considered to be primary magmatic cancrinite are described by Zhabin and Sveshnikova (1971). The cen-tral bodies of phoscorite conform to the concentric struc-ture of the complex. Their composition is highly variable with wide variations in the modal contents of the four principal rock-forming minerals, that is forsterite, magnetite, apatite and calcite. Amongst the phoscorites

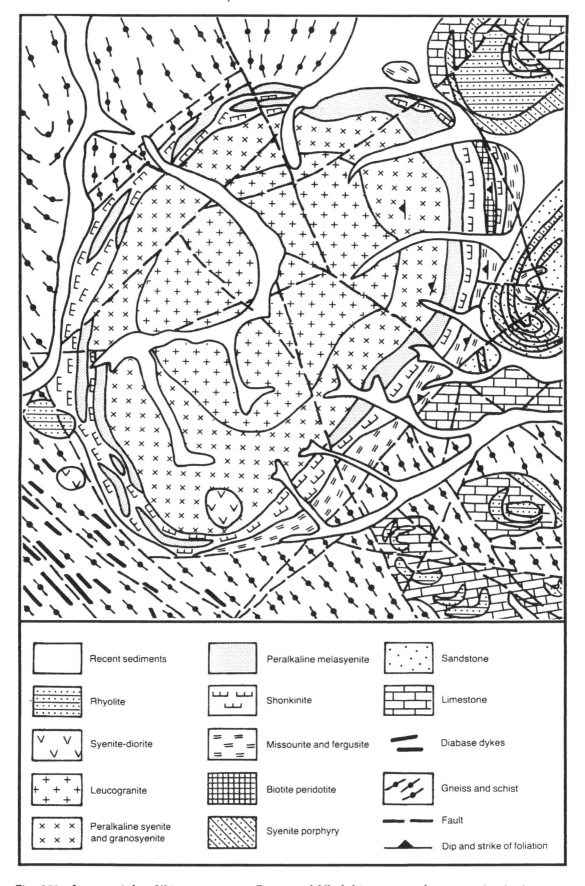

Fig. 232. Lomam (after Khitrunov, 1990, Fig. 2 and Vladykin, personal communication).

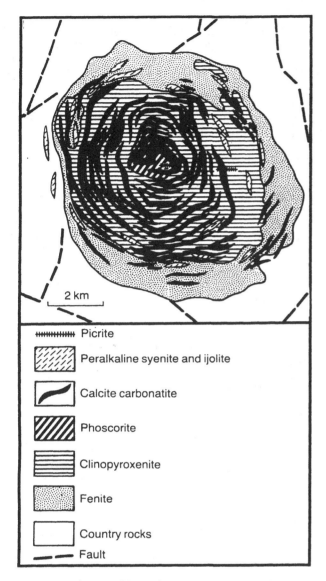

wwwwww Picrite

Peralkaline syenite and ijolite

Calcite carbonatite

Phoscorite

Clinopyroxenite

Fenite

Country rocks

— — Fault

Fig. 233. Arbarastakh (after Borodin et al., 1973, Fig. 4.1).

three groups can be distinguished which were generated in the following sequence: (1) magnetite–forsterite, (2) apatite–magnetite–chondrodite, in which the chondrodite develops at the expense of forsterite, and (3) apatite–calcite with magnetite, mica and chondrodite. Between the pyroxenites and phoscorites a zone of pegmatitic pyroxene rocks with magnetite is observed. Carbonatites are represented by a number of varieties. The earliest group are calcite carbonatites with alkali feldspar, cancrinite and zircon, the silicates comprising up to 45–60% of the rock. They are found near contacts with the alkaline syenites. For the second group of carbonatites the order of emplacement is calcite, calcite–dolomite, dolomite, and ankerite carbonatites. The calcite and calcite–dolomite carbonatites comprise 90% of the carbonatites of the complex. Mineralogically they include forsterite (chondrodite), magnetite, apatite and accessory pyrochlore, baddeleyite and zirconolite. These are medium- and coarse-grained rocks with spotted and linear textures and structures. Dolomite carbonatites are fine-grained, massive rocks which contain

a small quantity of phlogopite and alkali amphibole of the riebeckite type. They form small lens-like bodies and a series of small, gently dipping veins in fenites and fenitized gneisses. The ankeritic carbonatites contain burbankite, bastnaesite, parisite, huanghoite, strontianite, baryte, witherite, galena and sphalerite. They form narrow veins which cut the calcite carbonatites. Three groups of vein rocks are distinguished: (1) picrite porphyry, (2) syenite porphyry, tinguaite porphyry and cancrinite syenite porphyry and (3) ijolite porphyry, feldspathic ijolite porphyry and kalsilite ijolite porphyry.

Economic There is a potentially economic apatite-magnetite deposit (Frolov, 1984).

Age K–Ar determinations on phlogopite from pyroxene–mica rocks gave 690 ± 28 Ma, on apatite–magnetite rocks 720 ± 28 Ma and on carbonatites 690 ± 28 Ma (Glagolev and Korchagin, 1974). K–Ar on phlogopite from forsterite–magnetite rocks gave 625 Ma (El'yanov and Moralev, 1961).

References Borodin *et al.*, 1973; Frolov, 1984; El'anov and Moralev, 1961; Glagolev and Korchagin, 1974; Semenov *et al.*, 1974; Zhabin and Karchenkov, 1973; *Zhabin and Sveshnikova, 1971.

28 INGILI 58°45′N; 135°50′E
Fig. 234

The Ingili complex has an area of about 30 km² and is located within a dome-like structure in Archaean basement. It has a concentric form with the 3 kilometre diameter nucleus of the complex formed of schorlomite ijolite and melteigite with bands of urtite; these rocks have a taxitic structure. In the central part the ijolites and melteigites have a coarse-grained and giant-grained texture in which crystals of pyroxene and melanite may reach 10–20 cm, and sometimes 50 cm, across. Amongst the ijolites and melteigites are small xenoliths of pyroxenite which are greatly corroded by nepheline and which have gradual transitions to nepheline–pyroxene rocks. The nucleus of the complex is surrounded by a ring 2–3 km wide of rocks of widely diverse compositions. They include leucocratic nepheline gabbro, theralite, nepheline syenites including albite, cancrinite, aegirine, biotite and amphibole varieties, nepheline pyroxenites, nepheline–amphibole rocks and carbonatites. Very intensive alkaline and carbonate metasomatism has affected the rocks of this zone and has given them a homogeneous character. The sequence of development of the rocks of the ring zone is as follows: (1) emplacement of the stock of ijolite–melteigites which resulted in the formation of metasomatic 'theralites' (fenites) which were generated at the expense of basement gneisses. (2) Intrusion of carbonatites of the first generation, which are mainly located in the theralites. (3) Intrusion of numerous veins and small bodies of nepheline–cancrinite and albite–cancrinite syenites, tinguaite and ijolite porphyry. (4) Emplacement of carbonatites of the second generation, which are calcitic and dolomitic in composition.

Age K–Ar dating of hastingsite from ijolite gave 660 ± 18 Ma and from melteigite 655 ± 22 Ma; biotite from ijolite gave 648 ± 12 and nepheline from ijolite 704 ± 26 Ma (Orlova *et al.*, 1986a).

References Orlova *et al.*, 1986a; *Shadenkov, 1989; Zlenko *et al.*, 1966.

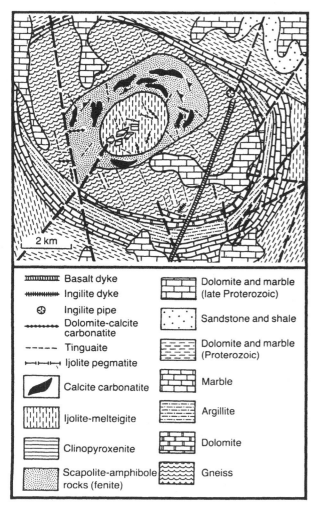

Legend:

- Basalt dyke
- Ingilite dyke
- Ingilite pipe
- Dolomite-calcite carbonatite
- Tinguaite
- Ijolite pegmatite
- Calcite carbonatite
- Ijolite-melteigite
- Clinopyroxenite
- Scapolite-amphibole rocks (fenite)
- Dolomite and marble (late Proterozoic)
- Sandstone and shale
- Dolomite and marble (Proterozoic)
- Marble
- Argillite
- Dolomite
- Gneiss

Fig. 234. Ingili (after Shadenkov, 1989, Fig. 1).

29 KONDER   57°30′N; 134°25′E
Fig. 235

Konder is an almost perfectly circular intrusion which cuts Archaean schists, gneisses and marbles that are overlapped by Sinian (Proterozoic to Lower Cambrian) sandstones and siltstones 450 m in thickness. The Sinian rocks are domed and Marakushev *et al.* (1991) discuss the formation of this dome and the emplacement of the early ultramafic rocks. Ultramafic rocks form the central stock and are surrounded by a ring intrusion of syenites, diorites and monzonites. There is a clear zonality to the massif (Fig. 235) with, from the centre to the margins, the following sequence: (1) Dunites dominate the central area and vary from coarse porphyritic varieties at the centre, with olivine phenocrysts of 2–3 mm diameter, to medium-grained dunites at the margin. Lenses of chromitite from 10 cm to 1 m thick are widespread in the dunites. In the southeastern part of the dunite core subparallel bodies of biotite pyroxenite with apatite (koswite) are concentrated, 24 such bodies having been identified in one 140 m section. (2) The outer part of the ultramafic stock consists of medium- and fine-grained olivinites. (3) An intermediate zone 8–10 m wide is composed of olivinites and clinopyroxenites. (4) Fine- and medium-grained pyroxenites. (5) The outer part of the complex is composed of essexites and shonkinites. The dunites are generally fine- and medium-grained rocks of

dark green and green hues; on weathered surfaces they are greenish-yellow. Compositionally they are rather homogeneous; in addition to olivine (80–96%, forsterite-chrysolite), they usually contain fine-grained and scattered chromite impregnation (up to 1–2% chrome picotite with Cr_2O_3 values up to 51.5%) and 3–20% serpentine. Pyroxenites are medium- and coarse-grained rocks which are dark green to almost black. They are built mainly of clinopyroxene (diopside–salite with 10% of hedenbergite molecule) and by a small amount of titanomagnetite (up to 5%). The koswites are fine- and medium-grained, massive rocks of clinopyroxene (56–77%), titanomagnetite (11–20%), hornblende (0–11%), biotite (0–21%) and accessory apatite, titanite and secondary chlorite. Apatite–biotite–titanomagnetite-pyroxene rocks form isolated occurrences amongst the dykes and veins of koswite and a large field in the centre of the complex; they are also recorded as small bodies within the dunites, pyroxenites and peridotites. The modal proportions of minerals vary widely, with clinopyroxene 20–50%, biotite 15–40%, titanomagnetite 20–40% and apatite 5–15%. The rocks are medium- to giant-grained with biotite crystals $10 \times 10 \times 3$ cm. In some areas occur veins up to 1 cm thick of chrysocolla. Diorites and monzosyenite of the external ring zone, with a width up to 700 m, are medium-and fine-grained rocks with a porphyritic textures. They consist of andesine (54–77%), microcline and orthoclase (0–20%), hornblende (6–23%), clinopyroxene (0–1.5%), biotite (1–5%) and quartz (0–7%), as well as accessory apatite, titanite and magnetite and secondary chlorite, sericite, and carbonate. There are dykes of nepheline syenite, nepheline pyroxenite, melteigite and nepheline syenite pegmatite. The following varieties of alkaline pegmatites have been identified: aegirine–arfvedsonite–feldspar with lamprophyllite, eudialyte, ramsayite and murmanite, as veins up to 6–7 m thick; aegirine–albite with ramsayite; arfvedsonite–zeolite with titanite; arfvedsonite–zeolite with ilmenite. As an example the composition of the first pegmatite type is aegirine (20–30%), arfvedsonite (20–30%), orthoclase (30–40%), eudialyte (1–2%), lamprophyllite (2–3%), murmanite (1%) and apatite (1–2%). A group of limestones at the margin of the complex were once considered to be carbonatites but these were later demonstrated to be skarns, a brief account of spinel- and perovskite-bearing examples of which is given by Orlova *et al.* (1981). Detailed accounts of the compositions of the rocks and minerals of Konder can be found in the paper by Andreyev (1987), and Lazarenkov and Malich (1992) give analyses, including platinum group elements, of the ultramafic rocks. Using iron–titanium oxides from a wide range of rocks Ardontsev and Malich (1989b) calculated temperatures and oxygen fugacities at the time of crystallization.

Age K–Ar on biotite from biotite–pyroxene rocks gave 650 Ma and K-feldspar from alkaline pegmatite 155 and 90 Ma (El'yanov and Moralev, 1961). According to Orlova *et al.* (1986a) K–Ar on olivine pyroxenite (whole rock) gave 113 and 124 Ma, shonkinite 120 ± 5 Ma and microcline from nepheline syenite pegmatite 110 ± 5 Ma. El'yanov and Moralev (1961) consider, from their K–Ar data, that the ultramafic core is Proterozoic in age but the ring of alkaline rocks Mesozoic. Orlova *et al.* (1986), however, suggest that the whole complex is Mesozoic.

References Andreyev, 1987; Ardontsev and Malich, 1989a and *1989b; Bogomolov, 1968; El'yanov and Moralev, 1961; *Lazarenkov and Malich, 1992; *Marakushev *et al.*, 1991; Orlova, 1991; Orlova *et al.*, *1981 and 1986a; Sheinmann *et al.*, 1961.

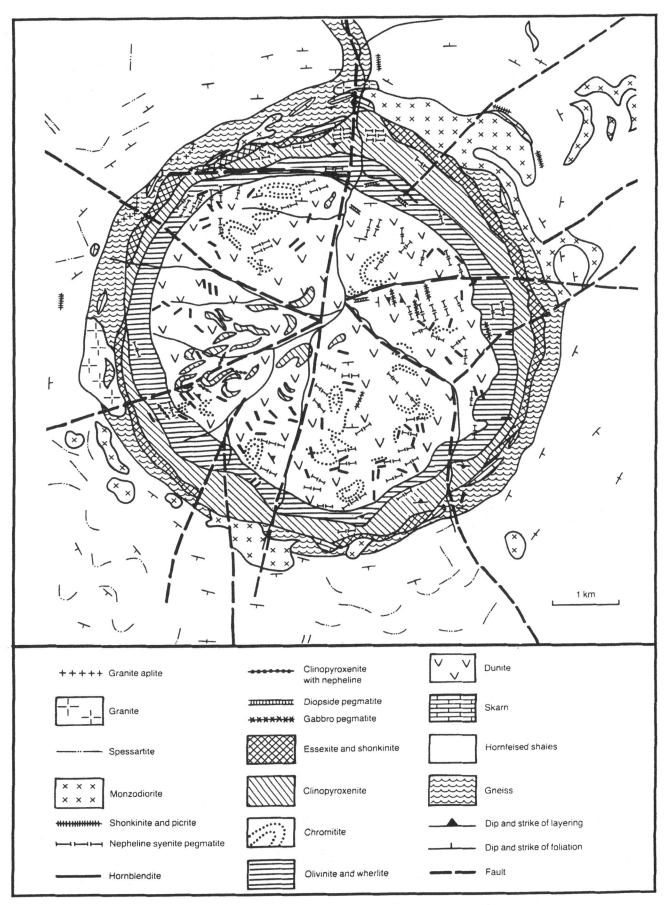

Legend:

- +++++ Granite aplite
- Granite
- —··—··— Spessartite
- ××× Monzodiorite
- Shonkinite and picrite
- ⊢—⊢—⊢ Nepheline syenite pegmatite
- —— Hornblendite
- Clinopyroxenite with nepheline
- Diopside pegmatite
- ××××××× Gabbro pegmatite
- Essexite and shonkinite
- Clinopyroxenite
- Chromitite
- Olivinite and wherlite
- V V V Dunite
- Skarn
- Hornfelsed shales
- Gneiss
- ▲ Dip and strike of layering
- ⊥ Dip and strike of foliation
- — — Fault

Fig. 235. Konder (after Orlova, 1991, Fig. 1).

30 SYBAKH   56°22'N; 135°00'E
                                              Fig. 236

This intrusion has an area of about 2 km², is lensoid in form and has a zonal structure. The central part is composed of serpentinite, dunite and olivinite, the intermediate part of clinopyroxenite and the margin shonkinite with an orientated fabric. Peralkaline syenite porphyry forms dykes.

Reference Ardontsev and Malich, 1989a.

---

31 CHAD   56°20'N; 135°00'E
                                              Fig. 237

A symmetrical intrusion, Chad is concentrically zoned with a nucleus 2 km in diameter of dunites which is surrounded by a ring of peridotite and clinopyroxenite. The peripheral part is formed by gabbroic rocks, syenites and diorites. Veins of peralkaline pegmatite, aplite and granite porphyry are present, particularly towards the centre of the intrusion. The dunites are medium-grained, compact, dark-green rocks of olivine with some serpentine, chrome spinel and platinum group minerals also present. The peridotites are formed predominantly of diopside and olivine, sometimes with a little titanomagnetite. The composition of the peridotites varies from the internal to the external contacts. In the zone adjacent to dunites they are coarse-grained bright green rocks. Closer to the periphery the peridotites become

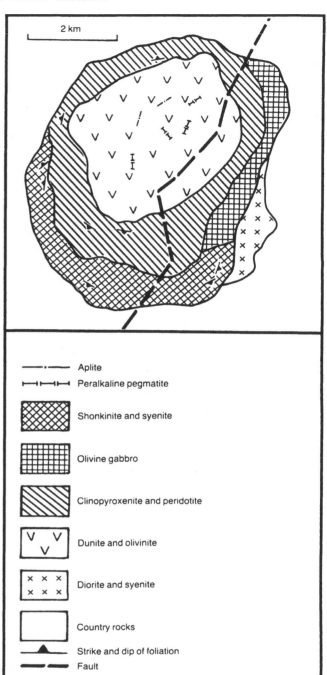

Aplite
Peralkaline pegmatite

Shonkinite and syenite

Olivine gabbro

Clinopyroxenite and peridotite

Dunite and olivinite

Diorite and syenite

Country rocks

Strike and dip of foliation
Fault

Fig. 237. Chad (after Bogomolov and Kitsul, 1964, Fig. 2).

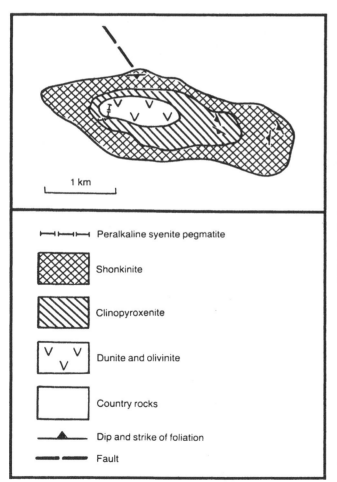

Peralkaline syenite pegmatite

Shonkinite

Clinopyroxenite

Dunite and olivinite

Country rocks

Dip and strike of foliation

Fault

Fig. 236. Sybakh (after Ardontsev amd Malich, 1989a, Fig. 1).

darker, finer-grained and titanomagnetite appears. Bogomolov and Kitsul (1964) regarded the peridotites as being generated by reaction around the dunite core during the introduction of the gabbroic intrusions. Moderately alkaline rocks of monzonitic type occur in the gabbroic intrusion, but olivine gabbro is also present. The moderately alkaline gabbros have poikilitic textures with large, up to 1 cm diameter, grains of K-feldspar with inclusions of pyroxene and olivine. In addition, plagioclase (An$_{50-55}$), titanomagnetite, biotite, hornblende and apatite are also present. The olivine gabbros are similar to the alkaline ones but K-feldspar is absent and the plagioclase has a more basic composition

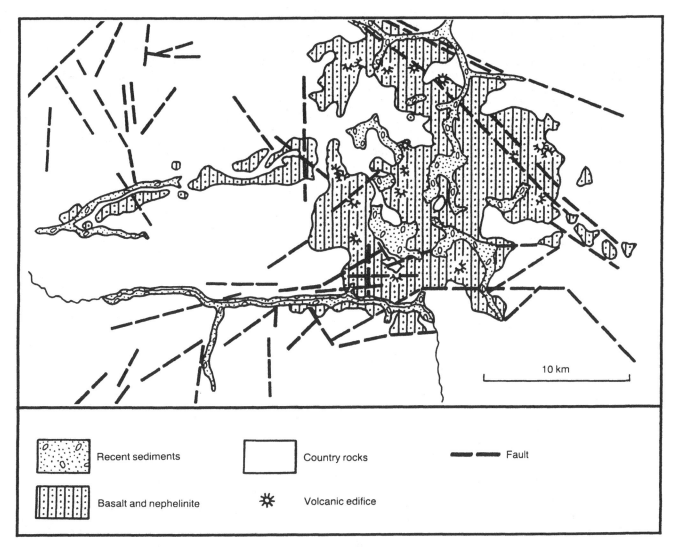

Fig. 238. Tokko (after Kost'uk, 1990, Fig. 8).

(An_{60-65}). Bodies of alkaline syenite and shonkinite form an arcuate body around the southern margin of the intrusion against the country rock limestones. The syenites consist of clinopyroxene, andesine and K-feldspar, while the shonkinites comprise clinopyroxene and K-feldspar.

Economic Platinum group minerals are present in river alluvials.

Age K–Ar on biotite from gabbro gave 160 ± 25 Ma (Bogomolov and Kitsul, 1964).

Reference Bogomolov and Kitsul, 1964.

---

32 TOKKO 55°36'N; 130°00'E
(Tokinskii) Fig. 238

Basaltic lava flows extend across the watershed of the Stanovoy ridge. They are up to some 280 m in thickness and cover about 200 km². The flows extend from the watershed along a late Quaternary river valley. The lower part of the section consists of massive pyroxene–plagioclase basalts and the upper sequence comprises peralkaline and weakly alkaline basalts including hawaiite, limburgite, nepheline and leucite basalts, olivine melanephelinite and phonolitic nephelinite.

There are also ash cones. There are abundant plutonic xenoliths in the alkaline rocks including spinel lherzolites, pyroxenites, wherlites and websterites, megacrysts of augite and sanidine and xenocrysts of olivine, orthopyroxene, spinel and clinopyroxene. Chemical compositions of igneous rocks and xenoliths are given by Shcheka *et al.* (1983) and Semenova *et al.* (1984 and 1987). $^{87}Sr/^{86}Sr$ values of peralkaline basaltoids and Ti-phlogopite, kaersutite and Al-augite of 0.7045, 0.7035, 0.7038 and 0.7045 respectively have been obtained by Solov'eva *et al.* (1983).

Age Quaternary.

References Semenova *et al.*, 1984 and 1987; Shcheka *et al.*, 1983; Solov'eva *et al.*, 1983.

Aldan References

ANDREYEV, G.V. 1987. *The Konder ultrabasic alkaline massif.* Nauka, Novosibirsk. 76 pp.

ARDONTSEV, S.N. and MALICH, K.N. 1989a. Geodynamic model of the formation of the massifs of the Konder complex. *Geologiya i Geofizika. Novosibirsk,* 7: 27–33.

*ARDONTSEV, S.N. and MALICH, K.N. 1989b.

Physicochemical conditions of iron–titanium oxide formation in rocks of the Konder alkalic ultramafic pluton. *Transactions (Doklady) of the USSR Academy of Sciences, Earth Science Sections*, 296: 225–7.

BILIBIN, U.A. 1958. *Selected publications* 1. USSR Academy of Sciences Publishing House, Moscow. 432 pp.

BILIBIN, U.A. 1974. *Petrology of the Yllymakh intrusive*. Gosgeoltekhisdat. Moscow and Leningrad. 235 pp.

*BIRYUKOV, V.M. and BERDNIKOV, N.V. 1993. The paragenetic relation between charoite mineralization and alkali metasomatism. *International Geology Review*, 35: 585–602.

*BOGATIKOV, O.A., YEREMEYEV, N.V., MAKHOTKIN, I.L., KONONOVA, V.A., NOVGORODOVA, M.I. and LAPUTINA, I.P. 1986. Lamproites of the Aldan and central Asia. *Transactions (Doklady) of the USSR Academy of Sciences, Earth Science Sections*, 290: 154–7.

BOGATIKOV, O.A., RYABCHIKOV, I.D., KONONOVA, V.A. *et al.* 1991. *Lamproite*. Nauka, Moscow. 320 pp.

BOGOMOLOV, M.A. 1968. Apatite-bearing rocks of the Konder massif. *In* O.H. Vorob'eva and V.P. Petrov (eds) *Apatite*. 224–7. Nauka, Moscow.

BOGOMOLOV, M.A. and KITSUL V.I. 1964. The Chadsky ultrabasic-alkaline massif at the eastern margin of the Aldan shield. *In* V.I. Kitsul (ed.) *Petrography of the metamorphic and igneous rocks of the Aldan Shield*. 156–65. Nauka, Moscow.

BORODIN, I.S., LAPIN, A.V. and CHARCHENKOV, A.G. 1973. *Rare-metal camoforite*. Nauka, Moscow. 176 pp.

*BULAKH, A.G., ZOLOTAREV, A.A., BOBROVA, I.P., GULIY, V.N. and VANDE-KIRKOV, Yu.V. 1985. Fundamental mineralogic features and origin of the Seligdar apatite deposit (Aldan crystalline shield). *International Geology Review*, 27: 144–56.

*BUTAKOVA, E.L. 1974. Regional distribution and tectonic relations of the alkaline rocks of Siberia. *In* H. Sorensen (ed.), *The alkaline rocks*. 172–89. John Wiley, London.

*CHEPUROV, A.I. 1973. Thermometric study of inclusions of melt in pseudoleucite minerals of the central Aldan district. *Transactions (Doklady) of the USSR Academy of Sciences, Earth Science Sections*, 213: 165–8.

EL'YANOV, A.A. and MORALEV, V.M. 1961. New data on the age of the ultramafic and alkaline rocks of the Aldan shield. *Doklady Akademii Nauk SSSR*, 141: 686–9.

EL'YANOV, A.A. and MORALEV, V.M. 1974. On the problem of the deep structure of the Yllymakh volcano-pluton (central Aldan area). *Izvestiya Akademii Nauk SSSR, Seriya Geologiya*, 1: 134–7.

ENTIN, A.R. 1986. High temperature mineral associations of apatite-bearing carbonatites of Seligdar type. *In* I.M. Frumkin (ed.) *Geology and geochemistry of the ore-bearing magmatic and metasomatic associations from the Maly BAM area*. 90–105. Yakutia Branch of the Siberia Division of the USSR Academy of Sciences, Yakutsk.

ENTIN, A.R., BELOUSOV, V.M. and GALKIN, G.F. 1977. New data on the geology of the Seligdar apatite deposits. *In* F.L. Smirnov (ed.) *Apatite of the Aldan Shield*. 5–18. Yakutia Branch of the Siberia Division of the USSR Academy of Sciences, Yakutsk.

*ENTIN, A.R., SOBOTOVICH, E.V., OL'KHOVIK, Yu.A., RUDNIK, V.A. and BELOUSOV, V.M. 1985. Conditions of formation of the Seligdar apatite deposit. *International Geology Review*, 27: 157–67.

EREMEYEV, N.V. 1987. Potassic alkaline magmatism Verhne-Yakokutsy graben (central Aldan). *Izvestiya Akademii Nauk SSSR, Seriya Geologiya*, 1: 126–30.

*EREMEYEV, N.V., ZHURAVLEV, D.Z., KONONOVA, V.A., PERVOV, V.A. and KRAMM, V. 1993. Source and age of the potassic rocks in the Ryabin intrusion, central Aldan. *Geochemistry International*, 30: 104–12.

FROLOV, A.A. 1984. Iron-ore deposits in the carbonatite-alkaline-ultrabasic central type massifs. *Geology of ore deposits, Moscow*, 26: 9–21.

GLAGOLEV, A.A. and KORCHAGIN, A.M. 1974. *Alkaline-ultrabasic massifs. Arbarastakh and Inagli*. Nauka, Moscow. 176 pp.

KANUKOV, B.V., MAKHOTKIN, I.L. and GOLOVANOVA, T.I. 1991. Petrology of the potassic volcanic series of the Yakokut volcanic–plutonic complex of the Central Aldan. *Izvestiya Akademii Nauk SSSR, Seriya Geologiya*, 12: 83–101.

KHITRUNOV, A.T. 1990. Petrology of the central type Mesozoic intrusions from the south-eastern part of the Aldan shield. *Geologiya i Geofizika. Novosibirsk*, 3: 62–71.

KOCHETKOV, A.Ya., IGUMNOVA, N.S. and KIM, A.A. 1986. Associations and mineralogical types of Mesozoic ores from Central Aldan. *In* I.M. Frumkin (ed.) *Geology and geochemistry of the ore-bearing magmatic and metasomatic associations from the Maly BAM area*. 20–31. Yakutia Branch of the Siberia Division of the USSR Academy of Sciences, Yakutsk.

KOCHETKOV, A.V., PAHOMOV, V.N. and POPOV, A.B. 1989. Magmatism and metasomatism at the Ryabinovy ore-bearing alkaline massif (central Aldan). *In* V.I. Sotnikov and A.A. Obolensky (eds) *Magmatism of the copper–molybdenum ore assemblages*. 79–110. Nauka, Novosibirsk.

KORCHAGIN, A.M. 1966. A vermiculite–phlogopite deposit at Inagli. *Izvestiya Akademii Nauk SSSR, Seriya Geologiya*, 8: 86–97.

KOST'UK, V.P. 1990. Geology and structural position of the potassic alkaline formations at the Baikal-Stanovoy rift system. *In* G.V. Polyakov and V.V. Kepezinskas (eds) *Potassic alkaline magmatism of the Baikal–Stanovoy rift system*. 65–123. Nauka, Siberian Division of the USSR Academy of Sciences, Novosibirsk.

*KRASILNIKOVA, N.A. and ILYIN, A.V. 1989. Igneous Proterozoic–Cambrian phosphate resources in eastern Siberia, USSR. *In* A.J.G. Notholt, R.P. Sheldon and D.F. Davidson (eds). *Phosphate deposits of the world. 2. Phosphate rock resources*. 510–17. Cambridge University Press, Cambridge.

KRAVCHENKO, S.M. 1972. Potassic-rich alkaline lava and ignimbrite of the Jurassic volcanics of the central Aldan. *Izvestiya Akademii Nauk SSSR, Seriya Geologiya*, 4: 24–34.

KRAVCHENKO, S.M., KAPUSTIN, Yu.L., KATAEVA, Z.T. and BYKOVA, A.V. 1982. Agpaitic mineralization of the Mesozoic metasomatic rocks of potassic alkaline complexes of the Central Aldan. *Doklady Akademii Nauk SSSR*, 263: 435–9.

*LAZARENKOV, V.G. and MALICH, K.N. 1992. Geochemistry of the ultrabasites of the Konder platiniferous massif. *Geochemistry International*, 29: 44–56.

MAKHOTKIN, I.I., ARAKEL'YANZ, M.M. and VLADYKIN, N.V. 1989. On the age of the lamproites of the Aldan Province. *Doklady Akademii Nauk SSSR*, 306: 703–7.

*MAKSIMOV, Ye.P. 1973. Mesozoic annular magmatic complexes in the Aldan Shield. *International Geology Review*, 15: 46–56.

MAKSIMOV, Ye.P. 1982. Mesozoic magmatism of the Aldan shield as the indicator of tectonic region. *Geologiya i Geofizika. Novosibirsk,* 5: 11–19.

MAKSIMOV, Ye.P. and UGR'UMOV, A.N. 1971. Mesozoic magmatic formations of the Aldan shield. *Sovetskaya Geologiya. Moskva,* 7: 107–19.

*MARAKUSHEV, A.A., YEMEL'YANENKO, Ye.P., NEKRASOV, I.Y, MASLOVSKIY, A.N. and ZALISHCHAK, B.L. 1991. Formation of the concentrically zoned structure of the Konder alkalic-ultrabasic pluton. *Transactions (Doklady) of the USSR Academy of Sciences, Earth Science Sections,* 311: 69–72.

*ORLOVA, M.P. 1988a. Recent findings on the geology of the Maly Murun alkalic pluton. *International Geology Review,* 30: 945–53.

*ORLOVA, M.P. 1988b. Petrochemistry of the Maly Murun alkalic pluton. *International Geology Review,* 30: 954–65.

ORLOVA, M.P. 1990. Mesozoic stage of magmatism. *In* G.V. Polyakov and V.V. Kepezinskas (eds) *Potassic alkaline magmatism of the Baikal-Stanovoy rift system.* 65–123. Nauka, Siberian Division of the USSR Academy of Sciences, Novosibirsk.

ORLOVA, M.P. 1991. Geologic structure and genesis of the Konder ultramafic massif (Khabarovsk Territory). *Pacific Geology,* 1: 80–8.

ORLOVA, M.P. and SHCHADENKOV, E.M. 1988. On the petrology of the Maly Murun alkaline massif (south-western Yakutia). *Geologiya i Geofizika,* 12: 77–86.

ORLOVA, M.P., ARDONTSEV, S.N. and SHCHADENKOV, E.M. 1986a. Alkaline magmatism of the Aldan shield and its specific mineralogical features. *In* I.M. Frumkin (ed.) *Geology and geochemistry of the ore-bearing magmatic and metasomatic associations from the Maly BAM area.* 4–12. Yakutia Branch of the Siberian Division of the USSR Academy of Sciences, Yakutsk.

*ORLOVA, M.P., BAGDASAROV, E.A. and SOSEDKO, T.A. 1981. Spinel and perovskite from rocks of the exoskarn zone of the Konder massif (Khabarovskiy Kray). *International Geology Review,* 23: 716–20.

ORLOVA, M.P., BORISOV, A.B. and SHCHADENKOV, E.M. 1992. Alkaline magmatism of the Murun area. *Geologiya i Geofizika. Novosibirsk,* 5: 57–70.

ORLOVA, M.P., KURANOVA, V.N., SOSEDKO, T.A., CHEREPANOV, V.A. and SHCHADENKOV, E.M. 1986b. Mineralogy and genesis of chrome-diopsides from the Inagli massif (Aldan). *In* J. Mincava-Stefanova (ed.) *Morphology and phase equilibria of minerals.* Proceedings of the 13th General Meeting of the International Mineralogical Association (IMA) in 1982. 449–60. Bulgarian Academy of Sciences, Sofia.

OSOKIN, E.D., LAPIN, A.V., KAPUSTIN, Yu.L., POHVISNEVA, E.A. and ALTUHOV, E.N. 1974. Alkaline provinces of Asia. Siberia-Pacific group. *In* L.S. Borodin (ed.) *Principal provinces and formations of alkaline rocks.* 91–166. Nauka, Moscow.

PANINA, L.I. 1990. P-T conditions of the potassic alkaline rocks formation according to the thermobarogeochemical data on minerals. *In* G.V. Polyakov and V.V. Kepezinskas (eds) *Potassic alkaline magmatism of the Baikal-Stanovoy rift system.* 150–73. Nauka, Siberian Division of the USSR Academy of Sciences, Novosibirsk.

PERVOV, V.A., KANUKOV, B.U. and ARAKEL'YANZ, M.M. 1991. New data on K-Ar age of the extrusive rocks of the Tommot volcanic–plutonic complex (Central Aldan). *Doklady Akademii Nauk SSSR,* 321: 349–52.

*POKROVSKII, B.G. and VINOGRADOV, V.I. 1991. Isotope investigations on alkalic rocks of central and western Siberia. *International Geology Review,* 33: 122–34.

*POKROVSKII, B.G. and ZHIDKOV, A.Ya. 1993. Origin of the ultrapotassic rocks of the Synnyr and Southern Sakun massifs (Transbaikal area) from isotopic data evidence. *Petrology,* 1: 195–204.

*PROKOF'YEV, V.Yu. and VOROB'YEV, Ye.I. 1992. P-T formation conditions for Sr–Ba carbonatites, charoite rocks, and torgolites in the Murun alkali intrusion, east Siberia. *Geochemistry International,* 29: 83–92.

SAMSONOVA, N.S. and DONAKOV, V.I. 1968. Kalsilite in the Murunskii alkaline massif (central Siberia). *Zapiski Vsesoyuznogo Mineralogicheskogo Obshchestva* 97: 291–300.

SEMENOVA, V.G., SOLOV'EVA, L.V. and VLADIMIROV, B.M. 1984. *Deep-seated inclusions from alkaline basaltoids from Tokinsky Stanovik.* Nauka, Siberian Division of the USSR Academy of Sciences, Novosibirsk. 119 pp.

SEMENOV, E.I., ESKOVA, E.M., KAPUSTIN, Yu.P. and HOM'AKOV, A.P. 1974. *Mineralogy of the alkaline massifs and their deposits.* Nauka, Moscow. 247 pp.

SEMENOVA, V.G., SOLOV'EVA, L.V., VLADIMIROV, B.M., ZAV'YALOVA, L.L. and BARANKEVICH, V.G. 1987. Glass and quenching phases from the deep-seated inclusions from alkaline basaltoids of Tokinsky Stanovik. *In* V.A. Zharikov and A.F. Grachev (eds) *Deep-seated xenoliths and structure of the lithosphere.* 73–95. Nauka, Moscow.

*SHADENKOV, Ye.M. 1989. Metasomatic assemblages in wall rocks of the Ingili pluton, eastern Aldan. *International Geology Review,* 31: 697–706.

SHCHADENKOV, E.M., ORLOVA, M.P. and BORISOV, A.B. 1989. Pyroxenite and shonkinite from Maly Murun massif are the intrusive equivalent of lamproite. *Zapiski Vsesoyuznogo Mineralogicheskogo Obshchestva. Moskva,* 118(6): 28–37.

SHCHEKA, S.A., LENNIKOV, A.M. and ROMANENKO, I.M. 1983. Extraordinary lherzolite inclusions from alkaline basalts from the Stanovoy Range. *Doklady Akademii Nauk SSSR,* 269: 915–18.

SHEINMANN, U.M., APELTSIN, F.R. and NECHAEVA, E.A. 1961. Alkaline intrusions, their distribution and related mineralization. *In* A.I. Ginzburg (ed.) *Geology of the deposits of the rare elements,* Gosgeoltekhizdat, Moscow. 177 pp.

SHNAI, G.K. and ORLOVA, M.P. 1977. New data on the geology and gold contents of the Yllymakh massif (central Aldan). *Geologiya i Geofizika,* 10: 57–65.

SILIN, I.I. and UGRYUMOV, A.I. 1972. Regularities of the distribution of Mesozoic alkaline rocks and gold deposits in the central Aldan region, south Yakutia. *In* Yu.P. Ivensen (ed.) *Ore formation and its connection with magmatism.* 275–82. Nauka, Moscow.

*SMIRNOV, F.L. 1976. New genetic type of commercial apatite deposits. *Doklady Earth Science Sections. American Geological Institute,* 230: 86–8.

*SMIRNOV, F.L. 1978. Geologic position and conditions of formation of the Seligdar apatite deposit. *International Geology Review,* 20: 1043–9.

SMIRNOV, F.L., ENTIN, A.R., UGR'UMOV, A.N. and BURNAIKIN, A.I. 1975. Precambrian apatite mineralization at the fault zones of the central part of the Aldan Shield. *Phosphorite of Yakutia.* 53–74. Yakutsk.

SOLOV'EVA, L.V., VLADIMIROV, B.M. and

SEMENOVA, V.G. 1983. Geochemistry of the deep-seated inclusions and alkaline basaltoids of Tokinsky Stanovik in view of the problems of their genesis. *Geologiya i Geofizika, Novosibirsk*, **2**: 75–82.

SVESHNIKOVA, E.V. and EREMEYEV, N.V. 1982. Ore-bearing polyformation magmatic complexes of the central part of the Aldan shield. *In* A.V. Sidorenko (ed.) *Ore-bearing structures of the Precambrian.* 105–16. Nauka, Moscow.

VASEILENKO, V.B., KUZNETSHOVA, L.G., HOLODOVA, L.D., EGIN, V.I., ENTIN, A.R., SUCHKOV, V.I. and BELOUSOV, V.M. 1982. *Apatite rocks from the Seligdar.* Nauka, Novosibirsk. 215 pp.

VAVILOV, M.A., BAZAROVA, T.Yu., PODGORNYKH, N.M., KRIVOPUTSKAYA, L.M. and KUZNETSOVA, I.K. 1986. Characteristics and conditions for the formation of potassic alkaline rocks of the Lomamsky massif. *Geologiya i Geofizika. Novsibirsk*, **3**: 40–6.

VLADYKIN, N.V. 1986. Crystallochemistry, paragenesis and genesis of charoite – a new precious stone. *In* J. Minceva-Stefanova (ed.) *Morphology and phase equilibria of minerals.* 387–94. Proceedings of the 13th General Meeting of the International Mineralogical Association (IMA) in 1982. Bulgarian Academy of Sciences, Sofia.

ZHABIN, A.G. and KHARCHENKOV, A.G. 1973. Arbarastakh carbonatite complex (south Yakutia province). *In* L.S. Borodin (ed.) *New data on the geology, mineralogy and geochemistry of the alkaline rocks.* 142–57. Nauka, Moscow.

*ZHABIN, A.G. and SVESHNIKOVA, Ye.V. 1971. Magmatic cancrinite. *International Geology Review*, **13**: 1269–74.

ZHIDKOV, A.Ya. 1990. The Palaeozoic stage of magmatism. *In* G.V. Polyakov and V.V. Kepezinskas (eds) *Potassic alkaline magmatism of the Baikal-Stanovoy rift system.* 32–64. Nauka, Siberian Division of the USSR Academy of Sciences, Novosibirsk.

ZLENKO, N.D., SHPAK, N.S. and EL'ANOV, A.A. 1966. Ultrabasic-alkaline intrusions of the Aldan anticlise. *In* I.I. Krasnov, M.L. Lur'e and V.L. Masaitis (eds) *Geology of the Siberian platform.* 233–42. Nedra, Moscow.

* In English

SETTE-DABAN

This province is situated in the southeast of Yakutia within the Sette-Daban Range and is just north of the eastern Aldan sub-province (see Aldan Fig. 220 for location of province and individual massifs). Four of the five occurrences of the province are located within a 90 km long zone which is related to a large, approximately north–south-trending fault. It has been suggested that the Sette-Daban zone of deep faults represents a system of rifts.

1 Voin   4 Ozernyi
2 Gek   5 Khamninskii
3 Provorotnyi

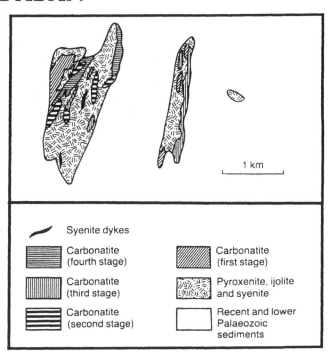

Syenite dykes

Carbonatite (fourth stage)

Carbonatite (third stage)

Carbonatite (second stage)

Carbonatite (first stage)

Pyroxenite, ijolite and syenite

Recent and lower Palaeozoic sediments

1 VOIN

60°45′N; 137°30′E
Fig. 239

This small 0.3 × 0.1 km stock cuts limestones of lower Ordovician age. It is formed mainly of nepheline and peralkaline syenites. A large xenolith of 130 × 70 m of utrabasic rock is preserved in the central part of the intrusion. It is cut by numerous, thin syenite dykes. The rocks of the intrusion are replaced by carbonate. Phlogopite–calcite carbonatite forms rare veins and nests of small size (1–10 cm diameter). Sulphides from calcite carbonatites gave $\delta^{34}S°/_{oo}$ values of $+0.6$ (Grinenko et al., 1970).

References Epstein et al., 1972; Grinenko et al., 1970.

2 GEK

60°41′N; 137°25′E
Fig. 239

In plan this massif is lens-shaped with an area of 2.3 × 0.4 km. It intrudes Middle Cambrian terrigenous carbonate deposits and is formed principally of alkaline syenites, but small bodies of ijolite and pyroxenite have been preserved amongst the syenites. The rocks of the intrusion have been extensively replaced by carbonate. Carbonatites are present and are dolomitic and ankeritic with rare-earth minerals abundant.

Reference Epstein et al., 1972.

3 POVOROTNYI

60°36′N; 137°23′E
Fig. 239

This occurrence, with an area of 3 km², takes the form of a stock which cuts limestones and mixed carbonate-terrigenous sediments of Lower and Middle Cambrian age. Judging from the nature of the bedding at the contacts it is presumed that the intrusion is lens-shaped and expands downwards, the pipe-like body being inclined steeply to the south. Carbonatites occupy more than 40% of the area of the intrusion with the rest formed by intensively carbonated silicate rocks. It is considered that the intrusion was formed initially of pyroxenites and alkaline rocks, including ijolites and peralkaline syenites. Calcite carbonatites are the most abundant but the composition depends on the nature of the surrounding rocks such that pyroxenite is replaced by augite–diopside–calcite carbonatite, ijolite by aegirine–diopside–calcite carbonatite, and peralkaline syenite by aegirine–calcite and biotite–calcite carbonatites. Secondary minerals, including epidote, albite and calcite, are widespread and form narrow veins and patches. $\delta^{34}S°/_{oo}$ values of sulphides in calcite carbonatite and ankerite

Fig. 239. Voin (right), Gek (centre) and Povorotnyi (left) (after Epstein, 1972, Fig. 9, II, III and IV). The relative positions on the map of the three occurrences are arbitrary.

carbonatite range from -0.5 to -2.0 (Grinenko et al., 1970).

References Epstein et al., 1972; Grinenko et al., 1970.

4 OZERNYI
(Gornoozerskii)

59°56′N; 136°53′E
Fig. 240

Oval in shape, Ozernyi has an area of 9.3 km² and cuts calcareous terrigenous rocks of Sinian age. About 85% of the area of the complex is composed of carbonatites. Silicate rocks, which have been preserved only in the form of relict blocks, which have diameters of up to 0.6 km, are represented by nepheline–cancrinite syenite, ijolite and pyroxenite. Four stages in the evolution of the complex have been established. Firstly a tube-like intrusion of ultramafic rocks was emplaced. This was followed by intrusion of nepheline–pyroxene rocks and dykes of ijolite porphyry and these in turn by nepheline syenites. Finally, carbonatites were intruded. The ultramafic rocks have been preserved mainly in the northern part of the complex where they form a body with sinuous margins which has an area of 0.12 km². Pyroxenites are formed mainly of diopside and are often recrystallized. The clinopyroxene is usually zoned from diopside cores to fassaitic rims. Ijolites and nepheline–pyroxene rocks are also found in the form of relict bodies within the carbonatites, the largest of which lie in the centre of the complex. The ijolites are generally uniformly grained, but are sometimes porphyritic, and may also be banded with alternating leucocratic and melanocratic layers. Usually the pyroxene in the ijolites contains up to 30% of aegirine component. Feldspathic ijolites with K-feldspar have been identified. Nepheline syenites comprise two irregularly shaped bodies with areas of 0.12 and 0.05 km² in the western part of the complex. In addition, they form veins and lenses within the ijolites. The nepheline syenites are fine- and medium-grained and generally have a trachytic texture. They comprise alkali feldspar, in which nepheline is included, clinopyroxene, with 40–45% of aegirine molecule, and biotite. The extensive carbonatites form large, lensoid and vein-like bodies which usually unite into continuous areas. Four types of carbonatite have been identified: (1) calcite carbonatites with forsterite, diopside, biotite, phlogopite, apatite and accessory perovskite, titanite, zircon and calzirtite; (2) calcite carbonatites with dolomite, diopside, biotite, phlogopite, tetraferriphlogopite, forsterite, magnesioarfvedsonite and accessory pyrochlore, hatchettolite, baddeleyite, rutile and ilmenite; (3) Calcite–dolomite carbonatites with aegirine, richterite, eckermannite, magnetite, pyrite, pyrrhotite and accessory apatite, zircon, pyrochlore, leushite and aeschynite; (4) Dolomite–ankerite carbonatites with aegirine, magnesioriebeckite, arfvedsonite, chlorite,

hydromica, albite, carbonate–apatite, baryte, strontianite, fluorite, zircon, sometimes quartz and with rare-earth minerals usually present including pyrochlore, fersmanite, columbite, monazite, burbankite, carbocernaite, parisite and bastnaesite. Sulphides from calcite carbonatite gave a $\delta^{34}S°/_{oo}$ value of -0.2 (Grinenko *et al.*, 1970).

Age K–Ar on biotite from nepheline syenite gave 280 and 300 Ma, and phlogopite from carbonatite 350 Ma (El'yanov and Moralev, 1961).

References El'yanov and Moralev, 1961; Ginzburg and Epstein, 1968; Pozharitskaya and Samoilov, 1972; Semenov *et al.*, 1974; Zlenko *et al.*, 1966.

---

5 KHAMNINSKII   59°45′N; 136°23′E

Stocks, of up to 0.7 km², and dykes of nepheline syenite and peralkaline syenites cut Sinian (Upper Proterozoic) terrigenous rocks. Many nests and veins of low-temperature carbonatites are known in the area. The alkaline rocks and carbonatites are concentrated along deep north–south-trending faults. Xenoliths of country rocks are preserved in the syenites and fenites are widespread. It is postulated that the intrusion has an area of 60 km² at a depth of 2 km below the present surface.

Reference Epstein *et al.*, 1972.

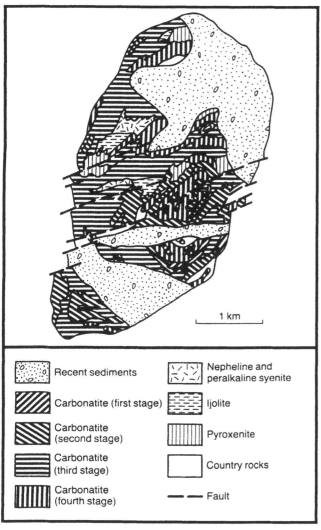

Recent sediments

Carbonatite (first stage)

Carbonatite (second stage)

Carbonatite (third stage)

Carbonatite (fourth stage)

Nepheline and peralkaline syenite

Ijolite

Pyroxenite

Country rocks

— — Fault

1 km

Fig. 240. Ozernyi (after Epstein et al., 1972, Fig. 9I).

Sette-Daban references

EL'YANOV, A.A. and MORALEV, V.M. 1961. New data on the age of the ultramafic and alkaline rocks of the Aldan shield. *Doklady Akademii Nauk SSSR*, 141: 687–9.

EPSTEIN, E.M., PANSHIN, I.P., MORALEV, V.M., VOLKODAV, I.G. 1972. On the vertical zonality of the massifs of the ultrabasic-alkaline rocks and carbonatites. *In* A.I. Ginzburg (ed.) *Geology of the ore deposits of the rare elements*. 35: 49–69. Nedra, Moscow.

GINZBURG, A.I. and EPSTEIN, E.M. 1968. Carbonatite deposits. *In* V.I. Smirnov (ed.) *Genesis of the endogenic ore deposits*. 152–219. Nedra, Moscow.

GOMBOEV, O.G., KONEVSEV, V.I. and SILICHEV, M.K. 1965. A new petrographic province of ultramafic-alkaline rocks in the north-east of the USSR. *Geologiya i Geofizika. Novosibirsk*, 5: 143–5.

GRINENKO, L.N., KONONOVA, V.A. and GRINENKO, V.A. 1970. Isotopic composition of sulphur of sulphides from carbonatites. *Geokhimiya. Akademiya Nauk USSR. Moscow*, 1: 66–75.

POZHARITSKAYA, L.K. and SAMOILOV, V.S. 1972. *Petrology, mineralogy and geochemistry of the carbonatites of eastern Siberia*. Nauka, Moscow. 267 pp.

SEMENOV, E.I., ESKOVA, E.M., KAPUSTIN, U.L. and HOM'AKOV, A.P. 1974. *Mineralogy of the alkaline massifs and their deposits*. Nauka, Moscow. 247 pp.

ZLENKO, N.D., SHPAK, N.S. and EL'YANOV, A.A. 1966. Ultramafic-alkaline intrusives of the Aldan anticline. *In* I.I. Krasnov, M.L. Lur'e and V.L. Masaitis (eds) *Geology of the Siberian Platform*. 233–42. Nedra, Moscow.

CHUKOTKA

Alkaline magmatism in the extreme northeast of Russia (Fig. 241) is variable in composition and facies (Lychagin, 1982). These alkaline rocks are known only superficially, because the region is both distant and almost inaccessible. A general review in English of Mesozoic volcanism and structural development in the northeast of Russia will be found in Belyy *et al.* (1989).

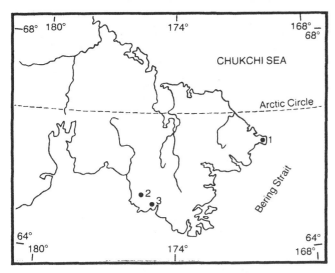

Fig. 241. The distribution of alkaline rocks in Chukotka (after Lychagin, 1982, Fig. 1).

1 Dezhnevskii	3 Nunyamuvem
2 Enmelen	

1 DEZHNEVSKII　　　　　66°05'N; 169°47'W
Fig. 242

The Dezhnevskii complex, covering an area of 125 km², lies within Carboniferous limestones and is of Palaeogene age. From the centre of the complex to the periphery there are a number of irregular concentric and semi-concentric zones. The central part is composed of granites which are succeeded outwards by quartz syenites, syenites and nepheline syenites. Within the nepheline syenites pseudoleucite shonkinites have been identified.

Age Geological evidence indicates a Mesozoic age.

References Perchuk, 1963; Zalishchak, 1978.

2 ENMELEN　　　　　65°07'N; 175°25'W

This volcanic area is represented by lava flows which are about 10 km long and up to 60 m thick; they fill a present day valley. As a rule the lavas are massive but the upper parts are typically scoriaceous in places. They include nepheline basalt (basanite), analcime basalt and limburgite, of which the nepheline basalt contains olivine and occasionally aegirine–augite and titanaugite. The content of nepheline varies between 5 and 10% and is present in the form of micro-poikilitic crystals. A little K-feldspar is occasionally present. Phenocrysts of aegirine–augite and olivine (30%) have been recorded in the limburgites in which augite, olivine and volcanic glass form the groundmass.

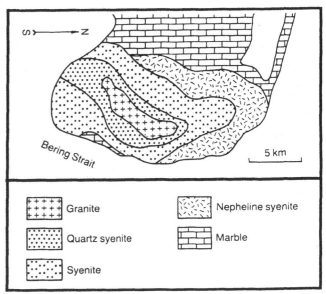

Fig. 242. Dezhnevskii (after Perchuk, 1963, Fig. 1).

Age Neogene to Quaternary, on stratigraphic evidence.

References Belyi and Migovich, 1971; Rabkin, 1954.

3 NUNYAMUVEM　　　　　64°50'N; 174°45'W

This volcanic field is similar to the Enmelen field described above (locality 2).

Age Neogene to Quaternary, on stratigraphical evidence.

Reference Belyi and Migovich, 1971.

Chukotka references

BELYI, V.F. and MIGOVICH, I.M. 1971. Neogene-Quaternary volcanic formations in the East Chukotka and Lower Penzhina. *Petrology of Neogene-Quaternary basaltoids in the North-West sector of the Pacific Mobile Belt.* 114–25. Nauka, Moscow.

*BELYY, V.F., GEL'MAN, M.L. and PARAKETSOV, K.V. 1989. Mesozoic volcanism and structural development in the northeastern USSR. *International Geology Review*, **31**: 455–68.

LYCHAGIN, P.P. 1982. Alkali basic rocks in the North-East of the USSR. *Pacific Geology*, **6**: 85–93.

PERCHUK, L.L. 1963. The magmatic replacement of the Carboniferous rock mass followed by the formation of the nepheline syenites and other alkali rocks in the Dezhnev massif considered as a model. *In* G.A. Sokolov (ed.) *Physico-chemical problems of rock and ore formation*, **2**: 160–81. Publishing House of the USSR Academy of Sciences, Moscow.

RABKIN, M.I., 1954. Alkali basic and ultrabasic effusive rocks in the southern part of the Chukotka Peninsula. *Proceedings of Scientific Research Institute of Arctic Geology*, **43**: 57–65.

ZALISHCHAK, N.L. 1978. Alkali magmatic rocks. *Geology of the Pacific Mobile Belt and the Pacific Ocean.* **2**: 66–82. Nedra, Leningrad.

* In English

KAMCHATKA-ANADYR'

In this province alkaline rocks occur in the south on the Kamchatka Peninsula and to the north near the Anadyr' and Apuka Rivers (Fig. 243). All of these occurrences lie within the same framework of Cenozoic folding, but the alkaline rocks are of only minor abundance in Kamchatka, forming but a small proportion of extensive non-alkaline volcanic sequences. In the central-southern part of the Kamchatka Peninsula the outcrops of volcanic rocks and intrusive complexes extend along the eastern slope of the Sredinny Ridge in an approximately north-south direction. They extend, discontinuously, for over 140 km in a 20–25 km wide zone.

1 Rarytkinskii
2 Vatyinskii
3 Western Kamchatka
4 Valaginskay Seria
5 Andriyanovka
6 Kirganic

1 RARYTKINSKII   63°45'N; 174°06'E

The lower part, about 600 m, of this volcanic structure is represented by pyroxene and olivine and pyroxene- and analcime-bearing basalts, hyalobasalts and andesites which are interbedded with tuffs and tuff breccias. The structure is also cut by dykes and sheets of teschenite, crinanite and trachydolerite. Ignimbrites of liparitic and dacitic composition, and occasionally andesitic, alternate with tuffs. The alkaline gabbroic rocks contain plagioclase of basic composition, titanaugite, barkevikite and analcime. A little olivine occurs in the crinanites.

Age Early Palaeogene, on geological evidence.

References Belyi and Migovich, 1966; Lychagin, 1982.

2 VATYINSKII   61°04'N; 169°25'E

Within siliceous volcanogenic rocks of Upper Cretaceous to Palaeogene age, basalts occur with compositions varying from normal alkaline to weakly alkaline. The former comprise about one-third of the sequence and include analcime-bearing varieties. Bodies of essexite gabbro have been discovered which contain plagioclase (labradorite–bytownite), clinopyroxene, olivine and analcime.

Age Assumed to be Quaternary.

References Lychagin, 1982; Rudich and Ustiyev, 1966.

3 WESTERN KAMCHATKA   57°30'N; 157°25'E

The alkaline rocks of Western Kamchatka occur within early Cretaceous, Palaeogene, and Neogene deposits and Lower Pliocene sedimentary rocks were contact metamorphosed by them. Generally the alkaline rocks form subvolcanic cupolas which vary from 10 m up to 1.5 km in diameter. Dykes and other intrusive bodies have thickness of 0.5 to 200 m and extend along the strike for about 2.5 km. Some small stock-like intrusions of alkaline syenites occupy areas of 1–2 km². Among these magmatic rocks are trachydolerites, alkali basalts, absarokites, alkaline syenites and crinanites. In most of them there is a small amount (0–10%) of analcime, but in crinanites and analcime basalts the analcime varies between 10 and 15%. The crinanites are holocrystalline, fine-grained, weakly porphyritic rocks composed principally of titanaugite (25–30%), analcime (10–15%), bytownite (40–45%) and olivine (10–15%). A little alkali feldspar (0–5%) and biotite (0–5%) are usually present and there is accessory apatite, magnetite and ilmenite. The analcime basalts are similar to the crinanite, but contain more olivine (15–20%) and less pyroxene and plagioclase. The alkaline syenites are leucocratic, medium-grained rocks of alkali feldspar (65–70%), pyroxene (5–10%), biotite (10–15%), analcime (0–15%) and accessory minerals (2–5%).

Reference Guznev, 1971.

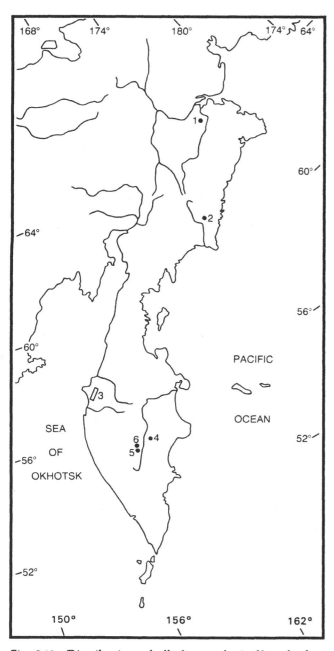

Fig. 243. Distribution of alkaline rocks in Kamchatka-Anadyr (after Guznev, 1971, Fig. 3).

4 VALAGINSKAY SERIA 54°52'N; 159°34'E

This volcanic series, formed at the beginning of a geosynclinal cycle, consists of ultramafic volcanics, volcanic breccias, basaltic lavas, trachybasalts and trachyandesites, which have a total thickness of 7.5 km and form dykes and sills as well as flows. Picrite lavas (SiO$_2$ 39.44%, Na$_2$O 0.07%, K$_2$O 0.08%), tuffs and breccias cover an area of about 7 km^2. Within the picritic tuffs are carbonatite bombs and blocks up to 1.5 m across. The carbonatites are light-grey to green-grey rocks which are porphyritic with a fine-grained groundmass. The phenocrysts are diopside–augite and garnet which are set in a matrix consisting of calcite microlites and brown volcanic glass. Accessories are apatite, baryte, strontianite, sulphides and pseudomorphs after olivine. The carbonatite is enriched in Ba, Sr and Zr but the concentrations of P and Nb are very low. Determinations of ^{87}Sr/^{86}Sr in the carbonatite gave 0.70365 ± 0.00010, 0.70342 ± 0.00010 and 0.70369 ± 0.00010.

Age Geological data indicate a late Cretaceous age.

Reference Rass and Frikh-Khar, 1987.

5 ANDRIYANOVKA 54°45'N; 158°30'E
Fig. 244

Trachybasalts are dominant at the Andriyanovka occurrence but alkali basalts, tephrites and trachyandesite-basalts are also present, with occasional occurrences of absarokite and leucitite. Tuffs as well as small stock-like bodies of monzonite also occur. The tephrites contain phenocrysts of augite, diopside, plagioclase, magnetite, biotite and apatite in a matrix of plagioclase, alkali feldspar, epidote, chlorite and glass. The leucitites contain phenocrysts of augite (20–25%), magnetite, occasionally olivine, which is replaced by iddingsite, and apatite. The groundmass of the rock is composed of epileucite, aegirine and magnetite. Analyses of the rocks are given by Flerov and Koloskov (1976).

Age K–Ar on leucite gave 20 ± 3 Ma.

Reference Flerov and Koloskov, 1976.

6 KIRGANIC 54°50'N; 158°30'E
Fig. 245

Volcanic rocks, including lavas and breccias, form the central part of a subvolcanic structure around the margins of which they pass into mixed volcanic and sedimentary and then into sedimentary deposits. There are dykes and sheets of orthoclase pyroxenite and porphyritic shonkinite and also small subvolcanic bodies (300–500 × 60–150 m) of porphyritic essexite. In the northern part of the field there is an area of carbonate-rich albite–orthoclase rocks. Pseudoleucite shonkinites are composed of alkali feldspar (40–45%) and augite (18–20%) together with pseudoleucite, biotite (7–9%), magnetite (5–8%), carbonate (18–23%), hastingsite (2–3%), sodalite, analcime, diopside and apatite. The pseudoleucites have isometric forms and are up to 5 mm in diameter. They are composed of radiating laths of feldspar and a fine-grained aggregate of clinozoisite, sodalite and sericite. The essexites contain phenocrysts of augite, diopside, magnetite and plagioclase set in a groundmass of clinopyroxene, plagioclase, biotite, magnetite, orthoclase, chlorite and epidote. Chemical analyses are available in Flerov and Koloskov (1976).

Age K–Ar on pseudoleucite shonkinite gave 44 ± 3 Ma.

Reference Flerov and Koloskov, 1976.

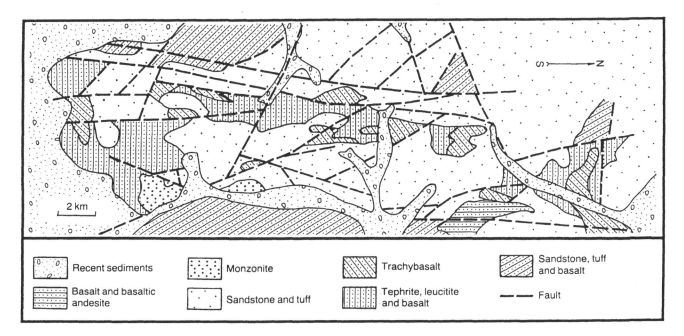

Recent sediments	Monzonite	Trachybasalt	Sandstone, tuff and basalt
Basalt and basaltic andesite	Sandstone and tuff	Tephrite, leucitite and basalt	—— Fault

Fig. 244. The volcanic field in the vicinity of the Andriyanovka and Zhupanka Rivers (after Flerov and Koloskov, 1976, Fig. 2).

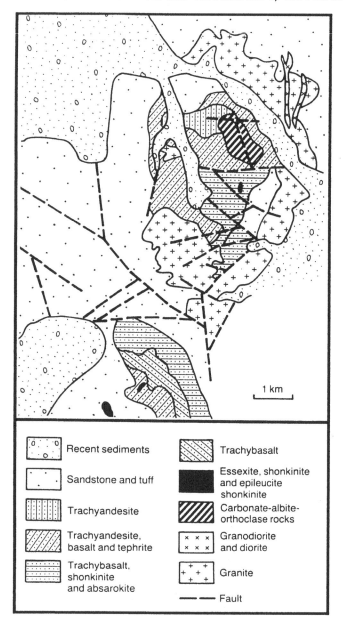

Recent sediments

Sandstone and tuff

Trachyandesite

Trachyandesite, basalt and tephrite

Trachybasalt, shonkinite and absarokite

Trachybasalt

Essexite, shonkinite and epileucite shonkinite

Carbonate-albite-orthoclase rocks

Granodiorite and diorite

Granite

— — Fault

1 km

Fig. 245. The volcanic field in the vicinity of the Kirganic River (after Flerov and Koloskov, 1976, Fig. 3).

Kamchatka references

BELYI, V.F. and MIGOVICH, I.M. 1966. Cenozoic volcanic formations in the Chukotka and Koryak highland. *Volcanic and volcano–plutonic formations.* 83–92. Nauka, Moscow.

FLEROV, G.B. and KOLOSKOV, A.V. 1976. *Alkali basaltic magmatism of central Kamchatka.* Nauka, Moscow. 147 pp.

GUZNEV, I.A. 1971. Neogene-Quaternary alkali basaltoids of Western Kamchatka. *Petrology of Neogene–Quaternary basaltoids of the North-West Sector of the Pacific Mobile Belt.* Proceedings of the All-Union Geological Institute. New Series, **174:** 107–13. Nedra, Moscow.

LYCHAGIN, P.P. 1982. Alkali basic rocks in the North-East of the USSR. *Pacific Geology* **6:** 85–93.

RASS, I.T. and FRIKH-KHAR, D.I. 1987. On the discovery of carbonatites within the upper Cretaceous ultramafic volcanics of Kamchatka. *Doklady Akademii Nauk SSSR,* **294:** 182–6.

RUDICH, K.N. and USTIYEV, E.K. 1966. The centres of Quaternary volcanism in the Mesozoic areas of the North-East of Asia. *Volcanic and volcano–plutonic formations.* 93–114. Nauka, Moscow.

OMOLON

In the northeast of Russia the greatest concentration of alkaline rocks is in the Omolon block (Fig. 246). This is a fault-bounded area of basement rocks preserved between Mesozoic fold belts.

1 Pyatistennyi	7 Omulevskii
2 Anyuiskii	8 Popovka
3 Aluchinskie	9 Khulichanskii
4 Bilibino	10 Omolonskii
5 Oloiskii	11 Anmandykanskii
6 Balagan-Tas	12 Kananyga

1 PYATISTENNYI   67°52′N; 161°39′E

This occurrence consists of dykes and lava flows of alkaline basaltic rocks which are overlain by tuffs and tuff breccias. The lavas are porphyritic with phenocrysts of zoned clinopyroxene, olivine replaced by chlorite, altered biotite, magnetite and accessory apatite; the presence of euhedral analcime has been demonstrated by Bazarova et al. (1981). Abundant micro-inclusions occur in the pyroxene, including leucite, analcime and glass, and homogenization temperatures of 1140–1240°C have been obtained from the inclusions (Bazarova et al., 1981). The leucite and analcime is readily apparent on weathered surfaces as circular spots. The groundmass consists essentially of brown glass, which is sometimes devitrified, in which are set acicular crystals of clinopyroxene and Ti–Fe oxide minerals; leucite and analcime coexist in the groundmass. The tuffs consist of lava fragments together with crystals of augite, leucite, orthoclase and plagioclase. Dykes of hauyne monchiquite are found.

Age Assumed to be Palaeogene

References Bazarova et al., 1981; Bilibin, 1958; Lychagin, 1982.

2 ANYUISKII   67°10′N; 165°50′E

Two stages of extrusive activity have been identified at this locality. The first stage is characterized by extensive lava flows of basalt, which ran along the valley of the

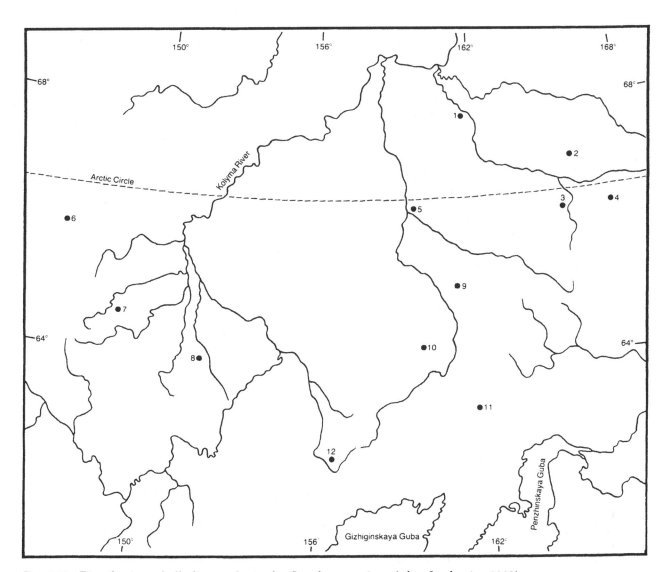

Fig. 246. Distribution of alkaline rocks in the Omolon province (after Lychagin, 1982).

215

Monka River; they have been traced over a region 53 km long by 2–4 km wide, cover an area of about 100 km² and have a total volume of 3 km³. The lavas were erupted from numerous fissures and consist of mildly alkaline basalts. The second stage of activity began with the explosive eruption of basaltic andesites which were then followed by further mildly alkaline trachybasalts, with normative nepheline, which form volcanoes and isolated lava flows. The most extensive lava flow in this final phase of eruption is about 12 km long, 1.5–2.2 km wide and has a thickness of 10–20 m. The total area of lava flows of the second stage is 124 km².

Age 0.4–0.5 million years.

References Dovgal and Chasovitin, 1965; Rudich and Ustiyev, 1966.

3 ALUCHINSKIE   66°18′N; 165°30′E

These volcanoes are accompanied by lava flows which extend for about 70 km. Two stages of effusive activity are separated by an interval that is indicated by the presence in the sequence of conglomerates. Compositionally the lavas are mildly alkaline basalts with about 5–6% normative nepheline (Dovgal and Chasovitin, 1965).

Age Holocene.

Reference Dovgal and Chasovitin, 1965; Rudich and Ustiyev, 1966.

4 BILIBINO   66°25′N; 167°12′E

The extinct Bilibino volcano reaches a height of 390 m above the adjacent river valley. It is of central type, forms a truncated cone which is elliptical in section and has a basal diameter of over 1 km. The structure of the upper part of the cone is clearly exposed and formed by alternating layers of dense, dark grey and reddish brown vesicular basalts and poorly cemented explosive products. The lavas are porphyritic with phenocrysts of olivine in a matrix of labradorite (An_{60-65}), clinopyroxene, Ti–Fe oxide minerals and glass. The chemistry indicates them to be nepheline normative (about 5–6% ne) mildly alkaline olivine basalts, but the occurrence has not been thoroughly studied and more alkaline types may occur. For a detailed mineralogical account see Dovgal and Chasovitin (1965).

Age Early Quaternary, from geological evidence.

Reference Dovgal and Chasovitin, 1965.

5 OLOISKII   66°23′N; 159°35′E

Peralkaline granites are found in a granitoid massif that occupies an area of 1500 km².

Age Geological evidence indicates a Late Cretaceous age.

Reference Lychagin, 1982

6 BALAGAN-TAS   66°00′N; 146°35′E

This volcano takes the form of an almost symmetrical cone about 190 m high which is composed of lavas similar in composition to those of peripheral lava flows. Explosive products are only found adjacent to the edge of the crater. The lava flows cover an area of 12 km². Chemically the lavas are the lowest in silica and most

alkaline (SiO_2 46%; $Na_2O + K_2O$ 5.5–8%) of all the Quaternary lavas in the Omolon province.

Age Upper Pleistocene.

References Dovgal and Chasovitin, 1965; Rudich and Ustiyev, 1966.

7 OMULEVSKII   64°40′N; 149°00′E

This occurrence consists of small and sparse dykes and stocks of porphyritic alkaline picrite and teschenite-picrite with segregations of more leucocratic teschenite. Camptonites and lamprophyres with amphibole also occasionally occur. The rock-forming minerals of the suite are plagioclase, barkevikite, biotite and occasionally olivine, clino- and orthopyroxene and analcime.

Age Late Cretaceous, on geological evidence.

Reference Lychagin, 1982.

8 POPOVKA   64°00′N; 152°10′E

Popovka consists of small, composite, differentiated intrusions in the form of stocks up to 100 m across. They are variable in composition and consist of tilaite (a gabbro with 80% of dark-coloured minerals), gabbro, dolerite, quartz-bearing syenites, nepheline syenites and alaskites.

Reference Zalishchak, 1978.

9 KHULICHANSKII   65°15′N; 161°02′E

This is a composite volcanic–plutonic complex in which the intrusive rocks are represented by alkaline gabbroids, which are located towards the centre, alkaline syenites, which occur around the periphery, and minor teschenites. The associated volcanic rocks include alkali olivine basalt and some trachytes. The mineralogy typically includes plagioclase, titanaugite and occasionally orthoclase and analcime.

Age Geological evidence indicates Early Cretaceous.

Reference Lychagin, 1982.

10 OMOLONSKII   64°16′N; 159°50′E

At Omolonskii a sequence of sedimentary volcanogenic rocks of Jurassic age is cut by numerous intrusive sheets, dykes, stocks and laccoliths, which extend over 20 km². These occurrences are composed of teschenite, crinanite, essexite–diabase, essexite, solvsbergite and tinguaite. The intrusive bodies are in some places associated with small outcrops of extrusive alkali basalt and augitite. The crinanites are composed of plagioclase (An_{62-63}), clinopyroxene, pseudomorphs after olivine, analcime, Ti–Fe oxide minerals and apatite. Accessory minerals include chlorite, calcite and biotite. The teschenites comprise considerable orthoclase (44%), hornblende (22%), analcime (7%), clinopyroxene of variable composition (4%), biotite (4%), plagioclase An_{47-48} (11%), apatite and Ti–Fe oxide minerals. The region is known for its numerous dykes of solvsbergite and tinguaite. The solvsbergites have trachytic textures and are characterized by the variability of the constituent aegirine and hastingsite. Alkali feldspar is the dominant phase (about 80%) and accessories are represented by apatite and zircon. Tinguaite comprises alkali feldspar (51%), nepheline (23%), analcime (6%) and aegirine

(20%), as well as riebeckite, apatite, zircon, rutile and magnetite.

Age Early Jurassic.

References Bilibin, 1958; Lychagin, 1975 and 1982; Shevchenko, 1975; Zalishchak, 1973.

---

11 ANMANDYKANSKII

63°12′N; 161°40′E
Fig. 247

The Anmandykanskii pluton occupies an area of about 310 km² and cuts Precambrian metamorphic rocks. The complex has a concentric zonal structure with quartz syenites in the central part while around the periphery are quartz-bearing syenite, syenite, peralkaline syenite and nepheline syenite. The nepheline syenites form sheet-like bodies with thicknesses up to 30 m. The peripheral rocks have layered structures. Thus, in the western part of the margin of the pluton, in a zone 2.5 km wide, there are up to 22 units formed of nepheline and alkaline syenites and microcline rocks. The nepheline syenites are medium- and coarse-grained porphyritic rocks composed of microcline–perthite (45–70%), nepheline or sericite pseudomorphs after nepheline (30–45%), biotite and aegirine (5–10%); there is also some albite, cancrinite, melanite and accessory minerals including titanite, apatite, zircon and Ti–Fe oxide minerals.

Age Palaeozoic.

References Fadeyev and Shpetny, 1978; Zalishchak, 1978.

---

12 KANANYGA

62°30′N; 156°39′E

This occurrence consists of just six outcrops of alkaline basaltic rocks. One of them is a small sheet, crowning the watershed and, apparently, directly linked with a feeder channel. The sheet is 70–80 m thick and consists of lavas with scoriaceous material at the base, while the upper section comprises massive basaltoids. The top of the section is composed of scoria together with volcanic bombs and lapilli of the same composition. Dykes and extrusive domes (0.1–0.4 km²) also occur and consist of dense and vesicular basaltoids with lherzolite inclusions. They are porphyritic rocks with olivine and titanaugite phenocrysts; the groundmass has a micropoikilitic texture and consists of oligoclase (An_{17-22}) and feldspathoids, which are mainly leucite but nepheline and analcime are also present, and abundant prisms of clinopyroxene, which are enclosed poikilitically by feldspathoid, apatite and Ti–Fe oxide minerals.

Age Quaternary.

Reference Ichetovkin *et al.*, 1970.

Omolon references

BAZAROVA, T.Yu., KOSTUK, V.P. and KHMELN-KOVA, O.A. 1981. Features of the formation of alkali basalts of Great Aniui (a tributary of the Kolyma). *Doklady Akademii Nauk SSSR*, 259: 1192–4.

BILIBIN, Yu.A. 1958. *Selected Works*. I. USSR Academy of Sciences, Moscow. 432 pp.

DOVGAL, Yu.M. and CHASOVITIN, M.D. 1965. The Bilibin volcano; a new Quaternary volcano in the North-East of Prikolymye. *Geologiya i Geofizika. Novosibirsk*, 6: 35–46.

FADEYEV, A.P. and SHPETNY, A.P. 1978. The peculiarities of geological structure and history of formation of the Anmandykanskii alkaline pluton. *Materials on the geology and mineral resources of the North-East of the USSR*. 24: 53–60. Magadan.

ICHETOVKIN, N.V., SILINSKY, A.D. and FADEYEV, A.P. 1970. Cainozoic alkaline basaltoids of the

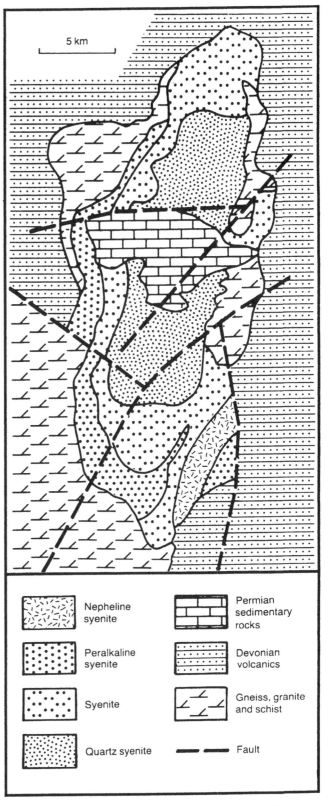

Fig. 247. Anmandykanskii (after Fadeyev and Shpetny, 1978, Fig. 1).

Legend:
- Nepheline syenite
- Peralkaline syenite
- Syenite
- Quartz syenite
- Permian sedimentary rocks
- Devonian volcanics
- Gneiss, granite and schist
- Fault

Kananiga River and Viliga River basins. *Geologiya i Geofizika. Novosibirsk,* **8**: 127–32.

LYCHAGIN, P.P. 1975. Early Jurassic alkali rocks in Omolonsky Massif. *Materials on the geology and mineral resources of the North-East of the USSR.* **22**: 62–9. Magadan.

LYCHAGIN, P.P. 1982. Alkali basic rocks in the North-East of the USSR. *Pacific Geology* **6**: 85–93.

RUDICH, K.N. and USTIYEV, E.K. 1966. The centres of Quaternary volcanism in the Mesozoic areas of the North-East of Asia. *Volcanic and volcano-plutonic formations.* 93–114. Nauka, Moscow.

SHEVCHENKO, V.M. 1975. The composition and volume of the Abkitsky granitoid complex in the Omolonsky Massif. *Materials on the geology and mineral resources of the North-East of the USSR.* **22**: 49–61. Magadan.

ZALISHCHAK, B.L. 1973. Syenite–crinanite complexes and the problem of alkali rocks genesis. *Magmatic rocks in the Far East. The works of the Far-Eastern Geological Institute.* 118–33. Vladivostok.

ZALISHCHAK, N.L. 1978. Alkali magmatic rocks. *Geology of the Pacific Mobile Belt and the Pacific Ocean.* **2**: 66–82. Nedra, Leningrad.

SAKHALIN

There are only two known areas of alkaline magmatism in Sakhalin, one on the Shmidt Peninsula at the northern end of the island, and the other on the western coast (Fig. 248).

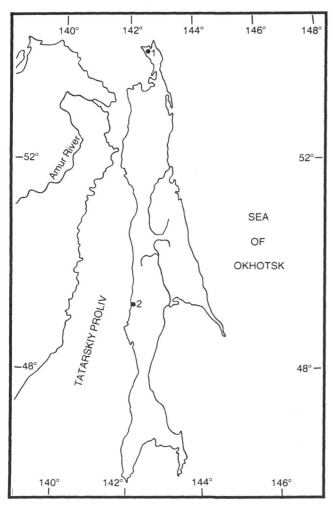

Fig. 248. *Distribution of alkaline rocks on Sakhalin (after Ganeshin and ZhiZhin, 1970, Fig. 2).*

1 Shmidt Peninsula	2 Nadezhdinka

1 SHMIDT PENINSULA   54°08′N; 142°30′E

Small stocks, sheets, laccoliths and dykes of gabbro, gabbro-diorite, essexite, crinanite and teschenite occur within the western (Zapadny) and eastern (Vostochny) ridges of the peninsula. The stocks do not exceed 100–120 m in diameter; the sills are 100–200 m thick and the dykes 2–3 m wide. Essexite dolerites are wholly crystalline fine- and medium-grained rocks composed of labradorite (30–35%), titanaugite (25–30%), olivine (10–15%), biotite (3–5%), analcime (10%) and anortho-

clase (10%) with accessory magnetite, ilmenite, titano-magnetite and apatite. The crinanites are rich in analcime (25%) and olivine (20%) and the teschenites include nepheline and natrolite.

Age K–Ar on crinanite gave 43 Ma and on porphyritic essexite 35 Ma (Semenov, 1970a).

References Erokhov and Shilov, 1971; Semenov, 1970a.

2 NADEZHDINKA   49°15′N; 142°10′E
(Morotu District)   Fig. 249

Dykes, sheets and laccoliths of alkaline rocks are intruded into Tertiary sediments which are metamorphosed, including coals that have been turned into coke. The margins of the larger sheets and laccoliths are of basic rocks but these grade inwards into syenites, although the basic rocks are also cut by syenite veins. Among the basic rocks are varieties which have been referred to by some workers as crinanites and by others as monzonite. The crinanites, which have ophitic textures, are composed of zoned andesine–labradorite (40–60%), analcime mixed with anorthoclase (10–20%) and dark-coloured minerals including titanaugite (up to 30%), olivine (up to 20%), biotite (up to 15%) and rare kaersutite, titanomagnetite and apatite. Among the alkali syenites biotite and analcime-bearing (up to 30%) varieties occur, and nepheline syenites have also been identified. They are composed predominantly of anorthoclase and microperthite (60–90%) and besides biotite aegirine–augite, aegirine, kaersutite, hastingsite and arfvedsonite are also present. The accessory minerals include aenigmatite, Ti–Fe oxide minerals, apatite, titanite and zircon. The chemistry and optical characteristics of the minerals have been studied in detail by Yagi (1953), including early classic work on the alkali pyroxene series; he also gives a range of rock analyses.

Age Early Pliocene.

Reference Semenov, 1970b; *Yagi, 1953; Zalishchak, 1973.

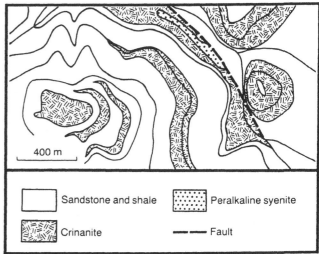

Fig. 249. *Nadezhdinka (after Yagi, 1953, Fig. 2).*

Sakhalin references

EROKHOV, V.F. and SHILOV, V.N. 1971. Volcanic formations in the Sakhalin and Kuril Islands. In G.M. Gapeeva (ed.) *Petrology of Neogene–Quaternary basaltoids of the North-West sector of the Pacific Mobile Belt*. Works of the all-Union Geological Institute. New series, **174**: 62–93. Nedra, Moscow.

GANESHIN, G.S. and ZHIZHIN, D.P. 1970. Physico-geographic and economic futures. *In* A.V. Siderenko (ed.) *Geology of the USSR*. **33**: 16–28. Nedra, Moscow.

SEMENOV, D.F. 1970a. The Shmidt Peninsula. *In* A.V. Siderenko (ed.) *Geology of the USSR* **33**: 285–9. Nedra, Moscow.

SEMENOV, D.F. 1970b. West Sakhalin Mountain. *In* A.V. Siderenko (ed.) *Geology of the USSR*. **33**: 298–303. Nedra, Moscow.

*YAGI, K. 1953. Petrochemical studies on the alkalic rocks of the Morotu District, Sakhalin. *Bulletin of the Geological Society of America*. **64**: 769–809.

ZALISHCHAK, B.L. 1973. The syenite–crinanite complexes and the problems of genesis of the alkaline rocks. *In* S.A. Shcheka (ed.) *Magmatic rocks in the Far East*. 118–34. Far East Scientific Centre, Far East Geological Institute, Vladivostok.

* In English

PRIMORYE

In the Primorye region (Fig. 250) there are alkaline rocks of a wide range of composition, age and facies type, including extrusive rocks of a basaltic association, alkaline granites, syenite–crinanite complexes and isolated intrusions of alkaline ultrabasic rocks.

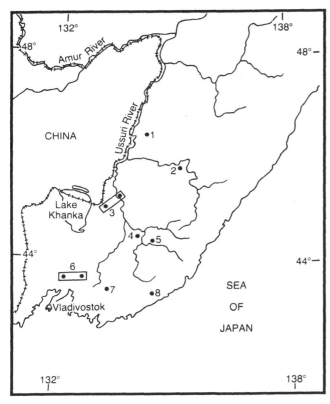

Fig. 250. *Distribution of alkaline rocks in Primorye (after Zalishchak, 1971, Fig. 1).*

1 Bikin
2 Sando-Vaku
3 Ussuri
4 Koksharovskii
5 Poginskii
6 Artem
7 Sitsinskii
8 Cape Orlova

1 BIKIN   46°26′N; 134°07′E

In the Bikin River basin volcanic fields of alkaline basaltoids form gently sloping hills and are present as lava flows and sills. They overlie lower Miocene coal-bearing strata and are overlapped by clays of Pleistocene age. The basaltoids are considerably metamorphosed. They contain analcime and zeolites.

Age Neogene.

Reference Kotlyur, 1968.

2 SANDO-VAKU   45°50′N; 135°12′E

Numerous stocks and dykes of alkaline syenites cut schists and siltstones of Upper and Middle Carboniferous age. The alkaline syenites are medium- and coarse-grained rocks with trachytic textures. They consist of K-feldspar, albite, aegirine and alkali amphibole. There

is a little biotite and accessory anatase, ilmenorutile and titanomagnetite. There is extensive secondary replacement of the alkaline syenites by albite, muscovite and sericite.

Age K–Ar on alkaline syenites gave 150 and 137 Ma.

Reference Kotlyur, 1968.

3 USSURI   45°08′N; 133°30′E

In the upper reaches of the Ussuri River more than 30 isolated outcrops of alkaline basaltoids are known. They traverse and overlie ancient crystalline Archaean and Proterozoic rocks and some outcrops cut through gravels of Pliocene age. The basaltic rocks form pyroclastics and lavas, dykes and breccia pipes up to 2 km² in area. Limburgite, leucite ankaratrite, nepheline basanite, nepheline basalt and essexite–dolerite are represented. The limburgites form small isolated outcrops overlying Archaean schists. They are black, dense rocks with brecciated structures and are often porous. They consist of volcanic glass in which are set microlites of titanaugite and olivine as well as xenocrysts and aggregates of olivine and clino- and ortho-pyroxene and spinel. The leucite ankaratrites have been found in one outcrop only, where they comprise part of the eruptive, fragmentary agglomerate of a breccia pipe. They have a fine-grained texture and consist of microlites of titanaugite, leucite, opaque minerals and plagioclase with interstitial volcanic glass. Xenocrysts of amphibole,

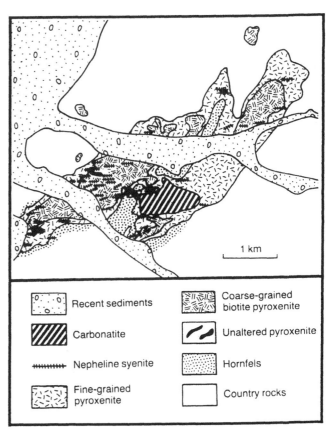

Fig. 251. *The eastern part of Koksharovskii (after Zalishchak, 1969, Fig. 1).*

biotite, olivine, pyroxene and spinel are also present. Ankaratrite–picrites are the most widespread rock type of this locality. They are fine-grained polycrystalline rocks consisting of microlites of titanaugite and small crystals of nepheline, olivine and ore minerals; apatite and analcime are present in accessory amounts. Larger xenocrysts of olivine, clino- and ortho-pyroxene and spinel surrounded by reaction borders of opaque minerals are also present.

Age Neogene

References Gapeyeva, 1971; Kotlyur, 1968.

4 KOKSHAROVSKII
44°30′N; 134°05′E
Figs 251 and 252

The Koksharovskii complex of alkaline and ultrabasic rocks is a faulted intrusion lying inside a complex tectonic block of Palaeozoic sediments. The complex has an area of 15 km² and the structure is different in the eastern (Fig. 251) and southwestern (Fig. 252) parts. The

complex can be divided into two phases, the first of which, covering about 90% of the area, is of pyroxenites which are variably replaced by biotite and contain hornblende, ilmenite and titanomagnetite in variable amounts; sometimes they contain garnet and sulphides. Titanaugite and aegirine–augite jacupirangites are rather less abundant. The second phase of alkaline igneous activity commenced with the emplacement of veins and dykes of foyaite, which are not very numerous. These are medium- and coarse-grained rocks consisting of alkali feldspar and albite (50–70%), nepheline (20–40%), aegirine–augite, some arfvedsonite and biotite (5–10%) and accessory zircon, magnetite, titanite, apatite and ilmenite. The foyaites are cut by a widespread system of alkaline dykes which traverse both the pyroxenites and the volcanic and sedimentary country rocks. The most abundant of the dykes are lujavrites, but rischorrites, mariupolites, cancrinite, zeolite and liebenerite syenites and syenite porphyries are also present. The lujavrites form zoned dykes 1.2 cm to 20–25 m thick which extend for up to 100–150 m. The central parts of the dykes are formed by even and medium-grained nepheline

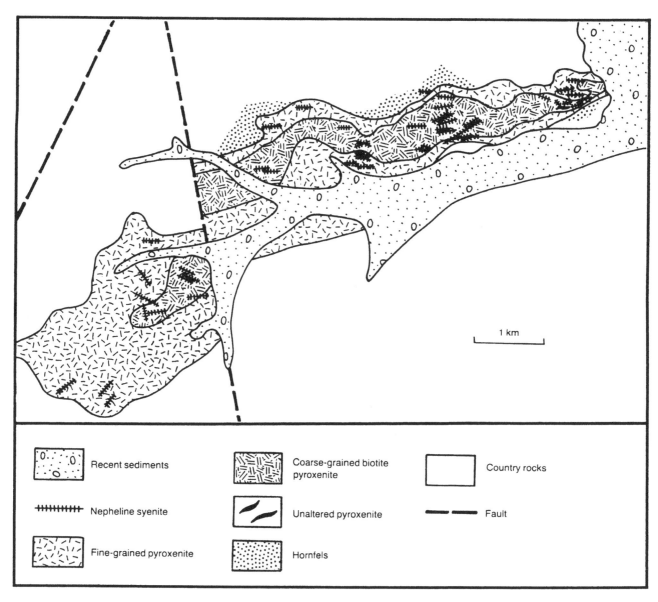

Recent sediments

Nepheline syenite

Fine-grained pyroxenite

Coarse-grained biotite pyroxenite

Unaltered pyroxenite

Hornfels

Country rocks

Fault

1 km

Fig. 252. The southwestern part of Koksharovskii (after Zalishchak, 1969, Fig. 4).

syenite containing nepheline (30–40%), feldspar (30–60%), aegirine (5–30%), biotite (1–5%), arfvedsonite, eudialyte, zirkelite, elpidite, lamprophyllite, titanite, apatite, Fe–Ti oxide minerals, zircon and pyrochlore. The outer parts of the dykes have a quenched texture and correspond compositionally to tinguaite. The intrusive sequence was continued by the introduction of veins, which are narrow and usually sinuous, of nepheline syenite pegmatites and syenite–aplites that cut pyroxenites and nepheline syenites. The igneous activity was completed by the emplacement of carbonatites. The carbonatite occupies an oval body of about 0.5 × 1 km at the eastern end of the complex (Fig. 251). It consists of calcite, up to 20% apatite, aegirine–augite, hastingsite, biotite, titanite, albite, alkali feldspar, Fe–Ti oxides and quartz. Apart from the apatite and titanite all the other minerals are intensively corroded by the calcite. Detailed information on the composition of rocks and minerals can be found in the paper by Zalishchak (1969).

Economic A weathered crust 10–20 m thick is developed over the complex and within it there are considerable reserves of vermiculite.

Age K–Ar on nepheline from tinguaite gave 135–145 Ma and on biotite from pyroxenite 139 Ma (Rub and Levitsky, 1962).

References Kotlyur, 1968; Rub and Levitsky, 1962; Zalishchak, 1969.

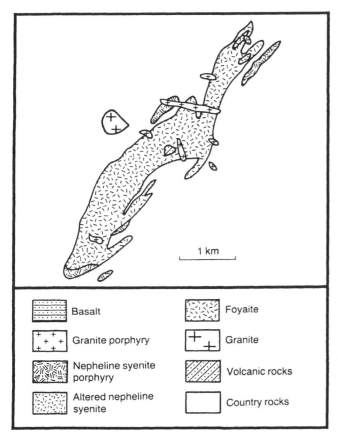

Fig. 253. *Poginskii (after Zalishchak, 1978, Fig. 17).*

Legend: Basalt · Granite porphyry · Nepheline syenite porphyry · Altered nepheline syenite · Foyaite · Granite · Volcanic rocks · Country rocks · 1 km

5 POGINSKII
44°08′N; 134°17′E
Fig. 253

One of the larger massifs in the Primorye region Poginskii is about 5 km in length. The intrusion cuts volcanic and sedimentary strata of Upper Palaeozoic age and is overlapped by Upper Cretaceous acid volcanics. It is composed predominantly of trachytic foyaites which are considerably metamorphosed, resulting in their replacement by aggregates of albite, zeolite, cancrinite and liebenerite. A series of veins is rather varied including as it does syenite porphyry, bostonite, nepheline syenite pegmatite and albitite. Alkaline basaltoids (essexite and teschenites) form extrusive bodies and numerous dykes.

Age Stratigraphic evidence indicates a Jurassic age (Kotlyur, 1968).

References Kotlyur, 1968; Zalishchak, 1978.

6 ARTEM
43°46′N; 132°28′E

In the Artem Depression, in the vicinity of the city of Artem, lavas and sills of trachydolerite, shoshonitic absarokite and leucite and pseudoleucite shoshonitic absarokites are widespread. They also form dykes, stocks of up to 4 km² and laccoliths. One of the laccoliths is formed by fergusites which are dark grey to black, porphyritic rocks with phenocrysts of biotite, sanidine, anorthoclase, pseudoleucite, a small amount of olivine and clinopyroxene of variable composition (diopside, aegirine–augite). Calcite is present in the fine-grained matrix and apatite is a usual accessory.

Age Neogene.

References Gapeyeva, 1971; Zalishchak, 1969.

7 SITSINSKII
43°29′N; 133°20′E

This symmetrically shaped intrusion, with an area of about 25 km², is formed predominantly of aegirine–riebeckite granites. In the outer parts of the intrusion the rocks are intensively albitized and alkaline pegmatites are present, while throughout there are numerous veins of grorudite and solvsbergite. The alkaline granites are formed of quartz (15–40%) and mainly perthitic K-feldspar with 20–25% of dark-coloured minerals, predominantly riebeckite and aegirine. Accessory minerals are aenigmatite, astrophyllite, titanomagnetite, zircon, xenotime, chevkinite and bastnaesite.

Age Late Cretaceous – from stratigraphic evidence (Kotlyur, 1968).

References Bevzenko, 1971; Kotlyur, 1968.

8 CAPE ORLOVA
43°28′N; 134°30′E

Two dome-like outcrops, respectively 1 km and 300 m in diameter, some 300 m apart, are formed by peralkaline granites which are cut by veins and dykes of peralkaline granite, granite porphyry and granite pegmatite. The peralkaline granites are formed of microperthite, quartz and riebeckite; albite and biotite are present in small amounts. Accessories include zircon, aenigmatite, astrophyllite, fluorite, garnet and magnetite.

Reference Kotlyur, 1968.

Primorye references

BEVZENKO, P.E. 1971. Types of granite intrusion associations from the south of the Far East. *In* S.S. Zimin (ed.) *Magmatic complexes of the Far East.* 29–53. Far East Scientific Centre of the Academy of Sciences of the USSR, Vladivostok.

GAPEYEVA, G.M. 1971. Petrology of the Neogene–Quaternary basaltoids from the northwest section of the Pacific belt. *Proceedings of the All-Union Geological Institute, New Series,* **174:** 126–46.

KOTLYUR, S.G. 1968. North-east and Far East. *In* Yu.I. Polovinkina (ed.) *Geological structures of the USSR.*

3 Magmatism. Nedra, Moscow. 640 pp.

RUB, M.G. and LEVITSKY, V.V. 1962. Petrological-geochemical features of the Koksharovka massif of ultrabasic-alkaline rocks with post magmatic products. *Alkaline rocks of Siberia.* 99–124. Publishing House of the USSR Academy of Sciences, Moscow.

ZALISHCHAK, N.L. 1969. *The Koksharovsky massif of ultrabasic and alkaline rocks (South Primorye).* Nauka, Moscow. 116 pp.

ZALISHCHAK, B.L. 1978. Alkaline magmatic rocks. *Geology of the Pacific Belt and Pacific Ocean.* **2:** 66–93. Nedra, Leningrad.

Locality index

Localities are indicated (a) by province (in *short form*) and occurrence number and (b) by page number, in bold type.

225

Printed in the United States
By Bookmasters